ANIMAL SCIENCE, ISSUES AND PROFESSIONS

LIVESTOCK

REARING, FARMING PRACTICES AND DISEASES

ANIMAL SCIENCE, ISSUES AND PROFESSIONS

Additional books in this series can be found on Nova's website
under the Series tab.

Additional E-books in this series can be found on Nova's website
under the E-books tab.

ANIMAL SCIENCE, ISSUES AND PROFESSIONS

LIVESTOCK

REARING, FARMING PRACTICES AND DISEASES

M. TARIQ JAVED
EDITOR

Nova Science Publishers, Inc.
New York

For permission to use material from this book please contact us:
Telephone 631-231-7269; Fax 631-231-8175
Web Site: http://www.novapublishers.com

Library of Congress Cataloging-in-Publication Data

Livestock : rearing, farming practices, and diseases / editor, M. Tariq Javed.
p. cm.
Includes index.
ISBN 978-1-62100-181-2 (hardcover)
1. Livestock--Diseases. 2. Feeds. 3. Livestock. I. Tariq Javed, M.
SF615.L58 2011
636--dc23

2011031633

Published by Nova Science Publishers, Inc. † New York

CONTENTS

PREFACE

This book presents current research in the study of livestock, with a particular focus on livestock rearing, farming practices and associated diseases . Topics discussed include a new generation of dietary supplements with microelements for livestock; pastoralism and the changing climate in northern Kenya; Bangladesh poultry sector; parasitic diseases in livestock under different farming practices; salmonella and salmonellosis in animals and humans; bovine tuberculosis at the human-animal interface and anthelmintic resistance.SChapter 1 - This chapter examines livestock farming practices of the Maasai pastoralists found in the semi-arid Laikipia North district and the Turkana pastoralists of the arid Turkana district, located in northern Kenya, the vulnerabilities, the adaptation strategies and responses to the changing climate. About 80percent of Kenya's total area is classified as arid and semi-arid land (ASAL) and over 60percent of pastoralism is practiced in the ASAL. Pastoralism accounts for about 10percent of the Gross Domestic Product (GDP) and 50percent of the agricultural GDP in Kenya. Livestock farming also accounts for 90percent of employment and more than 95percent of family incomes and livelihoods in the ASAL of Kenya. Laikipia, North district lies on the northern edge of the Laikipia Plateau at an altitudinal of 1600 to2000m above sea level with two major hills, the Mukogodo and Loldaiga located in the eastern part of the district. The district covers an area of over 1,000km^2 with annual rainfall ranging between 400 to 600 mm and temperatures ranging from 25 to 30^{0}C. The woody vegetation comprises of the Acacia specie,while the grass layer is predominantly Themeda, Cynodon, Eragrostis and Pennisetums. These are interspersed with a tree/shrub layer of Dodenea, Solanum, and Ipomoea. Turkana District is located to the west of Lake Turkana and covers an area of about 77,000km^2. Most of the district consists of low-lying plains with isolated mountains and hills; it has an annual rainfall ranging from 120 to 500mm and temperatures ranging from 24^0 to 38^{0}C. Vegetation in the district consists of scattered Acacia bush and annual herbaceous plants. Cattle are the dominant animals among the Maasai pastoralists, while camels are the most dominant livestock among the Turkana pastoralists. Sheep and goats (shoats) are kept in large numbers in both areas. The pastoralist's vulnerabilities to climate change include; frequent drought which cause lack of pasture, water and drought-related diseases, and the notable drought years include; 1995-96, 1999-2000, 2004-2006 and 2009-2010, and the one-off flood events which cause flood-related diseases and drowning of livestock. The impact has been poor livestock prices, livestock mortality and increased conflicts over scarce resources. The main coping strategies include; transhumance, pastoralists in Laikipia North District drive their livestock to Mt. Kenya, the Aberdare ranges and Mukogodo forest reserves, while the Turkana pastoralists take their livestock to Uganda and Sudan, use herbs to treat livestock during severe droughts. The pastoralists cope with floods by moving to high altitude lands and draining the floodwaters.

Chapter 2 - One of the most important problems in the animal production is trace-elements supplementation. A particular attention should be paid to animals in high-production systems. Micro- and macroelements demand is significantly higher in these animals than in extensive animal production. Resulting, deficiency of microelements has a dramatic consequence on health, production and at last economic balance. Commonly used preventive method is microelements supplementation.

Traditionally, trace-elements are introduced as inorganic salts or organic mineral products (e.g., amino acid chelates, proteinates and polysaccharides). However, several problems were encountered. Major disadvantage of inorganic salts supplementation is low microelements bioavailability and its quick alimentary tract transit, while chelates can be the cause of digestive tract irritation. Microelements supplemented in the form of inorganic salts possess low bioavailability and transit character, while chelates can cause alimentary tract irritation. Therefore, a new generation of dietary supplements produced by biosorption can be useful in solving this problem.

Contemporary results proved high bioavailability of Cr (III), Cu (II), Mn(II) and Zn (II) ions from preparations in which microelements are bound with a biological carrier. Preliminary studies showed significant increase of copper concentration in eggshell, blood and feather of hens. The investigations carried out on pigs revealed the positive influence on liver and higher copper level in meat. Positive influence of these dietary supplements on animal health was preliminarily proved, but the precise metabolic effects should be confirmed in the future. Very important is to increase bioavailability of micronutrients, reduce the problem of microelements segregation, through the production of less concentrated form of microelements and finally, improvement of health and performance of animals. These properties can also be related with digestive microflora modulation. Moreover, prohibition of nutritive antibiotics as one of the results of increased antimicrobial resistance in human and veterinary medicine caused subclinical and clinical infections.

Modern science states that the supplementation can play a crucial role in promotion of advantageous microflora. Native intestinal microfloras like Lactobacillus spp. are important inhabitants of the mammalian gastrointestinal system and it is emphasized that increasing their numbers improves gastrointestinal health. It seems that the new dietary supplementation method would have a beneficial influence on animal products, animal health, and environment, and can be used as prophylactic, metaphylactic and supportive treatment of the digestive system.

Chapter 3 - This chapter describes different soy protein products based on production processes, their nutritional value and application in animal feed.

There are several commercial soy protein products available in the market. These local products come from different origins, varieties and are produced differently. This causes variations in their nutritional value. For a systematic classification, appropriate terms need to be defined and used to describe different commercial products. First, the production process leads to different protein levels in the final product that can be used for their classification. Based on production process, soy protein products can be classified into two categories: 1) Basic processed soy protein including full fat soybean meal (36% CP), defatted soybean meal (44% CP) and high protein soybean meal (48% CP); 2) Highly processed soy protein including enzymatically/microbially fermented soy protein (52-56% CP), non-classically produced soy protein concentrate (60% CP) and classically produced soy protein concentrate (65-67% CP). Next to the protein level, other components are intrinsically linked to the products in above defined categories, which may have significant consequences for their usage in different animal species and age groups within a species.

Raw soybeans contain high amount of soy anti-nutritional factors (ANFs) that can be divided into heat stable and heat sensitive components. In animal nutrition, the most important heat labile ANFs are trypsin inhibitor activities (TIA) and lectins. The most important heat stable ANFs are antigens and oligosaccharides. High antigen levels can cause allergic response and intestinal damage in calves, young piglets and in some fish species. Oligosaccharides cannot be digested by young calves and monogastric animals, but can be fermented by bacteria, resulting in intestinal disturbance and diarrhea in young animals (such as weaning piglets).

These ANFs must be removed or inactivated by proper treatments before their applications in animal feed. Basic processed soy proteins are typically produced by heat treatment. The proper heating process reduces heat sensitive ANFs, such as trypsin inhibitor activity and lectins. However, large variation may exists for levels of trypsin inhibitor activity among differently processed soy protein products due to heat treatment conditions. The heat stable ANFs, in general, are not removed from these basic processed soy proteins. During the process of making of highly processed soy protein, both heat labile and heat stable ANFs are removed/eliminated, thus the final product have low levels of ANFs. In the process for soy protein concentrate, indigestible carbohydrates are also removed, leading to higher protein content and digestibility compared with basic processed soy products.The basic processed products are valuable sources of protein for poultry, growing and finishing pigs, ruminants and aquatic animals. However, due to the high amounts of heat stable soy ANFs, they have limited application in young animals` feed.Highly processed soy protein products have low soy ANFs; however, the ANF content and indigestible soy carbohydrates content are related to the processing methods and treatment conditions.

Production processes have direct impact on product quality and amino acids digestibility. Highly processed soy protein products have improved protein and AA digestibility when compared to basic processed soy products. The available literature demonstrates that the indigestible carbohydrates and other ANFs play an important role in digestibility of amino acids in soy protein, whereas the molecular size of the protein seems to have no or lesser importance.Due to the differences in production process, the quality assessment criteria are also different for basic processed soy proteins and highly processed soy proteins. For basic processed soy protein, TIA and PDI (Protein Dispersibility Index) are common methods for evaluating heat treatment efficiency. For highly processed soy protein, the following parameters should be considered: crude protein and essential AA content as percentage of crude protein, AA digestibility at different age of animals, indigestible carbohydrates content, soy ANFs level including TIA, antigens, oligosaccharides and soluble non-starch polysaccharides.

The composition and nutritional value of soybeans as well as soybean meal are not only related to production process, but also varies among origin and year of production. These variations affect feed formulations. Awareness of the variation in soybeans and processed soybean meal composition as well as variation on digestibility is important for nutritionists. In the past, animal feeds have been formulated with excess nutrients to meet/exceed the nutritional requirement of animals. Economic pressure and environmental regulations have made modern feed formulation to meet the nutritional requirements more precisely. This optimizes performance and minimizes waste output. However, this concept is not working without precise knowledge of ingredients used in feed formulation. Since soy products are the overwhelming supply source of dietary crude protein and amino acids, the quality assessment of soy products with respect to amino acid composition and digestibility and ANFs disserves utmost attention. Additionally, many different processing technologies are around and may affect soy quality as well.

For basic produced soy protein, the use of table values to formulate animal feed is not advisable seeing the existing variations in composition, ANF content and amino acid digestibility. It is recommended that nutritionists use producers' data sheets wherever available to formulate feed or make own analysis. The use of table values is only advisable in case there are no other sources of data available. Due to the advanced production process, highly processed soy protein can provide more consistent product quality, as these products are tightly controlled to meet nutritional specifications.

Chapter 4 - Since the 1960s, the world poultry meat production has been growing faster than any other meat, indicating its rising performance. Bangladesh poultry sector has emerged with great potential during the past two decades. Similar to other developing countries worldwide, Bangladesh also has a long historical record of poultry rearing under the scavengingsystem. The poultry sector in Bangladesh has not yet been fully industrialized and/or transformed from scavenging to a commercial system, although the sectorhas potential to generate income, employment as well as to fill meat consumption deficiency. The present chapter examines thetrends in production and consumption ofpoultry meat over a 35 year time period (1971–2005) in Bangladesh. Next, it examines the responses ofthe country's poultry and related input marketsto the poultry development policies ofthe 1990s. The chapter also examines the comparative cost competitiveness of poultry meat in relation to world's three major poultry meat suppliers (USA, Brazil and Thailand). Results reveal that transformation of the sector is slow although it has great potential. Demand for poultry meat is strong and the income elasticity of demand is the highest (0.65) as compared to other substitutes (meats and fish). Policy recommendation includes speeding the pace of the transformation process of the sector towards commercialization through various means (e.g., by favoring contract farming) and increase production in order to meet the growing demand for poultry meat.

Chapter 5 - Three are the main rearing practices classically developed worldwide, intensive, semi-extensive and extensive, on the basis of time the animals graze in the fields. Infection by parasites can be influenced by their management. Most of helminths (cestodes, trematodes, nematodes) present a life-cycle with an external phase in the environment, where resistant forms passed by feces (ova, larvae, cysts) develop to reach the infective stages (larvae, metacercariae) and infection occurs in livestock when they feed on contaminated pastures. Protozoa affecting the digestive apparatus are released to the environment as oocysts and infection is improved in indoor management systems, as occurred with certain ectoparasites (mange and lice). This seems to point that animals under an intensive regime could be mainly exposed to protozoan. By opposite, the livestock under a semi-extensive or extensive management should be infected by helminths and ectoparasites living in the environment. In spite of the trueness of this statement, several aspects related to the rearing of the animals must be taken into account. The administration of herbage to animals maintained indoors might increase the risk of infection by different helminth parasites (*Moniezia*, *Fasciola*, *Paramphistomum*, gastrointestinal nematodes). The supplementation of animals reared on extensive systems by using feeders placed in the grounds, the exposition to protozoan parasites could be enhanced. The main internal parasitism affecting livestock in respect to the type of farming have been analyzed. Fecal samples belonging to livestock (cattle, goats, sheep, horses and pigs) under intensive, semi-extensive and extensive regimes were collected and analyzed. The possibilities for controlling the parasitic diseases were discussed.

Chapter 6 - Trypanosomosis is an important constraint to the livestock production in many parts of Africa, Asia and Latin America. Tsetse-transmitted trypanosomosis (nagana) is a disease complex caused particularly by *Trypanosoma vivax*, *T. congolense* and *T. brucei brucei*. Non tsetse-transmitted trypanosomosis, on the other hand, is principally caused by *T. evansi* (surra), a widely distributed pathogenic trypanosome affecting livestock, but *T. equiperdum* and *T. cruzi* are also relevant pathogenic trypanosomes. From an economic viewpoint, it has been estimated that trypanosomosis reduces the cattle population between 30percent and 50percent and the production of milk and meat by at least 50percent in those infected areas of Africa. The presence of *Trypanosoma vivax* in and out of Africa transmitted by mechanical vectors rather than tsetse flies or the recent descriptions of *T. evansi* and *T. equiperdum* in European countries could pose a new threat for animal production in those territories. The purpose of this chapter is to review the current knowledge of the pathogenic trypanosomes that affect livestock, including the economic impact and control programs

Chapter 7 - Surveillance for tsetse and cattle trypanosomiasis distribution aimed at instituting control measures was undertaken in parts of Kauru Local Government Area in Kaduna State, Nigeria between late February to early March, 2007 and October, 2008 following claims of severe trypanosomiasis which had disrupted farming activities, devastated communal cattle industry and flight of 32 cattle herders with over 70 herds. A total of 964 cattle selected at random in nine sampling sites were each bled from the jugular vein and blood collected. Dark ground buffy coat and Giemsa stained thin film were methods used in diagnosis of trypanosomes. Anemia was quantified by packed cell volume assessment using the hematocrit method. Traps were set for tsetse catch. The knowledge, attitude, perception and treatment seeking behaviors (KAPTSB) and economic losses were assessed using questionnaire administered on 36 cattle herders, focus group discussions held with selected cattle herders and in-depth interview with the Head, Agriculture Department. Overall, 210 cattle were found positive with an infection rate of 22percent. The infection rates in the dry and wet seasons were 23.6 and 20.1percent, respectively, indicating increased trypanosome prevalence in the dry season. We detected 203 *Trypanosoma vivax* and 7 *T. congolense* infections and these differed significantly ($P< 0.05$) across the sampling sites. Out of 964 cattle, 395 (41%) were found anemic with significant variation across the sampling sites ($P < 0.05$). The traps captured few *Glossina palpalis palpalis* and *G. tachinoides* with lots of Tabanidae. The tsetse flies did not harbor trypanosomes. From the qualitative data, it became apparent that the cattle herders, though very much aware of trypanosomiasis problem, were faced with the challenge of inappropriate health care seeking behavior due to ignorance about where to seek for the service as a result of unavailability of veterinary clinics in the study districts. The 61percent of the respondents procured veterinary drugs from the open markets and resorted to quacks, while the rest said they relied on traditional remedies. The Local Government on its part lacked reliable information for planning, management and decision making against the disease. This research provided data on the magnitude of trypanosomiasis and linkage between community and Government. The Government treated cattle in all the districts and brought down the incidence of trypanosomiasis. It recommended the strategies for sustainable control to enhance livestock productivity, which is critical to poverty reduction and creation of wealth.

Chapter 8 - Antibiotics are in wide use for the control of microbial infections and as growth promoters in food animals. The indiscriminate use of these antibiotics has given rise to the menace of resistance in bacteria against these antibiotics. In similar fashion, wide spread use of anthelmintics resulted in resistance among parasites which has become a challenge for the clinician. The anthelmintic resistance in parasites has wide implications in terms of animal health and their productivity. Resistance is said to be present when there is a large frequency of microbes within a population able to tolerate doses of a compound than in a normal population. The resistance against a particular antimicrobial agent is also heritable. Among the parasites, nematode infestations are of major economic concern, particularlyin domesticated animals throughout the world. Anthelmintics have been developed with objective to control the parasites and finally reduce the production loss. Unplanned and injudicious use of anthelmintics has resulted in the development of resistance. Anthelmintic resistance has been noted against all classes of anthelmntics. Parasites (nematodes) have developed resistance against most prescribed groups of anthelmintics like benzimidazole through changes in primary target β-tubulin,whereas, resistance against levamisole, morantel, and pyrantel occur through loss or changes in nicotinic-acetylcholine receptors (nAChR).The P-glycoprotein, a drug efflux protein, was over expressed in avermectin resistant nematodes. Main goal of clinician is to minimize the resistance against the anthelmintics in planned ways which include correct dose administration, cyclical changeover of anthelmintics, regular monitoring on development of resistance by *in vivo* as well *in vitro* tests, controlling of livestock density on particular pasture area and use of combination of anthelmintics.

Chapter 9 - Salmonellosis is one of the most common and widely distributed foodborne diseases. It is a major public health problemin many countries and requires a significant amount of money to deal with. Millions of human cases are reported worldwide every year and the disease results in thousands of deaths. Salmonellosis is caused by the bacteria Salmonella. According to contemporary classification, the genus Salmonella contains two species; Salmonella bongori and Salmonella enterica, but there are more than 2,500 serotypes of S. enterica. Salmonella serovars can be divided into host restricted, host adapted, and ubiquitous serotypes with important implications for epidemiology and public health. Most of these cause acute gastroenteritis characterized by a short incubation period and a predominance of intestinal over systemic symptoms. Only a small number of serotypes typically cause severe systemic disease in man or animals, characterized by septicemia, fever and/or abortion, and such serotypes are often associated with one or few host species.

Since the beginning of the 1990s, antimicrobial-resistant Salmonellastrains have emerged and are threatening to become a serious animal and human health problem. This resistance results from the use of antimicrobials, both in humans and animals. The global spread of multi-drug resistantstrains to critically important antimicrobials;including the first-choice agents used for the treatment of humans are of great concern.

Chapter 10 - The situation of bovine tuberculosis in Pakistan and the role of *M. bovis* as cause of human tuberculosis have been presented. The study in 1969 indicated a prevalence of 6.72percent in animals in Faisalabad. In 1972, it was 2.9, 1.6 and 8.6percent in buffaloes,

Australian and Sahiwal cattle, respectively; in 1974 in Quetta, it was 0.53percent in buffaloes. In 1989, it was 7.3percent in cattle and buffaloes being slaughtered at Lahore Abattoir. In 2001, a study reported 1.76 percent prevalence in buffaloes in Faisalabad. In 2003, the respective prevalence of bovine tuberculosis in buffaloes and cattle was 6.91 and 8.64percent, respectively in Lahore. In 2006, the prevalence in buffaloes at two farms was 2.45 and 8.48percent. A study during 2006 and 2007 reported 2.2percent prevalence of tuberculosis in buffaloes at 11 farms. Another study reported a prevalence of 2.4percent in goats and 0.9percent in sheep at 3 and 7 livestock farms, respectively. A study during same period in buffaloes reported three percent prevalence involving 14percent farms around two cities of Punjab Pakistan. Recently, in cattle a prevalence of 7.6percent in cattle at 100percent farms has been reported. An overall prevalence of 3.3percent has been reported in zoo animals with 3.6percent in Bovidae, 3.2percent in Cervidae and 0percent in Equidae families.It is quite possible that some of the sheep and goat breeds are genetically resistant to tuberculosis. The stronger risk factors in cattle at 11 farms were the age of cattle, number of calving, total milk produced and lactation length. Certain possible risk factors identified in buffaloes were the lactating status of the buffalo, the presence of cattle at the farm, total number of cattle at the farm and the total number of animals at the farm. Results in zoo animals suggested that odds of tuberculosis were 1.2 times higher when animal number was less than 10. Because of the growing trend in developing livestock farms in the private sector in Pakistan, there is likelihood that the situation of bovine tuberculosis will worsen in coming years. Therefore, it is important to develop a strategy to control the disease right at the beginning rather than to think about it when it is already late. Certain strategies are suggested to control the disease in animals and humans.

Chapter 11 - Paratuberculosis (Johne's disease) is a chronic inflammatory disease of the gastrointestinal tract caused by Mycobacterium avium subspecies paratuberculosis(M. a. paratuberculosis). Paratuberculosis affects mainly ruminants (domestic and wild), but it has also been reported in monogastric animals and the main route of transmission is the fecal-oral.

There are no clinical signs that are pathognomonic of the disease, but it is very characteristic to observe progressive weight loss in the infected animals. The disease hasbeen divided into four stages: silent infection, subclinical, clinical and advanced clinical stage, where in the latter stage animals present intermandibularedema due to hypoproteinemia, emaciation, profuse diarrhea, generalized muscle atrophy and in some cases, death.

The diagnosis of the disease can be made by the direct detection of the agent, M. a. paratuberculosis(staining, culture –conventional, radiometric, non-radiometric, or direct DNA extraction), the detection of the host immune response by serological tests (complement fixation test, agar-gel immunodifusion or enzyme-linked immunosorbent assay) or by tests that measure the cell mediated immune response (gamma interferon assay or delayed-type hypersensitivity).Moreover, some studies have reported M. a. paratuberculosisas the etiological agent of Crohn's disease in humans. This hypothesis shows the possible zoonotic potential of this organism.

Chapter 12 - Consumer demand for additional information on food labels has been accompanied by legislative and judicial edicts that are expected to affect livestock production in the United States.One development involves legislation enacted by the state of Ohio that limited labels on milk products concerning recombinant bovine somatotropin, commonly called rbST.Dairy processors challenged the regulations because they wanted to be able to tell consumers more about whether products contained milk from cows are treated with rbST.Because of potential concerns for dangers to human health, some consumers are willing to pay more for milk produced from cows that were not treated with rbST.An appellate court found some of the Ohio labeling restrictions to be unconstitutional.This decision should facilitate more labeling, a reduction in market share for milk from cows treated with rbST, and a corresponding need for more dairy animals and high yielders.

The second issue involves the labeling of livestock products with attributes claiming that they are "natural" or "naturally raised."Firms have successfully convinced consumers that products with these labels are healthier and more environmentally friendly than their "non-natural" counterparts.Over the past 20 years, consumer demand for "natural" food products has steadily increased.Accompanying labels is consumer confusion of the actual meaning of the terms.In 2007, the U.S.D.A's Agricultural Marketing Service proposed a Naturally Raised Marketing Claim standard for producers that want to identify their livestock as "naturally raised."A voluntary marketing claim standard for naturally raised livestock and meat products would preclude the use of promotants (hormones), antibiotics (except to prevent parasites), and animal by-products as feed.Further regulatory developments may define these terms to reduce the confusion and alter producers' marketing strategies.

In: Livestock: Rearing, Farming Practices and Diseases
Editor: M. Tariq Javed
ISBN 978-1-62100-181-2

Chapter 1

PASTORALISM AND THE CHANGING CLIMATE IN THE ARID NORTHERN KENYA

Julius M. Huho **[*]** ***and Josephine K.W. Ngaira*** **[**]**
Geography Department, Maseno University, Kenya

ABSTRACT

This chapter examines livestock farming practices of the Maasai pastoralists found in the semi-arid Laikipia North district and the Turkana pastoralists of the arid Turkana district, located in northern Kenya, the vulnerabilities, the adaptation strategies and responses to the changing climate. About 80percent of Kenya's total area is classified as arid and semi-arid land (ASAL) and over 60percent of pastoralism is practiced in the ASAL. Pastoralism accounts for about 10percent of the Gross Domestic Product (GDP) and 50percent of the agricultural GDP in Kenya. Livestock farming also accounts for 90percent of employment and more than 95percent of family incomes and livelihoods in the ASAL of Kenya. Laikipia, North district lies on the northern edge of the Laikipia Plateau at an altitudinal of 1600 to2000m above sea level with two major hills, the Mukogodo and Loldaiga located in the eastern part of the district. The district covers an area of over 1,000km2 with annual rainfall ranging between 400 to 600 mm and temperatures ranging from 25 to 300C. The woody vegetation comprises of the Acacia specie,while the grass layer is predominantly Themeda, Cynodon, Eragrostis and Pennisetums. These are interspersed with a tree/shrub layer of Dodenea, Solanum, and Ipomoea. Turkana District is located to the west of Lake Turkana and covers an area of about 77,000km2. Most of the district consists of low-lying plains with isolated mountains and hills; it has an annual rainfall ranging from 120 to 500mm and temperatures ranging from 240 to 380C. Vegetation in the district consists of scattered Acacia bush and annual herbaceous plants. Cattle are the dominant animals among the Maasai pastoralists, while camels are the most dominant livestock among the Turkana pastoralists. Sheep and goats (shoats) are kept in large numbers in both areas. The pastoralist's

[*]Corresponding author: jhuho2003@yahoo.com.
[**] E-mail: ngaira06@yahoo.co.uk.

vulnerabilities to climate change include; frequent drought which cause lack of pasture, water and drought-related diseases, and the notable drought years include; 1995-96, 1999-2000, 2004-2006 and 2009-2010, and the one-off flood events which cause flood-related diseases and drowning of livestock. The impact has been poor livestock prices, livestock mortality and increased conflicts over scarce resources. The main coping strategies include; transhumance, pastoralists in Laikipia North District drive their livestock to Mt. Kenya, the Aberdare ranges and Mukogodo forest reserves, while the Turkana pastoralists take their livestock to Uganda and Sudan, use herbs to treat livestock during severe droughts. The pastoralists cope with floods by moving to high altitude lands and draining the floodwaters.

1. Introduction

Pastoralism is a form of agriculture that involves keeping of large numbers of livestock mainly for subsistence purposes although social and cultural functions are also important. Pastoralism is practiced in arid zones where rainfall is too low and temperatures too high for cropping to serve as the base for subsistence [1]. Pastoral activities, which cover about 25percent of the world land area [2], support over 120 million pastoralists of which about 50 million live in theSub-Saharan Africa [3]. In North America, Australia, and parts of South America, pastoral production system is extensiveand enclosed, while in Africa, Andes, Asia and Siberia, it is an open access system [2]. The preferred livestock type varies amongst pastoralists. Reindeers are the preferred livestock type in northern Scandinavia and northern Mongolia, cattle in East Africa, sheep and goats (shoats) in the mountainous regions of Southwest Asia, horses in Mongolia andCentral Asia and camels in the more arid lowlands of Southwest Asia and North and East Africa [3, 4].

On the basis of frequency of movement, Jahnke[1] and independent source [5] identified three forms of pastoralism namely nomadic, transhumance and agro-pastoralism. Nomadic pastoralism is characterized by long-range migration, while transhumant and agro-pastoralism are characterized by medium and short-range migrations, respectively. Nomadic pastoralism predominates in the deserts such as the Thar and Deccan Plateau of India and Pakistan as well as in the Himalaya in the north. In southern India, pastoralism is transhumant in nature [6]. The large herds of livestock kept by pastoralists in Africa serve both cultural and economic purposes and therefore, play an integral part of their livelihoods [7]. Pastoralism in Africa ranges from nomadic to agro-pastoralism. According to Jahnke[1]nomadic pastoralism is practiced in the very arid zone where total annual rainfall range from 0 to 200 mm.Camel and goats are the preferred livestock type under nomadic pastoralism and there is the use of oasis products as supplements to the livestock products. In areas where annual rainfall range from 200 to 400 mm, transhumant pastoralism is practiced. This form of pastoralism is based on regular seasonal migration from permanent homesteads, which lacks in nomadic pastoralism. Transhumant pastoralism is characterized by medium to long-range migration, a mixture of grazer and browser and the use of wildlife as supplements to the livestock products. Agro-pastoralism is practiced in areas where rainfall range from 400 to 600 mm and is characterized by short-range migration.Cattle and sheep are the preferred livestock type and there is use of grains as supplement to the livestock products.

In Kenya, pastoralism is practiced in the arid and the semi-arid lands (ASALs) which cover over 80percent of the landmass of the Kenya. Over 60percent of the national herd is held by pastoralists. Pastoralism accounts for a half of the agricultural Gross Domestic Product (GDP) and 10percent of the national GDP. Overall, livestock raised by pastoralists in the ASALs of Kenya are worth US$800 million each year [8].Based on [1] classification of pastoralism in Africa, pastoralism in Kenya can be placed into 2 categories: transhumant and agro-pastoralism. Transhumant pastoralism is commonly practiced in the arid northern Kenya where annual rainfall is less than 500 mm and pastoralists have permanent households. Agro-pastoralism is practiced in the semi-arid areas where annual rainfall is less than 750 mm. The pastoral communities in Kenya include the Turkana (2.56% of the Kenya population), the Maasai (2.18%), the Samburu (0.61%), the Pokot (0.54%), the Kalenjin (12.87%), the Rendile (0.16%), the Gabra (0.23%), Borana (0.42%) and the Orma (0.17%). Although all pastoralists in Kenya mainly keep cattle, shoats, donkeys and camels, variations occur in the preference of livestock type depending on the environmental and cultural factors such as dowry payment and signs of wealth and social status. Among the Turkana and Rendilepastoralists,who live in the very arid areas, camels are the preferred livestock type. The Maasai, the Samburu and the Kalenjin pastoralists who live in the semi-arid to arid areas keep cattle as the preferred livestock type.

2. Study Area

This chapter focuses on Turkana and Laikipia North Districts that lie in the arid northern Kenya. Turkana district lies between longitudes $34^0 00'$ and $36^0 40'$ east and between latitudes $10^0 30'$ and $5^0 30'$ north. The District is located to the west of Lake Turkana and covers an area of about 77,000km^2. Most of the district consists of low-lying plains with isolated mountains and hills. According to GoK[9], the district is inhabited by 855,399 people who mainly belong to the Turkana tribe. Laikipia North District (formerly Mukogodo Division of Laikipia District) lies on the northern edge of the Laikipiaplateau at an altitudinal range of 1600 to 2000 m above sea level between longitudes $36^0 50'$ and $37^0 24'$ east and latitudes $0^0 15'$ and $0^0 33'$ north (Figure 1). It has two major hills; the Mukogodo and Loldaigathat are located in the eastern part of the district. The district covers an area of over 1,000 km^2 with a population of 32,762 people who mainly belong to the Maasai tribe.

2.1. Climate

The climate of Turkana and Laikipia North Districts is hot and dry with erratic and unreliable rainfall. Thesedistricts lie in the low corridors of northern Kenya where climate is strongly influenced by dry northeast monsoon which originates from the dry Arabian continent and the Harmattan air mass which originates from the Sahara desert [10, 11]. Rainfall is the most important climatic parameter in thesedistricts since the Turkana and Maasai pastoralists entirely depend on it for the regeneration of pasture and replenishment of water sources for their livestock. The concentration of the district’s population is thus determined by rainfall, water and browse. Turkana District receives annual rainfall with range between 120 to 500 mm. The district has two rainfall seasons,the "long rains" occurring between March and July and the "short rains" occurring between October and November. Laikipia North District receives annual rainfall with range between 184 to 784 mm with a bimodal distribution.The "long rains" occur between March and May,while the "short rains" occur between October and December. Temperatures range from 24 to 38°C and from 20 to 28°C in Turkana and Laikipia North Districts, respectively. These districts fall in arid and semi-arid area.

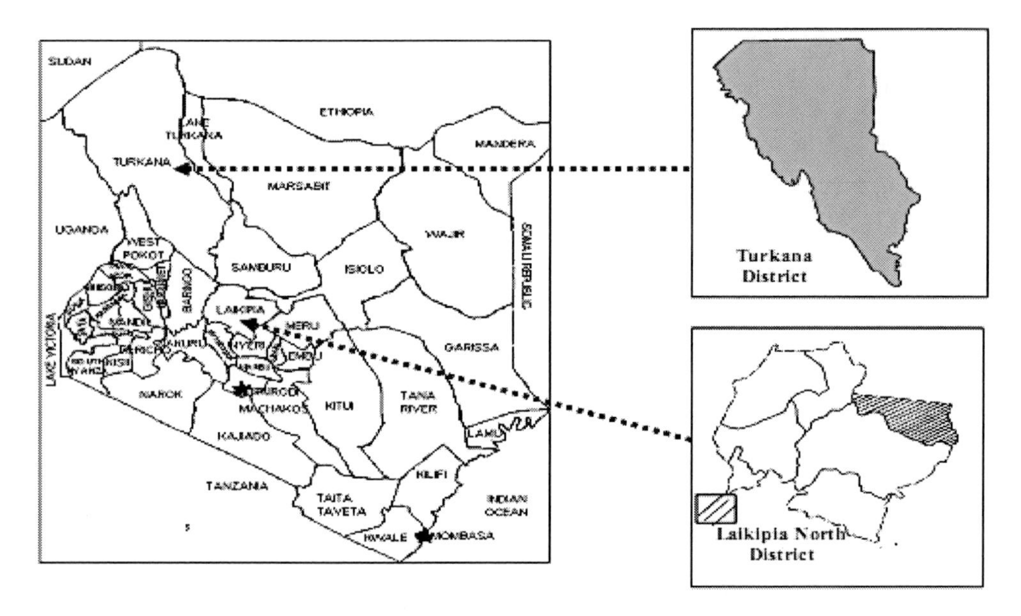

Figure 1. Location and sizes of Turkana and LaikipiaNorth Districts.

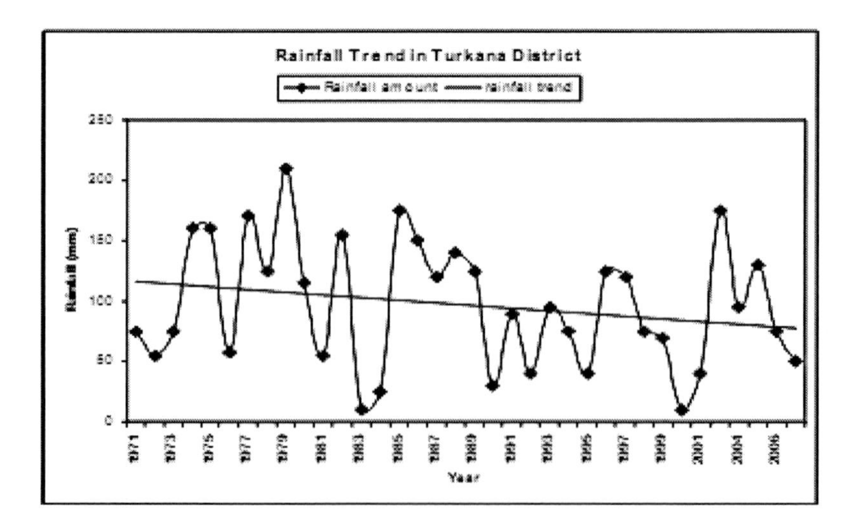

Figure 2. Rainfall trend in Turkana District.

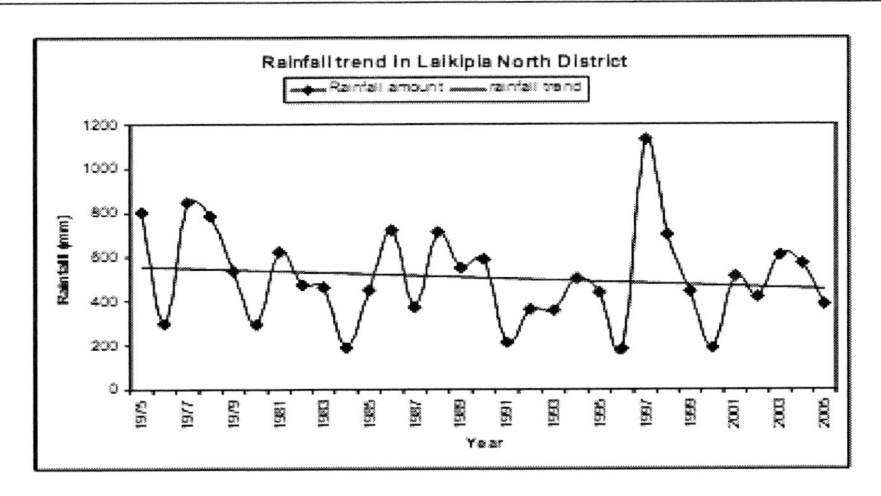

Figure 3. Rainfall trend in Laikipia North District.

Climate change in the study areas has been witnessed through the declining annual rainfall and through the occurrence of extreme climatic events, particularly droughts and floods. In Turkana District,e.g.,the mean annual rainfall decreased from 217 mm between 1950 and1974 to 204 mm between 1975 and 1999 [12](Figure 2). In Laikipia North District, decadal mean annual rainfall decreased from 374.9 mm in the 1980s to 357.7 mm and 332.1 mm in the 1990s and 2000s,respectively (Figure 3). As a result, the probability of drought events persisting in Laikipia North District increased from 25percent in two consecutive years to 26 and 27percent in three and four consecutive years, respectively [13].

Droughtsin the arid northern Kenya have been increasing in frequency from once in every 10 years in the 1970s; once in every 5 years in the 1980s; once in every 2 to 3 years in the 1990's and now a norm since the year 2000 [14].Since 1980s, episodes of moderate and severe droughts have increased in frequency. The most notable severe and prolonged droughts occurred in the 1982 to1984, 1990 to 1994, 1999 to2003 and 2008 to 2009 in Turkana Districts and during 1982 to 1985, 1991 to 1996, 1999 to 2001, 2004 to 2006 and 2008 to2009 in Laikipia North District [13,12]. During the last two decades, the 1992 to 1993, 1995 to 1996, 1999 to 2001, 2004 to2006 and 2008 to 2009 droughts were declared as national disasters by the Kenyan Government [15]. In between these years, droughts of lesser magnitudes were experienced. Floods in the arid northern Kenya mostly occurred immediately after the droughts. In the recent past, flash floods occurred in 1997 to1998, 2002, 2006, 2008 and 2010. The 1997 to1998 and 2002 floods were declared as national disasters.

2.2. Economic Activity

The main economic activity for the inhabitants of Turkana and Laikipia North Districts is pastoralism.Over 90percent of the inhabitants rely heavily on extensive livestock production as their principal livelihood and source of income. Cash income is largely obtained from either livestock sales or livestock products [16]. The most common livestock types kept by the Turkana pastoralists are goats, sheep, cattle, donkeys and camels. With regard to livestock population, goats had the highest population followed by sheep, camel, cattle and donkeys. However, goats and camels were the most important livestock type with regard to survival of the family [8]. The Maasai pastoralists of Laikipia North District keep goats, sheep, cattle, donkeys and camels. Their herds are dominated by goats, followed by sheep, cattle, camels and donkeysin that order. With regard to family survival, sheep and cattle were the most important livestock.

The pastoralists in northern Kenya sell their livestock in organized livestock sale yards/ markets within and outside their districts. The main livestock market in Turkana District is Kakuma livestock sale yard, which is under the Livestock Marketing Association (LMA). However, an informal livestock market is conducted at the Kakuma Refugee Camp [17]. In Laikipia North District, livestock sales takeplace every two weeks in livestock sale yards located at DolDol (on Fridays), Kimanjo (on Wednesdays) and OlDonyiro (on Tuesdays).During dry seasons, livestock sales are not confined in the sale yards because of the extensive livestock movements.

Although the livestock product forms the main source oflivelihood and food securityfor the pastoralists, some pastoral households practice small-scale irrigation or rain-fed crop farming.In Turkana District, irrigation infrastructures such as Katilu, Nakuamoru, Morulem and Lokubae irrigation schemes have been developed where maize, sorghum and millet are grown. Other economic activities in the Turkana Districtinclude charcoal burning, selling firewood, and small-scale business. In Laikipia North Division, the Maasai pastoralists engaged in non-pastoral economic activities such as bee-keeping, charcoal burning, small-scale businesses such as selling of local artifacts such as beads, while some pastoralists had formalemployment in the cultural centers in the district.

2.2.1. Vulnerability of Pastoralists

The vulnerability here refers to people's exposure to risks, shocks and stresses caused by the fragile arid ecosystem, the changing climate registered in extremes of droughts and unpredictability of the rainy seasons and their ill preparedness to adapt to the impacts due to lack of climate information [39]. Vulnerability in the districts had been exacerbated by high poverty levels, severe, frequent and prolonged droughts such as during 1991 to1992, 1999 to2001, 2005 to2006 and 2009 to2011, resulting in water and pasture scarcity, marked environmental degradation and frequent resource based conflicts which correlate with hunger, lack of shelter, illiteracy, joblessness and political insecurity. The harsh arid and semi-arid climate of the districts renders the pastoralists very poor due to low productivity of the rangelands. Inadequate rainfall and frequent droughts dwindlethe pastoral economy through livestock mortality [13] leading to high poverty levels of more than 60percent. The frequent movements in search of pasture and water coupled with high poverty levels in the districts resulted in low literacy level of about 31percent leading to un-employment. Lack of government policies on sharing of the scarce water and pasture resources also exacerbated the pastoralist's vulnerability through poor marketing of agricultural and livestock products, poor infrastructure, especially roads, inadequate health facilities, insecurity and poor water supply systems. In Turkana District, e.g., the actions of abstracting the waters of Rivers Omo and Turkwell, the main tributaries of Lake Turkana for purposes of irrigation and hydroelectric power generation by the Ethiopian and the Kenyan Government, respectively increased the vulnerability of the inhabitants in the districts [18].It is observed that blocking of Rivers Omo and Turkwell resulted in the reduction of the size of Lake Turkana by 10 m and fish production by 40percent over the last 10 years. In addition, the blocking of the rivers also resulted in drying of riverine indigenous vegetation which forms part of the grazing areas in dry season.

2.3. Water Resources

The main sources of water include; traditional hand dug shallow wells, boreholes, rivers, lakes, dams & water pans, springs and rock catchments. In Turkana District, most of the households (41%) relied on traditional hand dug shallow wells for domestic water. About 19percent of the households depended on rivers, 11percent on boreholes, 6percent on Lake Turkana, 5percent on springs, 5percent on rock catchments and 3percent on water pans and dams [19]. In Laikipia North District, most of the households depend on the water from hand dug wells, rivers, boreholes, dams & water pans, rock catchments and springs in that order.

2.4. Vegetation

Vegetation in Turkana District consists of scattered Acacia bush and annual herbaceous plants. Perennial grasses include*Aristida spp.* and *Cenchrusciliaris.* The northern half of Turkana District is open grassland dominated by*Confines,Eragrostisracemosa*, *Echinochloa-haploclada* and *Chlorisvirgata.* In the southern half of the district, areas along the River Turkwelare dominated by *Acaciatortilis*, *Hyphaenecoriacea*and*Salvadorapersica.* Areas away from the river are dominated by *A. reficiens*, *Digitariavelutina*, *Aristidamutabilis, Dicomatomentosa, Pupalialappacea, Zaleyapentandra* and *Eragrostiscilianensis.*In Laikipia North District, the woody vegetation comprises of the *Acacia spp.*,while the grass layer is predominantly *Themeda, Cynodon, Eragrostis* and *Pennisetums*. These are interspersed with a tree/shrub layer of *Dodenea, Solanum*, and *Ipomoea.*

3. Effects of Extreme Climatic Events on Pastoralism

3.1. Droughts

The frequency of severe droughts in Kenya has increased from once in every 10 year to once in every three tofour years [20]. In the arid northern Kenya, droughts are now a norm rather than an exception occurring every year but with varied severities [14]. Frequent droughts have resulted in massive loss of rangeland resources incapacitating livelihood, particularly those of the pastoralists [20]. Droughts affect the level of primary production of plants which support livestock and human populations. Changes in micro-ecology due to frequent droughts coupled with overgrazing have favored abundant regeneration of non-palatable plant species in Turkana and Laikipia North Districts. It is observedthat the perennial grasses in Turkana District are replaced by non-palatable species like *Tribulusterrestris* and *Portulacaoleraceae*and in areas near Kakuma camp, a high density and cover of the non-palatable shrubs such as *Cissus qua-drangularis*, *Euphorbiacuneata*, and *Cadabarotundifolia*dominate[21].

Plate 1. Non-palatable *Sansavellia specie* near DolDol town in Laikipia North District.

In Laikipia North District, reduction in primary biomass production of the rangelands pasture was caused by the replacement of palatable forage by non-palatable plant species such as *Sansavellia specie* (Plate 1) and *Opuntiamegacantha*. Some sections of the district were devoid of pasture due to the invasive species [4].It is observed that the woody vegetation got depleted through charcoal burning, while *Harpachneschimperi* and *Microchloakunthii*is replaced by palatable species such as *Themedatriandra*inLaikipia North District [22]. The overall effect is the decline in rangeland carrying capacity.

Drought is the main threat to pastoral economy in Kenya due to massive livestock losses. It is observed that the 1979 to 1980 drought in northern Kenya led to 90percent loss of cattle, 80percent shoats and40percent camels, while the 1991 drought led to 28percent loss of cattle and 18percent loss of shoats [7]. Livestock mortality rates in the Turkana and Laikipia North Districts rose during drought periods depending on the severity of the drought and the mobility of people and livestock. Table 1 shows the number of livestock lost during droughts of varying severities in Laikipia North District.

During the 1992 to 1993 and 1999 to 2000 droughts in Laikipia North District, a decline in growth rate of cattle to 12 and 17.9percent, respectively has been observed [26]. Livestock mortality was exacerbated by increased livestock diseases such as Contagious Caprine Pleuropneumonia (CCPP), mange, foot and mouth disease and tick borne diseases [38]. Livestock movement and congestion at watering points enhanced the spread of these diseases. Tick borne diseases were also associated with livestock coming in contact with wild animals. Similarly, in the Greater Horn of African countries, droughts have been the major hindrance to agricultural activities due to frequent crop failure and livestock losses. In Ethiopia, for example, the 1995 to1997 drought led to a decrease of 45and 78percent in camel and cattle herd sizes in the Somali and Borena areas, while during the 2000 drought, cattle herds decreased by 15 to 45percent in Afar, 30 to 80percent in Borena Zone, and 40 to80percent in Somali Region [27].It has been reported that livestock losses due to mortality, poor quarantines and diseases and missed trade opportunities in the ASALs of Kenya leads to losses amounting to about US$ 2 billion annually[23]. Livestock losses during the 1999 to

2002 drought was valued at US$0.1 billion [24] while in north-eastern parts of Kenya livestock worth more than US$0.9 billion were lost during the 2005 to 2006 drought [25].

Table 1. Number of livestock lost during droughts in Laikipia North District

Drought years	Drought severity index	Drought category	Number of livestock lost			
			Cattle	Sheep	Goats	Total
1983	-0.27	Mild	199	907	670	1,776
1984	-2.10	Extreme	4,050	4,100	4,600	12,750
1985	-0.39	Mild	1,489	1,660	1,397	4,546
1987	-1.00	Moderate	3,000	3,270	2,875	9,145
1991	-1.93	Severe	3,300	3,350	3,100	9,750
1992	-0.97	Mild	500	540	460	1,500
1993	-0.98	Mild	170	185	150	505
1994	-0.03	Mild	230	300	210	740
1995	-0.45	Mild	500	600	480	1,580
1996	-2.13	Extreme	4,100	4,325	4,000	1,425
1999	-0.43	Mild	540	770	460	1,770
2000	-2.11	Extreme	4,760	6,200	6,000	16,960
2002	-0.58	Mild	150	190	125	465
2005	-0.78	Mild	180	210	170	560
Total			23168	26607	24697	74,472

Reduction of livestock prices was a common feature during droughts because of poor livestock health and the increased number of livestock offered for sale. The average price for shoats during the year 2002 drought in Turkana District reduced from US$ 12 in September to US$ 9 in October [28].In Laikipia North District, the cattle prices were 36.6, 52.4 and 40.5percent below the mean price between 1975 and 2005 during the 1991, 1996 and 2000 droughts, respectively. During the same period, shoat prices reduced by 54.5, 63.6 and 49.5percent during the 1991, 1996 and 2000 droughts in that order [15,39].The sale of livestock to avoid more losses during droughts resulted in very low prices, while on the other hand the price of grains rose sharply; reducing the purchasing power of the pastoralists forcing themto gather wild fruits and depend on food handouts from the government. It is reported that during dry season in Nakururm area in northern parts of Turkana District, wild fruits formed 42percent of the diet in contrast to 25percent during normal period [17]. Chronic food insecurity caused by droughts forced thepastoralists to depend entirely on relief food for the Government. For example, livestock deaths and food shortages during the 1979 to 1980 drought caused a severe famine in Turkana District that led to the creation of relief camps [29]. During the 2002 drought, some 69,000 people required urgent relief food, while 13,000 children under five years required supplementary feeding, a situation that called for urgent humanitarian intervention [30]. In Laikipia North District, the distribution of relief food during severe and extreme droughts occurred after every two weeks. During the 1999 to2000 drought, about 75percent of the Maasai pastoralists in Laikipia North District survived on relief food.

Scarcity of rangeland resources in the study area persistently threatened the security of the pastoralists. The scramble for resources triggered conflict among the neighboring pastoral communities. In addition, the post drought recovery of herds through raids heightened the conflicts. In February 2008 for instance, the Pokot pastoralists raided Katilu area of Turkana District and escaped with 98 cows and 108 shoats and also set the pastures on fire. The burning of grass by the Pokot raiders after every attack was meant to destroy pastures to make the opposing group helpless. Two people were killed during the raids [31]. During the 2009drought, the Samburupastoralists from Samburu District raided about 600 heads of cows from Baragoi village in Turkana District and in August 2010, they raided Laikipia North District in an attempt to recover the herd decimated by the droughts. Another cause of insecurity in the study area emanated from human/ wildlife conflict. During droughts, hyenas and leopards attackedand killedlivestock,whileelephants took over most of the watering points hindering livestock and human access forcing residents to search for alternative sources of water. Elephants also destroyed crops.

Drying up of surface water sources was a direct effect of droughts in the arid northern Kenya. Pastoralist travelled greater distances to secure water during droughts. In West Pokot District, water pans and dams dried up during the 2010 drought increasing the average distance travelled by households in search of water from 2.4 km in October to 4.2 km in January 2011. The longest travelled distances recorded were at Nyangaita and Pserumand that was 8.5 km each [32]. During the 2003 drought in Turkana District, most surface water sources dried up forcing pastoralist communities to migrate. Similarly, during the 2008 drought, surface water sources dried up in most parts of Turkana District increasing the average travelled distance from an average of 3.2 km in May to 3.8 km in June. Ground water sources and Rivers Turkwel and Kerio yielded little amounts of water [33]. In Laikipia North District, the 1984 extreme drought led to complete drying up of Ngaboli and Ildupata springs in Murupusi group Ranch. The waters of Ilpuduk spring reduced to very low levels during the 1999 to2000 prolonged extreme [34]. Droughts of mild and moderate severities led to drying up of rock water catchments, streams and water pans.

3.2.Floods

Extremely high rainfall which follows immediately after the droughts cause flash floods in the ASALs of Kenya leading to loss of human and livestock lives. Flash floods in the arid northern Kenya affected pastoralism throughlivestock drowning, increase in livestock diseases and poor livestock markets. The 1997 and 2006 floods for example, caused the Rift Valley Fever (RVF) disease which affected livestock. It is reported that in Lagdera District, about 50 households were separated from their livestock by a distance of about 10 km by 2006 flash floods, while in Garissa District, the flash floods led to loss of livestock through drowning in Balambala and Saka areas[35]. In Turkana District, the 2008 flash floods caused livestock diseases such as pestes des petitis Ruminants (PPR), which affected goats and RVF which affected cattle [17]. Death toll due to flash floods rose to 4,362 for shoats, 235 for cattle, 60 for donkeys and 193 for camels during the March 2010 flash floods in Turkana District. In May 2010, River EwasoNyiro and its tributaries flooded causing livestock and human deaths in Laikipia North District. Flash floods also covered grazing and browsing areas and pollutedthe water sources [36].

Road transport was cut to both Turkana and Laikipia District by flash floods in 2008 blocking livestock traders fromaccessing the districts. The overall effect was poor livestock prices. Flash floods destroyed other sector of economy in the districts particularly crop farming. The December 2009 and January 2010 flash floods destroyed irrigation infrastructure in Katilu and Nakuamoru irrigation schemes and inundated 800 acres of sorghum and maize in Turkana District. The bursting of Kerio River at Lokori in Turkana East in March 2010 destroyed irrigation infrastructure in Morulem and Lokubae irrigation schemes and in April 2010, 42 acres of land were destroyed in Jonuk Location in southern Turkana District [36].

Flash floods refilled water pans and dams in the districts. In Laikipia North District for example, the May 2010 flash floods reduced the average travel distance to water sources from 4 km to 2 km. However, flash floods caused the problems of water pollution and increased siltation of the water pans and dams reducing their holding capacity. During droughts, silted water pans and dams dried more quickly than un-silted water sources.

3.3. Other Challenges Faced by Pastoralists in Arid Northern Kenya

The livelihoods of the pastoralists arechallenged by insecurity and conflicts. Migration outside the traditional grazing lands lead to big challenges such as establishment of new administrative boundaries, changing land use and settlement, Human-Wildlife conflicts, violent conflicts with the indigenous pastoral communities such as the Turkana of Kenya and Toposa of Somalia, Karamojong of Uganda [42].

Food aid dependency in pastoral communities dominates the aid agenda of the Government leading to sustained dependency year by year. Lack of Government working policy to invest in pastoral production system has left pastoralists at the mercy of climate change vagaries and at the margins of development. The current crisis management where the Government only acts during droughts or floods by providing food and drugs is unsustainable and inadequate [39].

Although the pastoralists are well aware of the changing climate as registered in extremities of droughts and floods,they lack climate information to help them adapt and mitigate against the impacts; the severe and prolonged droughts adversely affect their livelihoods make them keep vary their coping strategies from year to year.

4. Coping and Adaptive Strategies

The relationship between droughts and the organization of pastoral production system was strongly positive. In the pastoral system therefore, drought is an element within the production system rather than an external event [27]. The coping and adaptive strategies adopted by the pastoralists are determined by the severity of the drought. It is noted that under traditional pastoral systems the coping and mitigation strategies are concerned with risk reduction than maximizing the individual gains [27]. The most common strategies adopted by pastoralists in Turkana and Laikipia North Districts include:

4.1. Out Migration

Negative changes in the availability of pasture and water resources directly affect livelihood security of the pastoralists through declining livestock body conditions, livestock reproduction and milk production [27]. It has been observed that the arid conditions of the pastoral environments, the fragile ecosystem and the vulnerability of pastoralists to drought make the mobility of people and livestock essential [7,12]. In Turkana District, pastoralists moved to the neighboring countries of Uganda and Sudan. During the 2008 drought for example, the Turkana pastoralists moved to areas along the Kenya-Sudan, Kenya-Ethiopia and Kenya-Uganda boundaries [17]. In Mukogodo Division, pastoralists moved to Mukogodo, Mt. Kenya and Aberdare forest reserves. During the 1999 to2000 drought for example, the Maasaipastoralists moved their livestock to Mt. Kenya and Aberdare forest reserves [4].

4.2. Diversification of Livestock Type

Diversification of animals is one of the most common drought adaptive strategies employed by pastoralists in the arid northern Kenya. Pastoralists keep several species of livestock which include a mix of grazers and browsers [38]. Different species of livestock have varied adaptability to water stresses, grazing scarcity and diseases. In addition, a mix of grazers and browsers make maximum use of rangeland resources than a single species alone. In Turkana and Laikipia North Districts, the grazers kept were cattle, sheep and donkeys, while the browsers were camel and goats. However, due to resilience of shoats to the vagaries of climate, the number of shoats had been on the increase as the climate getdrier. The number of camels in Laikipia North District also increased from 25 in 1983 to 3500 in 1998 [4]. Today, majority of the pastoral households keep indigenous chicken for egg production and meat. Though men dislike chicken meat, womenand children benefited from it, especially during dry season when livestock move to dry season grazing areas. Eggs formed basic source of proteins for the children in the absence of camel and cattle milk/meatduring droughts.

4.3. Increase of Livestock during Inter Drought Periods

Efforts to improve pastoral livelihood in arid northern Kenya were geared towards herd maximization since large herds guaranteed subsistence and provided insurance against the impact of droughts. All members of the pastoral societies, regardless of the social status, continued to invest in livestock during favorable weather conditions [7, 37]. It has been observed that pastoralists maximize their herds during wet seasons through restricted sales, natural increase and purchasing of livestock [38]. Similar practices are replicated in Turkana and Laikipia North Districts.

4.4. Diversification of Livelihood Sources

Diversification of livelihood sources was triggered by the dwindling of the pastoral economy due to frequent droughts and occasional floods.Pastoralists lost part or all of their herds, while young adults were unable to acquire and accumulate large herds of livestock [16]. As a result, pastoralists engaged in small-scale businesses, small-scale crop farming, charcoal burning, selling of cultural artifacts such as beads, small-scale basketry such as weaving of mats and baskets, trading livestock and livestock productsand waged employment. These activities were based on the region's natural resource base and majority of these activities were meant to generate cash income [39]. In Turkana District, women pastoralists engaged in collection of wild fruits from doum palm (*Hyphaenecompressa*) for home use or sale. The fruits are sold to traders in Kalokol, who extract oil from the seeds for frying fish. Women also gathered and sold building materials, wove baskets and mats for sale, collected and sold gum Arabica, Aloe Vera and honey [40,39,16]. In Laikipia North District, some Maasaipastoralists engaged in selling firewood and charcoal to hoteliers in Nanyuki town, entertaining tourists in eco-lodges such as Ilngwesi, sand harvesting, bee keeping and selling of gum Arabica. Tassia Lodge, which was managed by members of Lekurruki group ranch in the district, had 90percent of its staff being the members of the Maasaipastoralist community. In 2003, pastoralists from Musul group initiated a bee-keeping project, Musul Bee-keeping and Environmental Conservation group (MBEKEC), where about 50 members were engaged in bee keeping.

4.5. Moving to Higher Land during Flood

To cope with floods, pastoralists move to high altitude lands together with their livestock, drain water from their dwellings and build gabions along the riverbanks to prevent future flooding.

4.6. Homemade Arms

To safeguard their livestock attacks and stealing by other pastoral communities such as the Pokots from Baringo, the Karamojong from Uganda and the Toposa from Somalia, Turkana pastoralists have acquired homemade arms leading to proliferation of illegal arms in Kenya. Cattle rustling once seen as a "cultural" practice used for restocking has become more violent and frequent.

4.7. Alteration of Traditional Mobility

Due to the severe droughts experienced in the 21st Century, Turkana pastoralists have altered their traditional mobility patterns such that they now move further and spend long time in "foreign" territories and countries outside the traditional grazing areas. In 2009, Turkana pastoralists went as far as Juba in Somalia [42].

5. Conclusion

Evidence indicates that there has been weather that is more erratic in the 21st Century, and the most vulnerable areas are the arid and semi-arid lands where pastoral communities live. The livelihood of these communities is on the verge of collapse unless the Government of Kenya urgently develops and implements policies to support and invest in pastoral economy. The current crisis risk management of providing food to pastoralists in the wake of adverse weather makes these communities more vulnerable to risks of climate change.

References

[1] Jahnke HE. *Livestock production systems and livestock development in Tropical Africa.* KielerWissenschaftsverlagVauk, Kiel. 1982.

[2] Blench R. *Pastoralism in the new millennium. Available online at:* http://www.odi.org.uk/work/projects/pdn/eps.pdf. 2001.

[3] Rass N. Policies and strategies to address the vulnerability of pastoralists in sub-Saharan Africa. *PPLPI Working Paper Number* 37, 2006.

[4] Huho JM, Ngaira JKW, Ogindo HO. Living with drought: the case of the Maasai pastoralists of northern Kenya. *International Research Journals,*2011; 2:779-789.

[5] http://courses.washington.edu/anth457/pastoral.htm. Pastoralism. 2011.

[6] Sharma VP, Kohler-Rollefson I and Morton J. Pastoralism in India: a scoping study. Available online at: http://www.dfid.gov.uk/R4D/PDF/outputs/ZC0181b.pdf. No date.

[7] Barton D, Meadows N and Morton J. Drought losses, pastoral saving and banking: a *Review. DFID Advisory and Support Services Commission Project* ZW0027, NRI Project L0114. Natural Resources Institute. June 2001.

[8] Bett B, Jost C and Mariner J. Participatory investigation of important animal health problems amongst the Turkana pastoralists: Relative incidence, impact on livelihoods and suggested interventions. *Discussion Paper No. 15.* International Livestock Research Institute. Addis Ababa. 2008.

[9] GoK. Kenya population and housing census 2009. Government printers, Nairobi. 2010.

[10] Okwany VO. Climate and vegetation relationships in southwestern Kenya and the potential impacts of a warmer climate. *A PhD thesis,* University of East Anglia. 1991.

[11] Ojany RB and Ogendo FB. Kenya: *A Study in physical and human Geography.* Longman group, Nairobi. 1973.

[12] Obando J,Ogindo H, Otieno W, Macharia A, Ensord J, Tumuising W, Kisiangani E,Muchiri L. and Amollo N. Reducing vulnerability of pastoralists communities to climate change and variability in Northern Kenya. *Paper presented in International Conference on Climate, sustainability and development in Semi Arid Regions.* 2010.

[13] Huho JM, Ngaira JKW, Ogindo HO. Climate change and pastoral economy in Kenya: a blinking future. *ActaGeologicaSinica (English Edition),*2009; 83:1017-1023.

[14] Howden D. The great drought in East Africa; No rainfall for three years. Available at: http://www.infiniteunknown.net/2009/10/03/the-great-drought-in-east-africa-no-rainfall-for-three-years/ gov/fews/africa/index.php. 2009.

[15] Huho JM, Mugalavai EM. The Effects of Droughts on Food Security in Kenya. The International Journal of Climate Change: *Impacts Resp.*,2010; 2:61-72.

[16] Watson DJ, and Binsbergen J. Livestock market access and opportunities in Turkana, Kenya. International Livestock Research Institute, *Report No. 3*. Addis Ababa. 2008.

[17] ASARECA. Livestock situation analysis zone 1 (Turkana District*). Pastoralist bulletin: Community based livestock early warning systems - CB-LEWS.* July,2007.

[18] Ngaira JKW. Challenges of water resource management and food production in a changing climate in Kenya. *Journal of Geography and Regional Planning,*2009; 2: 097-103.

[19] GoK. Drought monitoring bulletin for May 2010 for Turkana Districts. *Arid lands resource management project II.* 2010.

[20] Orindi VA, Nyong A, and Herrero M. Pastoral livelihood adaptation to drought and institutional interventions in Kenya. Human Development Report 2007/2008. *Fighting climate change: Human solidarity in a divided world.* 2008.

[21] Okoti M,Ng'ethe JC,Ekaya WN and Mbuvi DM. Land use, ecology, and socio-economic changes in a pastoral production system. *J. Hum. Ecol.*, 2004;16: 83-89.

[22] Olang MO. Classification of Kenya rangeland. In Dzowela B. H. (ed) African forage plant genetic resources, evaluation of forage germplasm and extensive livestock production systems. *Proceedings of the third workshop held at the international conference centreArusha,* Tanzania, 27-30 April, 1987.

[23] USAID. Drylands livestock development program. Available at: http://kenya.usaid.gov/ programs/economic-growth/412. 2010.

[24] Akillu Y and Catley A. Livestock exports from pastoralist areas: An analysis of benefits by wealth group and policy implications. *IGAD LPI Working Paper No.* 01 - 10. 2010.

[25] IRIN. Kenya: *Act now to mitigate drought effects, say agencies.* Available online at: http://www.irinnews.org/PrintReport.aspx?ReportID=91666. 2011.

[26] Kairu E. Poverty, target groups and governance environment in Laikipia District. *A report for SARDEP.* 2002.

[27] Ahmed AGM,Azeze A, Babiker M and Tsegaye D. Post-drought recovery strategies among the pastoral households in the Horn of Africa: A *review. Development research report series no. 3.* No date.

[28] Reliefweb. *ACT appeal Kenya: West Pokot/ Turkana famine – AFKE 31.* Jan 15, 2003.

[29] Critchley W. *Looking after our land: Soil and water conservation in dryland Africa.* Oxfam publication, Oxford. 1991.

[30] IRIN. Kenya: *Food crisis looms in the west and north.* Available online at: http://irinnews.org/PrintReport.aspx?ReportID=40079. 2011.

[31] Ekar S. *Conflict early warning alert Turkana District. RiamRiam Turkana peace network,* Lodwar. 2008.

[32] GoK. Drought monitoring bulletin for West Pokot County. Ministry of State for Development of Northern Kenya and other Arid Lands: *Arid lands resource management project II*, January, 2011.

[33] GoK. Drought monitoring bulletin for May 2010 for Turkana Districts. *Arid lands resource management project II.* 2010.

[34] Gitau MK. Participatory rural appraisal (PRA) report for Marupusi resource management area (RMA), Mukogodo GCA: *Natural resources trends of Maru*pusi. Unpublished report. 2002.

[35] OCHA. Kenya: Floods and landslides. *Situation report No 4*, Nov, 2008.

[36] KRCS. Kenya: *Floods. Preliminary emergency appeal no MDRKE012*. 14 May 2010.

[37] Kerven C. Customary commerce: *A historical reassessment of pastoral livestock marketing in Africa*. ODI, London. 1992.

[38] Ngaira JK. *Environmental impact of rainfall variability in the semi arid areas: A case study of Baringo District, Kenya*. Unpublished PhD thesis, Moi University, Kenya. 1999.

[39] Ngaira JK. The impact of climatic variations on agriculture in semi-arid regions of Kenya, 1960-2000.In Power Play and Policy in Kenya: *An interdisciplinary Discourse*. 2006.

[40] Watson DJ and Binsbergen J. Life beyond pastoralism: Livelihood diversification opportunities for pastoralists in Turkana District, Kenya. *ILRI briefs, Nairobi*. 2006.

[41] Eriksen S and Lind JA.The impacts of conflict on household vulnerability to climate stress: Evidence from Turkana and Kitui Districts in Kenya. Paper presented in human security and climate change: *An International Workshop in Asker*. 21–23 June 2005.

[42] UNEP. Advocating for safe movement as a climate change adaptation strategy for pastoralists in the Horn and East Africa. Available online at: http://reliefweb.int/sites/reliefweb.int/files/resources/8E4A32AF1BCEE254492577580006583l-Full_Report.pdf. 2010.

In: Livestock: Rearing, Farming Practices and Diseases
Editor: M. Tariq Javed
ISBN 978-1-62100-181-2

Chapter 2

NEW GENERATION OF DIETARY SUPPLEMENTS WITH MICROELEMENTS FOR LIVESTOCK - POSSIBILITIES AND PROSPECTS

Maciej Janeczek,[1*] Katarzyna Chojnacka,[2] Aleksander Chrószcz,[1] and Agnieszka Zielińska[2]

[1]Department of Biostructure and Animal Physiology, Faculty of Veterinary Medicine, Wrocław Environmental and Life Sciences, Wroclaw, Poland
[2]Institute of Inorganic Technology and Mineral Fertilizers, Wroclaw University of Technology, Wroclaw, Poland

ABSTRACT

One of the most important problems in the animal production is trace-elements supplementation. A particular attention should be paid to animals in high-production systems. Micro- and macroelements demand is significantly higher in these animals than in extensive animal production. Resulting, deficiency of microelements has a dramatic consequence on health, production and at last economic balance. Commonly used preventive method is microelements supplementation.

Traditionally, trace-elements are introduced as inorganic salts or organic mineral products (e.g., amino acid chelates, proteinates and polysaccharides). However, several problems were encountered. Major disadvantage of inorganic salts supplementation is low microelements bioavailability and its quick alimentary tract transit, while chelates can be the cause of digestive tract irritation. Microelements supplemented in the form of inorganic salts possess low bioavailability and transit character, while chelates can cause alimentary tract irritation. Therefore, a new generation of dietary supplements produced by biosorption can be useful in solving this problem.

Contemporary results proved high bioavailability of Cr (III), Cu (II), Mn(II) and Zn (II) ions from preparations in which microelements are bound with a biological carrier. Preliminary studies showed significant increase of copper concentration in eggshell,

* Department of Biostructure and Animal Physiology, Faculty of Veterinary Medicine, Wrocław Environmental and Life Sciences, Kożuchowska 1/3, 51-631 Wroclaw, Poland, e-mail: janeczekm@poczta.onet.pl/ phone +48713205746/ fax +48713205744.

blood and feather of hens. The investigations carried out on pigs revealed the positive influence on liver and higher copper level in meat. Positive influence of these dietary supplements on animal health was preliminarily proved, but the precise metabolic effects should be confirmed in the future. Very important is to increase bioavailability of micronutrients, reduce the problem of microelements segregation, through the production of less concentrated form of microelements and finally, improvement of health and performance of animals. These properties can also be related with digestive microflora modulation. Moreover, prohibition of nutritive antibiotics as one of the results of increased antimicrobial resistance in human and veterinary medicine caused subclinical and clinical infections.

Modern science states that the supplementation can play a crucial role in promotion of advantageous microflora. Native intestinal microfloras like Lactobacillus spp. are important inhabitants of the mammalian gastrointestinal system and it is emphasized that increasing their numbers improves gastrointestinal health. It seems that the new dietary supplementation method would have a beneficial influence on animal products, animal health, and environment, and can be used as prophylactic, metaphylactic and supportive treatment of the digestive system.

1. Introduction

Microelements are essential constituents of animal's diet. Deficiency of either of these nutrients, according to Liebig's law of the minimum causes the phenomenon of microelement hunger, also termed 'hidden hunger'. In contrast to conventional hunger (meaning the lack of coverage of required quantity of energy), in the case of hidden hunger, the requirement for nutrients is not covered. The problem of microelement deficiencies is difficult to diagnose, because clinical symptoms are not obvious and frequently postponed and are frequently manifested by hindered growth. Therefore, microelements as essential nutrients play a significant role, in particularly in intensive breeding systems.

To cover animal's requirements for microelements, diets are supplemented with feed supplements, usually in the form of inorganic salts. However, several problems have been encountered, including non-uniform distribution in feed due to segregation during transportation, low availability of microelements from inorganic salts. Consequently, new preparations are being elaborated. In modern supplements, microelement ions are bound (chelated, complexed) with an organic molecule or with the biomass in biosorption process. This assures supplementation of the form, which resembles mostly the microelements, which occur in natural feeding materials.

In the present chapter, various aspects related with microelements, their role in animals and problems with their imbalance (mainly deficiencies) is discussed. In addition, information on various preparations that are used to supplement animal's diet with microelements is provided.

2. The Biological Function of Trace-Elements

Trace-elements have a crucial role in animal's heath, especially in ensuring efficient growth, reproduction and immunological response [1]. The role of Zn, Cu, Se, Fe and Mn in animal feed is significant. These microelements are necessary for a number of physiological processes, including synthesis of various proteins and activation of enzymatic systems. The lack of microelements can have fatal consequences.

2.1. Zinc

It is a structural component of more than 300 enzymes. The mechanism of its influence is based on regulation of enzymatic activity, gene expression of proteins and on a transduction of mitogenic hormone signal, transcription of gene and RNA. Zinc plays important role in hormones like testosterone, osteocalcin, insulin, growth hormone and thyroid hormones, in their production, secretion and function. Zn plays importantrole at many places including keratin synthesis, maintenance of epithelium, bone matrix, nucleic acid synthesis, protein synthesis, enzyme function, metabolism of lipids and proteins, spermatogenesis, immune system function and central nervous system function [2, 3]. Zn plays another important role as anti-oxidant being integral part of the superoxide dismutase (SOD). It is proved, that Zn has an influence on leukocytes like T- and B-cells function [4]. These cells are the important player of immune system. Immunostimulation effect of Zn is a result of its wide influence on various components of immune system like lymphocyte function, natural killer cells function, neutrophil function, lymphokine production and antibody-dependent cell-mediated cytotoxicity [1, 5, 6]. Thus, Zn is important for animal health. This microelement is a significant factor in reproduction, too. The role of zinc in fetal development and growth was proved. It has an influence on the fetal expression of Cu/Zn SOD, glutathione peroxidase activity and cell membrane protection from oxidative damage by these enzymes [7]. Active regulation of insulin-like growth factors is also known role of Zn during pregnancy [8, 9].

2.2. Copper (Cu)

Similar to zinc, copper is integral part of many metolloenzymeslike cytochome oxidase, lysyl oxidase, ceruloplasmin, tyrosinase and superoxide dismutase (SOD) [10]. The main functions of this trace-element are anti-oxidant; participate in cellular respiration, cardiac muscle metabolism, immune system function, connective tissue maintenance, keratinization process, neuronal development, etc. Copper has also influence on other microelements metabolism, like iron. Copper is an activator of enzymes associated with iron metabolism and functioning. Ceruloplasmin is required for iron oxidation process, from ferrous to ferric state transition. In this form (ferric), iron can be utilized for erythrocyte formation. Being a part of SOD, copper is an important element in anti-oxidative defense system [11].

2.3. Iron

The other essential trace-element is *iron (Fe)*. Iron performs several vital functions and participates in many biochemical reactions. Most of the iron is present in erythrocytes as hemoglobin and as myoglobin in syncytium and cells of muscular tissue. The hem moiety binds oxygen. Iron containing enzymes are cytochromes, which contain hem, too. These enzymes participate in the oxidative metabolism that supports energy transfer within the cell, especially in the mitochondria. The iron-containing enzymes are responsible for synthesis of steroid hormones and bile acids too. Iron is also part of detoxification processes of various substances in the liver and takes part in controlling the signal in some neurotransmitters, like the dopamine and serotonin systems in the brain. Iron has a crucial role in erythropoiesis especially for hemoglobin synthesis by the maturing erythroblasts [12]. One of the most important iron-binding proteins, which does not contain hem is lactoferrin (Lf). Because of multiple functions, the lactoferrin is a very interesting iron-binding protein. This glycoprotein is found in colostrum, milk and external secretions of the body like saliva, bile, tears or sperm. Lf is secreted from secondary neutrophil granules in response to inflammatory stimulation, too. Lf is a part of the inborn immune system, but also participates in specific immune reactions [13]. Because of its strategical location on the mucosal surfaces, lactoferrin is an element of the first line of defense against microbial agents attacking the organism through mucosa. Lf produced by the mammary gland presumably plays a double role, protecting the mammary gland and the intestine from infections in perinatal period. Lf has broad-spectrum antimicrobial activity against Gram-positive bacteria, Gram-negative bacteria, viruses and fungi [14, 15]. Its activity against *Toxoplasma gondii, Eimeriastiedai, Pneumocystis carinii*washas been proved [16, 17]. On the other hand, lactoferrin may be an efficient anabolic factor influencing osteocytes. Lactoferrin stimulates proliferation of osteoblast, enhances incorporation of thymidine into osteocytes, and reduces osteoblasts apoptosis. A similar effect was also observed in cartilaginous tissue [18].

2.4. Chromium

The next essential trace-element important for animal physiology is chromium (Cr). This microelement takes part in metabolic functions of carbohydrates, lipids, proteins and nucleic acids. It is well known that chromium is a structural element of the glucose tolerance factor (GTF) and is synergistic with insulin promoting cellular glucose uptake [14]. The investigations confirmed the relationship between chromium and the metabolism during various stress occurrences. The demand of chromium increases during exposition to higher stress factors [19]. During the stressor action, there is an increase in cortisol secretion and it acts as an insulin antagonist through increasing concentration of blood glucose and glucose utilization reduction by peripheral tissues. Increased levels of blood glucose stimulate process of chromium mobilization. In this situation, Cr is irreversibly excreted in urine [20, 21]. It is also possible, that chromium contributes in maintaining the configuration of RNA molecule, since this trace-element has been demonstrated to be especially effective as a cross-linking factor for collagen [22].

2.5. Selenium (Se)

It is particularly required for numerous biochemical functions in living organisms. Selenium is important component of several major metabolic pathways, e.g., metabolism of thyroid hormone, antioxidant defense systems or immunological function. This trace-element is an integral part of many selenoproteins including enzymes like glutathione peroxidase, thioredoxinreductase and iodothyroninedeiodinase. It is necessary, to have a complete understanding about the role of selenium in an organism. Here we describe some of the most important selenoproteins. Glutathione peroxidase 1 (GPx1) is the first discovered selenoprotein. One of the biological functions is its participation in antioxidation defense system. Gastrointestinal glutathione peroxidase 2 (GPx2) protects organism against the toxic effect of ingested lipid hydroperoxides [23]. The next selenoprotein, extracellular seleno protein glutathione peroxidase mitochondrial RNA (GSHPx mRNA), appears in the renal proximal tubular epithelial cells, and because GSH concentrations are high in kidney, glutathione peroxidase 3 (GPx3) presumably plays a unique antioxidant function in renal tissue [24]. GPx3 was found in other cells, too. Cellular membranes associated phospholipid hydroperoxide glutathione peroxidase (GPx4) plays an important role in biological membranes defense from oxidative stress. This enzyme is accountable for the lipid hydroperoxide reductive destruction [25]. Thioredoxinreductase contains enzyme catalyzing the NADPH dependent reduction of thioredoxin and play regulatory role in its metabolic activity [26]. Selenoprotein P primarily serves to transport selenium inside the organism [27]. About 40% of the blood plasma selenium is incorporated in to selenoprotein P [28]. It is possible, that selenoprotein P plays a role in antioxidant processes, too. It was demonstrated that GPx4 plays a role of oxidative stress sensor and cell death signals transducer [29]. The iodothyroninedeiodinase enzymes are another important group of selenoproteins. The role of this is to catalyze the 505- mono-deiodination of the prohormone thyroxin (T_4) to the active thyroid hormone 3, 305-triiodothyronine (T_3) and the inactive reverse T_3 conversion to three to thirty diiodothyronine. It is commonly known that the thyroid hormones exhibit a crucial role in numerous physiological catabolic processes. It can be stated that the selenoproteins have a direct influence on T_3-T_4 balance. It was proved, that selenoprotein W is necessary for correct muscle metabolism especially during cell differentiation [30, 31]. Selenoprotein W interacts with glutathione, which points that it plays a role of an antioxidant in organism [32]. The metabolic function of other selenoproteins is miscellaneous. Selenoproteins H and K have a strong antioxidant activity, especially in embryonic period. Selenoprotein I has influence on cells and development of various tissues. Other selenoprotein types (selenoprotein M, Sep15, O, R, S) are the essential part of oxidoreductive processes. Selenoprotein N is important for the differentiation of myoblasts and subsequently for muscle physiology. Selenoprotein T influences on calcium mobilization in embryos and selenoprotein V is important for male reproductive biology [33].

2.6. Manganese

Another essential mineral for normal activity of an organism is manganese (Mn). This microelement is an integral part of many enzymes, e.g., glutamine synthetase, arginase, phosphoenolpyruvate decarboxylase, and the mitochondrial superoxide dismutase [34, 35]. As a part of superoxide dismutase, manganese plays a role in oxidative stress protection. Manganese is especially important for central nervous system homeostasis, presumably followed by neural activity alteration. Another function is its participation in catalytic urea synthesis reaction in the liver. Manganese is also important for cartilaginous tissue function and injury prevention. The significance of manganese in physiological growth, osteogenesis, embryonic development, reproduction and for lipid and carbohydrate metabolism is well known [36].

3. Trace-Elements' Deficiencies

The modern animal herding relies on intensive animal production. Thus, various factors like environment, feeding and other stress factors can contribute in load of organism. In this situation, supply of microelements is especially important, because of their important physiological role. Low intake of trace-elements leads to their deficiency. This metabolic disturbance has a subclinical and chronic character in most cases without clear clinical symptoms. Thus, deficiency of trace-elements results in large economic losses.

One of the most important negative aspects of *selenium* deficiencies have been noted on different components of female and male reproductive system. The negative influence of selenium deficiency on ovulation rate was proved [37]. The determined results of Se deficiency include weak or silent heat periods, delayed conception, poor fertilization and cystic ovaries [38]. In the male, the insufficiency of this trace-element results in lower sperm mobility [39]. Segerson and Libby [40] observed reduced uterine activity in the Charoiles cattle with Se deficiency. The retained fetal membrane (RFM) occurs during Se deficiency in cattle. In this disease, quick directional veterinary intervention is necessary. Olson [41] stated that in 10.3% animals with retained fetal membrane, deficiencies of selenium and vitamin A were detected. This relatively large number suggests the important role of this microelement in prevalence of RMF in cattle. In boar, the low status of selenium has negative influence on sperm concentration, mobility and amount of sperm with a high occurrence of cytoplasmic droplets [42, 43]. It is clear, that poor status of sperm is connected with low percent of fertilization. The other results of selenium deficiency in the cattle can range from ill-thrift, diarrhea, poor growth rates in young cattle, decreased milk production, calves born prematurely and selenium-responsive muscular dystrophy (white muscle disease) [44]. The influence of selenium status on muscular system is especially significant in hyper muscled beef cattle breeds like Belgian Blue or Charolaise. The "shoulder lameness" disease resulting from the deficiency of selenium or vitamin E in the beef cattle herd was described [45]. The bilateral dorsal scapular displacement flying (flying scapula) is observed in this disorder. This situation is commonly preceded by the rupture of serratusventralis muscle caused by degenerative myopathy.

3.1. Hypocuprosis

It occurs often in cattle. This is the second most commonly distributed mineral deficiency observed in grazing cattle [46]. The aftermaths of copper deficiency include copper metaloenzymes failure, which were described for various animal species. Lower Cu/Zn SOD system activity in rats, mice, chickens and cattle with hypocuprosis was observed [47, 48, 49, 50, 51]. Based on these results, it can be stated, that the failure in antioxidant system occurs with copper deficiency. The positive correlation between hypocuprosis and DNA damage was proved [52, 53, 54, 55]. Copper deficiency is the cause of immune system dysfunction. The signs of this negative influence include the absolute decrease in T lymphocytes number and especially T-helper lymphocytes subpopulation as well as a marked decrease in the T and B-lymphocytes mitogens on splenic lymphocytes [56, 57, 58]. The decrease in the response of antibody cell with increased susceptibility to infection was noted in animals with hypocuprosis [59, 60]. In such situation, the possibility of adequate immune system reaction on destructive activity of microbial, viral and other agents is lower than normal. Ahmed at al. [61] observed the hair discoloration, ovarian inactivity, low level of serum progesterone, especially in luteal phase of estrus, anemia and low level of ceruloplasmin in buffalo/cows with hypocuprosis. However, the mechanism by which copper deficiency affects reproduction is not clear, these investigations indicate the potential role of hypocuprosis in reproductive problems in buffalo breeding, which is very important in some countries like Egypt. In general, it can be stated that copper deficiency gives many different clinical signs, like pale coat, poor quality of sheep fleece, anemia, spontaneous fractures, poor capillary integrity, myocardial degeneration, spinal cordhypomyelinization, infertility, low pregnancy rate, lowered resistance to infectious diseases, diarrhea and general ill-health [62, 63]. Thus, the copper deficiency can cause a large economical loss.

3.2. Zinc (Zn)

It belongs to the group of trace-elements involved in many enzymatic functions, especially as metaloenzymes [64, 65]. Zinc deficiency causes reproduction problems in cattle. It is well established, that zinc is important for a normal physiology of male genital system [66]. It is proved that this microelement plays crucial role in fertility, because its deficiency in bulls is often the cause of the late stage alterations in spermatozoa formation and androgen synthesis. The metabolism in female is less dependent on zinc concentration, but infertility and estrus abnormalities were observed [38, 64]. Zinc and selenium supplementation to ram can be important for semen production according to its polygamy and short breeding season [67]. Moreover, it is proved, that zinc has an influence on the growth of stocker/feedlot cattle and immune system function that can be significantly disturbed by zinc deficiency [64]. Zinc deficiency induces lymphoid atrophy and T-cell-dependent antigen response, reduction in IgM and IgG plague-forming cells per spleen, loss of humoral immune capability and reduction in thymus weight [68, 69, 70]. Skin lesions in American camelids were described according to zinc and copper deficiency [71]. In calf with lethal trait A 46, zinc deficiency manifests with diarrhea, lethargy, depression in motor skills and external stimulation. Symmetrical skin lesions (skin peripheral to nares, oral rim commeasures and intermandibular space) with tendency to spread, suckling inability and decrease in growth rate have been reported [72]. Newborn animals might be influenced to a higher degree when compared to mature ones, since the young will earlier demonstrate the nutritional effects of a nutritional stress and will not be able to prevent diseases from past immunological activity when the nutrition was proper [68, 69, 70]. The National Research Council [73] suggests 20-40 mg/kg as the recommended dietary level of zinc. Zinc deficiency cannot be easily diagnosed with blood chemical analysis. The most important is serum level interpretation [64, 74]. Zinc deficits in plasma are observed quickly during deficiency, but it can be easily manifested by some processes that increase the internal demand like bacterial infections. In the latter, false-low zinc level will be accompanied by high Cu and haptoglobin levelsin serum as the consequence of Zn-Cu antagonism [74]. In cattle and birds, zinc supplementation can be successfully provided with zinc oxide, zinc carbonate, zinc sulfate and zinc methionine [64, 75]. Zinc sulfide is poorly utilized [75]. Cattle zinc intoxications occurs rarely, but levels above 500 mg/kg can lead to occurrence of clinical disease [76]. The metabolic function in domestic birds is also strongly related to zinc levels. In birds, zinc is crucial for normal growth, development of skeleton and epithelial tissues and for egg production. Zinc deficiency in birds result in growth retardation and poor development of feathers. The hock joint enlargement and long bones shortening can also be observed [75]. Footpad skin becomes dry and thickened which is accompanied by hyperkeratosis and fissures development. Decrease in egg production and hatchability, together with micromyelia or amyelia, spine shortening, hyper curvature and fusion of thoracic/lumbar vertebras in poultry embryos have been proved[77, 78]. Elemental zinc prevents and heals the thickening and hyper keratinization of esophagus and skin epithelium in pigs and chicken. In pigs, appetite loss, decrease in growth rate and parakeratosis accompanies zinc deficiency [75]. In young birds, poor growth, dermatitis, poor feather development, respiration problems, skeletal dysfunctions, leg weakness and ataxia are common results of deficiency [75, 78].

3.3. Manganese (Mn)

It is well known because of its strong relationship with birth defects of congenital and teratogenic character in rabbits, guinea pigs, pigs, cattle, mice and rats [79]. Deficiency symptoms usually are caused by its deficiency in the diet, whiletheteratogenic changes are associated with binding of Mn to DNA and RNA molecules [80]. Manganese belongs to the group of microelements, which can influence the production including reproduction in grazing cattle. The recommendations for this trace-element are very difficult due to the metabolic and nutritional factors and its role in microelement complex formation. The latter strongly influences nutritional availability for domestic cattle [64]. Manganese is less accessible in alkaline soils [74]. Manganese is one of the least under valuatedtrace-element and its absorption equals to only 14-18% of ingested from diet [81]. Manganese absorption can be inhibited with excessive calcium and phosphorus amounts [75]. The iron rich milk supplementation provokes reduced manganese absorption [82]. This trace-element can be hyper absorbed during sow's pregnancy and in chicken with coccidiosis [83]. In cattle, the manganese deficiency can cause the lower conception rate, delay in estrus, abortion and malformations in calves (knocked over fetlock joint calf) [64, 84]. In perinatal period, poor growth can occur, hair color loss in both cow and calf and chances of cystic ovaries increases [38, 85]. The pathogenesis of these symptoms is still unknown, but it seems to be true that enzymatic system plays an important role [64]. Cases of the chondrodystrophy in newborn calves have been reported [74]. There is no proof of direct influence of manganese on immunodeficiency [38]. The skeletal deformations in animals and shell defects in laying chicken were caused by inadequate manganese intake [75, 78]. The manganese deficiency causes lameness, hock joint swelling, short limbs in pigs. Chicken can be affected with perosis and slipped tendons, moreover the nutritional chondrodystrophy in chick embryos can occur [75]. There are some reports of manganese deficiencies in laboratory animals, too [86]. Sometimes, nervous symptoms like ataxia and mood changes or abnormal otolith (statoconium or otoconium) formations can be observed [75]. Birds are usually more susceptible to the lack of manganese when compared to mammals, due to higher requirements and poor intestinal absorption [78]. Therefore, the exact trace-element supplementation is very important in poultry and mammals fed with high-corn diet [75]. Manganese inaccessibility is deteriorated in dry seasons of the year (autumn, winter and transition to spring) according to low quality and quantity of dry matter and thus there is a need of supplementation. Adding mineral supplementation, including manganese, reduced the birth of calves with skeletal malformations [79]. Prevention and treatment of manganese deficiency is based on evaluation and monitoring of manganese concentration in the diet [76]. Usually, the treatment should be based on reducing the manganese deficiency using the manganese oxide [64]. There is a lack of manganese toxicity proofs [73].

3.4. Chromium (Cr)

It is a crucial trace-element in humans and in animals as well [87]. This element plays an important role in maintaining the configuration of RNA molecule [88]. It was proved that the trivalent chromium is a component of glucose tolerance factor [89, 90]. In experimental animals, chromium deficiency leads to decrease in removal rate of ingested glucose (low insulin sensibility of peripheral tissues), growth rate, life span, corneal lesions and diabetic state [91, 92]. Hyperinsulinaemia follows chromium deficiency caused by a decreased level of chromium-dependent glucose tolerance factor in goats [93]. Chromium toxicity causes kidney, liver, nervous system and blood disturbances and it usually leads to death [75, 94].

4. Microelement Supplements in the Market and Their Bioavailability

The biosorption method using dietary addition of supplements can have a crucial role in animal health and production. The regulation of trace-elements in animal products is important for human health [95]. The precise regulation of the trace-elements level in the diet is valuable for metaphylactic, too. In the trace-elements deficiencies, the exact and effective dietary supplementation would be valuable. Because of widespread microbial antibiotics resistance, all alternative methods of pathogenic microbe's elimination should be revised and improved. One of these methods is probiotics promotion in animal health preservation. Probably, the supply of probiotics can have an influence on reduction of pathogens [97]. Other aspects of influence of biosorbant dietary supplements on animal health should be proved in detail.

Demands of animals for minerals are different for different species. The composition of feed additives is based on age, needs and direction of use [98, 99]. Numerous investigations have shown that appropriate supplementation of minerals in the animal's diet increases breeding efficiency and consequently high productivity [100, 101].

Table 1 summarizes seven key micronutrients, along with the role which they play in the body of the animal, also the list of names of the feed additives approved by the European Commission [102, 103], which may be used as the source of trace-elements [98, 102, 104-108], as well as the standards of nutrition [95,109].

Table 1. Micronutrients and their functions in the animal body, with the names of feed additives and their components

	Function	Name of feed additives [102, 103]	Chemical formula [99]	Standard of nutrition mg/kg feed [108]	
				Laying hens	Piglets
Zn (zinc)	Immune system Endocrine system Enzyme Reproductive System	Zinc lactate trihydrate Zinc acetate dihydrate Zinc carbonate Zinc chloride monohydrate Zinc oxide Zinc sulfate heptahydrate Zinc sulphate monohydrate Zinc chelate	$Zn(C_3H_5O_3)_2 \cdot 3H_2O$ $ZnCH_3COO)_2 \cdot 2H_2O$ $ZnCO_3$ $ZnCl_2 \cdot H_2O$ ZnO $ZnSO_4 \cdot 7H_2O$ $ZnSO_4 \cdot H_2O$ $Zn(X)_{1\text{-}3} \cdot nH_2O$	50-60	70-150
Mn (manganese)	Energy Production Enzyme Reproductive System	Manganese carbonate Manganese chloride tetrahydrate Acid phosphate, manganese trihydrate Manganese Oxide Manganese oxide Manganese sulfate tetrahydrate Manganese sulphate monohydrate Manganese chelate Manganese oxide Manganese	$MnCO_3$ $MnCl_2 \cdot 4H_2O$ $MnHPO_4 \cdot 3H_2O$ MnO Mn_2O_3 $MnSO_4 \cdot 4H_2O$ $MnSO_4 \cdot 1H_2O$ $Mn(C_3H_5O_3)_2 \cdot 3H_2O$ $MnO \cdot Mn_2O_3$	60-70	30-40
Cu (copper)	Immune system Endocrine system Enzyme Reproductive System	The complex of copper sulfate and lysine Copper acetate monohydrate Copper carbonate monohydrate Copper chloride dihydrate Methionate copper Copper oxide Copper sulfate pentahydrate Copper chelate of amino acids	$Cu(C_6H_{13}N_2O_2)_2SO_4$ $Cu(C_3COO)_2 \cdot H_2O$ $CuCO_3 \cdot Cu(OH)_2 \cdot H_2O$ $CuCl_2 \cdot 2H_2O$ $Cu(C_5H_{10}NO_2S)_2$ CuO $CuSO_4 \cdot 5H_2O$ $Cu(x)_{1\text{-}3} \cdot nH_2O$	5-6	20-165
Co (cobalt)	Production of vitamins Biosynthesis of vitamin B12	Cobalt acetate tetrahydrate II, Cobalt II carbonate monohydrate Cobalt chloride hexahydrate II Nitrate hexahydrate Cobalt II Cobalt II sulfate monohydrate Cobalt sulfate heptahydrate II	$Co(CH_3COO)_2 \cdot 4H_2O$ $2CoCO_3 3Co(OH)_2 \cdot H_2O$ $CoCl_2 \cdot 6H_2O$ $Co(NO_3)_2 \cdot 6H_2O$ $CoSO_4 \cdot H_2O$ $CoSO_4 \cdot 7H_2O$	-	0.5
Fe (iron)	Immune system Endocrine system Blood-forming Enzyme	Ferrous-amino acid chelate Ferrous II Carbonate Ferric chloride tetrahydrate II Ferric chloride hexahydrate III Ferrous citrate hexahydrate II Ferrous fumarate II Ferrous II lactate trihydrate Ferrous Oxide III Ferrous sulphateheptahydrate II Ferrous sulphate monohydrate II	$Fe\ (x)_{1\text{-}3}nH_2O$ $FeCO_3$ $FeCl_2 4H_2O$ $FeCl_2 6H_2O$ $Fe_3(C_6H_5O_7)_2 6H_2O$ $FeC_4H_2O_4$ $Fe(C_3H_5O_3)_2 3H_2O$ Fe_2O_3 $FeSO_4 7H_2O$ $FeSO_4 H_2O$	50-55	90-100

Presented in Table 1, forms of feed additives differ with availability of microelements. It is important, in which form agiven mineral is applied. The consequences of introduction of the element in the unavailable form or those are not easily digestible, may result into low production yields with additional environmental problems associated with the accumulation of these components in the soil [109, 110]. The modern trend of increasing livestock productivity through genetic selection bring forth the need to use microelements in highly bioavailable form in animal feeding to meet their requirements and to ensure the rapid growth and reproductive potential. Biotechnology provides many new opportunities to increase the productivity by improving the utilization of nutrients, improving the quality of animal products and increasing the safety and animal health. Additionally, significant is minimizing the quantity of Cu, Fe, Mn and Zn that is transferred to manure, through supplementation of animals diet with preparations of higher bioavailability. The concentration of microelements in manure, especially Cu, Fe, Mn and Zn, is an important problem of the modern animal wastes neutralization. Therefore, higher microelements bioavailability in animal diet can significantly decrease its transfer to manure and environment contamination. An example of the application of biotechnology in animal nutrition is the supplementation of amino acids and endogenous enzymes as well as hormones, probiotics and antibodies in order to enhance such availability of nutrients, reduction of pollution load in animal manure and increasing resistance to diseases. Mineral management is of particular importance. Currently, tendency to replace inorganic compounds that demonstrate the limited bioavailability in the digestive system and cause environmental concerns, by the organic sources (micronutrients) that have better availability, has been observed. Additionally, literature reports that the use of traditional inorganic sources of micronutrients causes oxidation of vitamins [111] and perforation of digestive tract [111, 112].

5. Biological Microelement Feed Additives

The solution to low productivity of farm animals can be the introduction of micronutrients in the available formin the diets of animals. Supplementation of trace-elements in the form bound to the well-absorbedmatrixcan provide efficient transportation of bound components and hence better utilization.

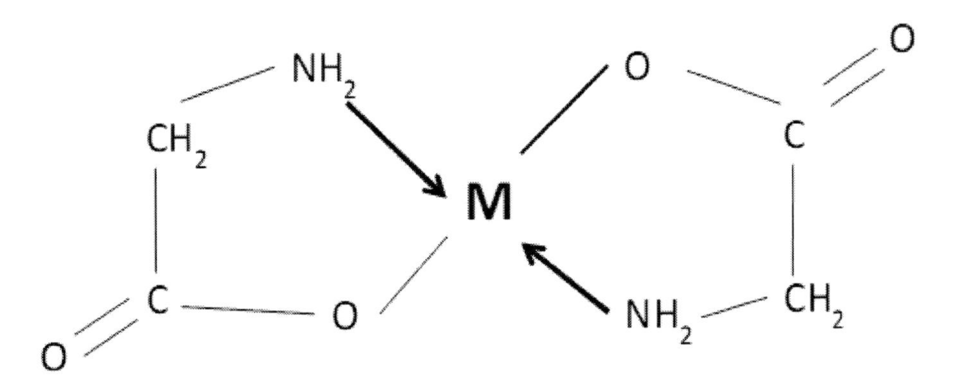

Figure 1. Structural formula of Fe glycine chelate [113].

Figure 1 shows the structural formula of glycine chelate. The carboxyl group (-COOH) of the ligand forms an ionic bond with the cations, whereas the alpha-amino group (=NH_2) donates an electron pair back into a vacant-orbital of the metal ion, thus forming a coordinate covalent bond. The tetrahedral metal conformation provides a degree of protection to iron by limiting its reactivity with dietary components [113].

Among the organic microelement supplements listed by Association of American Feed Control Officials [114, 115] the following forms can be distinguished:

- *Amino acid chelates*– the product of soluble metal salt reaction with an amino acid in a molar ratio of1:2(3). Metal ion is connected with each of the amino acids at least by one bond to form heterocyclic rings (Figure 1). Positively charged metal ion is neutralized by the free electron pairs from α-carboxyl and α-amino group, resulting in the coordination bonds. The mean molar mass of the hydrolyzed amino acids has to be 150 g, and the entire weight of chelate should not exceed the level of 800 g. The heaviest of the amino acid tryptophan is at 204g, the heaviest of microelement is molybdenum 96 g. The mass of chelate of molybdenum ion with three molecules of tryptophan is 709 g. Chelate molecular weight should not exceed the limit of 800 g.That limit was introduced to prevent the use of *amino acid chelates* term to define product produced by the reaction of metal ion with the products of proteins hydrolysis, which are more often peptides than amino acids [114, 115].
- *Amino acid complex* – the product of the complex formation of metal ion in the form of a soluble salt with the corresponding amino acid. Metal ion is connected to only one site of an amino acid [114, 115].
- *Polysaccharide complexes*– the product of the reaction of metal ion complex formation in the form of soluble salts in the solution of the polysaccharide. Metal ion is not bound chemically with polysaccharide [114, 115].
- *Metalproteinate*– the product of the reaction of metal ion chelation in the form of a soluble salt with amino acids or in the form of partially hydrolyzed protein. Partially digested protein is too large to be absorbed. This step must be preceded by hydrolysis of the protein into amino acids, before the metal ion in the digestive system will be released [114, 115].

Chelates create a ring structure around the ion of metal and form coordination bond with free electron pair derived from nitrogen (amino group) or oxygen (carboxyl group) from amino acid or protein. Gluconate, fumarates, and citrates can form complexes with metal ions by ionic bonds too, but are less stable when compared to the chelates. All the chelates are complexes;however,notall complexes are chelates [116, 117].

In the literature, many reports about the effectiveness of the use of microelements in organic form can be found. Apparent absorption is defined as one of the measures to express the microelements availability for digestive intake from animal diet. It is often presented as a percentage of the difference of amounts taken and mineral that was excreted (Equation 1).

$$Apparentabsorption = \frac{intake - totalfecalexcretion}{intake} \cdot 100\ \% \qquad (1)$$

The difference between the amount of mineral collected and excreted shows only the final disappearance of the ingredients from the digestive tract. It does not explain the origin of the component, which is present in animal manure (or it is due to abrasion of muscle cells or secretion of the digestive tract). This method of assimilation is reserved only for the components for which the gastrointestinal tract (animal manure) is the main route of excretion (Ca, P, Zn, Mn and Cu) [118].

Table 2 shows the effects of using these preparations (apparent availability), where the first position for each micronutrient represent the availability of sulfate to each ion. In addition, the remaining part of the bioavailability of other forms of microelement was assessed in correlation to the sulfate. Data reported are the averages collected from the literature and rounded.

Table 2. Comparison of apparent availability of different sources of Cu, Fe, Mn and Zn, %

Microelement	Animal species	
Cu	Poultry	Pigs
$CuSO_4$	100	100
Cu- lysine	105 [119, 120]	-
Cu- methionine	90 [121, 122]	110 [126]
$CuCO_3$	65 [123, 124]	85 [127, 128]
$CuCl_2$	145 [125, 121]	-
Fe		
$FeSO_4$ $5H_2O$	100	100
Fe-proteinate	-	125 [132]
Fe- methionine	-	185 [133]
$FeCO_3$	90 [129]	22 [129]
$FeCl_2$	60 [130, 131]	-
Mn		
$MnSO_4 \cdot 5H_2O$	100	100
Mn - proteinate	110 [134]	-
Mn- methionine	120 [135]	-
$MnCO_3$	90 [136, 137]	-
$MnCl_2 \cdot 4H_2O$	100 [136, 138, 139]	-
Zn		
$ZnSO_4$	100	100
Zn- lysine	-	100 [144]
Zn- methionine	150 [140]	100 [144]
$ZnCO_3$	115 [141, 142]	-
$ZnCl_2$	100 [143]	-

According to the information presented in Table 2,regarding the effect of the use of organic forms of micronutrients, many applications [145-148] and production methods that allow to obtain the product characterized by improved properties [149] were submitted for patent protection.

However, not all agree according to the properties of metal amino acids chelates. Davidson [150] as well as Miles and Henry [151] in their work claimed that in the popular scientific literature or advertising material, different ideas about organic additives were presented.

The following features can be included:

(i) Ring structure protects the metal ion from undesirable reactions in the digestive system,
(ii) Chelates can easily pass the intestinal barrier and reach the bloodstream in the intact form,
(iii) Organic forms are absorbed by other metabolic pathways than their inorganic forms,
(iv) Provide minerals in a form similar to the naturally occurring in the living organism, since the chelates are negatively charged, they are more easily absorbed and metabolized,
(v) Chelation increases the solubility, transport through cell membranes and stability at low pH,
(vi) Chelates are absorbed by the route of transport of amino acids.

Table 3. Forms of biological trace-elements. Commercial products of JHBiotech, USA company [153]

Product	Content, %	Comments /Dose	Price US$/kg	Costs of enrichment 1 Mg feed	
				Laying hens	Pigs
Instamin® *Copper*	Glycine - 25 Cu - 10	0.02 - 0,1 %	19.0	1.14	31.5
Instamin® *Iron*	Glycine - 20 Fe - 15	0.02 - 0,2 %	17.0	6.23	11.3
Instamin® *Zinc*	Glycine - 20 Zn - 15	0.02- 0.2 %	14.8	5.43	9.87
Buffermin® *AF Liquid*	Ca– 0.05 Co – 0.0012 Cu – 0.011 Fe – 0.026 Mg – 0.5 Mn – 0.055 Zn – 0,16	Brown liquid with a density of 1.2 g/cm^3, added to water or milk	No data	-	-
Buffermin® *Chromium Yeas*	Crude protein–35.0 Cr – 0.1	0.20-0.40 mg/kg - cattle, pigs, sheep, 0.05-0.3 mg/kg - smaller animals. It is produced by a fermentation process with chromium saltswith *Saccharomyces cerevisiae*	22.6	-	-
Buffermin® *Copper Proteinate*	Crude protein - 35 Cu - 10	0.02 – 0.2 %	9.2	0.552	15.2
Buffermin® *Iron Proteinate*			8.0	2.94	5.33
Buffermin® *Zinc Proteinate*	Crude protein – 25.0 Zn – 15.0	0.02 – 0.2 %	34.0	12.5	22.8
Buffermin® *Zinc Methionine (20% Zn)*	Methionine – 40.0 Zn – 20.0	0.02 – 0.1 %	19.0	5.7	14.2
Buffermin® *Selenium Yeast (0.1% Se)*	Se - 0,1 Crude protein - 35	0.1 mg/kg-cattle, pigs, sheep, horses, 0.05 mg/kg - poultry. Brown powder with a slight odor of yeast	11.7	1.40	31.4

However, none of these reports were confirmed or rejected in the scientific literature, with one exception of zinc and copper chelates that were not stable at pH 2[152]. In Table 3, metal amino acids chelates manufactured by JHBiotech (USA), along with prices, as well as the approximate cost of enrichment at1 Mg of feed for laying hens and pigs are given.

There is a possibility to improve the biomass nutritional value by biosorption and bioaccumulation, which is confirmed by literature. It was reported that copper-enriched yeast has solved the problem of micronutrient deficiency in the diet of humans and animals [154]. Among the products offered by JHBiotech (USA), (Table 3) yeast enriched with Se and Cr can be found. Tompkins et al. [155] proved that Zn bound with the yeast cells is better absorbed than zinc gluconate. However, the use of yeast in the diet as a dietary supplement creates the risk of candidiasis [156, 157].

6. Biosorption Application in the Production of New Generation Feed Supplements

6.1. Biosorption - General Information

Biosorption is defined as passive binding and removal process of metal ions or metalloid species by biological material [157-161].

It has been demonstrated that many types of materials can be used as biosorbents for the removal of metals or organics. The analyzedbiosorbents can be classified according the following categories: bacteria [162-164], fungi [165-168], yeast, algae [169] and microalgae [169-172]. Their biosorption properties have been used in many techniques related with metal pollution control. Moreover, its potential application and future has also been discussed in the literature [160, 164, 174, 175, 176].

Using the biosorption process, a method used to improve the nutritional value of the components of the animal fodder which results in higher Enrichment Coefficient (EC), it is referred to as the ratio of the content of microelements in enriched and natural biomasses (Table 4).

Table 4. The comparison of biosorption EC (Enrichment Coefficient) of green plants

Enrichment Coefficient (EC)	*Spirulina maxima*[177]	*Pithophoravaria*Wille [178]	*Lemna minor*[179]
Cr	-	-	1857
Co	60877	8172	25513
Cu	3218	1486	8379
Fe	37.5	-	-
Mn	768	37.2	-
Zn	1435	84.6	567

- No information.

Literature describes the method of chemical modification of binding sites on the cell surface, in order to enhance their concentration or activity, by pre-treatment, increasing the number of binding sites, their modification and polymerization. For example, adding a long chain polymer on the surface of the biomass and/or monomer polymerization creates additional binding sites on cell surface. Additional binding sites on cell surface can be created with a long chain polymer adding or monomer polymerization on the surface of biomass. Another way to develop new and better biosorption properties is by genetic engineering, which results in higher selectivity and the biosorption capacity of the biomass cells [180-182].

6.2. Biosorption – Mechanism

The postulated mechanism of biosorption is ion exchange and complex formation process in the system of donor-acceptor of electrons coming from functional groups derived from polymers and macromolecules that are the element of the cell wall. The formation of complexes involves organic compounds containing atoms with free electron pair, a coordination bond that is formed by oxygen, nitrogen and sulfur.

The literature states that the ion exchange is the dominant biosorption mechanism [159, 171, 183-186]. The natural biomass contains light metal ions like K^+, Na^+, Ca^{2+} and Mg^{2+}that are bound to the functional groups on the cell wall. In the biosorption process, these ions are exchanged for other metal ions and their concentration in the solution after biosorption is higher than before the process. At the time light metal ions and protons are released, their binding sites are taken by the microelement ions from the solution (after biosorption pH is lower than before) [185, 186]. Metal ions binding to the cell wall of biomass on the basis of physical adsorption, occurs due to electrostatic interactions and Van der Walas forces. Extracellular precipitation of metal ion may also occur by the reaction with the components of the cell wall [187].

The main factor responsible for the biosorption is cell wall composition. Physical and chemical interactions between the components of the cell wall with the metal ions allow binding of ions to the cell surface. The cell wall is composed principally of polysaccharides, lipids and proteins that in its structure have many sites where metal ions can be bound [183, 184].

6.3. Biosorption – Effects of Application in Animal Feeding

The utilization of new mineral feed additives based on the biomass that is enriched with microelements in biosorption process was confirmed in animal breeding experiments. New preparations of microelements were examined on laying hens and swine, and the assessment of production parameters, the quality of eggs and carcass was performed for laying hens and swine, respectively. Animals were fed with fodder where inorganic forms of microelements were replaced with biomass enriched with microelements via the process of biosorption.

As a result of the use of biomass enriched with Cu (II) and Fe (II) ions, in the laying hens diet as a source of micronutrients, the following results were achieved: increased shell strength, lower breakage, egg mass was higher; egg whites mass lower; parameter Lform CIE color model of yolk was lower, and the parameter A higher; eggs fortified with Fe, Zn and Mnwere obtained [188]. Eggs from the experimental group were richer in all the examined micronutrients and macronutrients when compared to eggs from the control group [188]. In the case of the application of microalgae in swine diet, the following results were obtained:

improvement of liver profile; LDL cholesterol and total cholesterol level decreased; carcass slaughter value improved according to the SEUROP classification; increase of the parameter A and reduction of natural leakage, which improved technological assessment of carcass grade from class PSE (pale, soft, exudative) to RFN (red, firm, normal) [188].

7. Conclusions

Recently, a particular attention is paid to the role of microelements and their proper doses in the diet of both human and animals. To fulfill the nutritional requirements, various supplements are used. This is particularly significant in highly intensive animal production system, because the proper doses of nutrients in feed facilitate, e.g., in combating stress [189]. In the latter case, the dose should be even higher than the recommended [189]. Microelements of the highest significance are copper and zinc [190].

The major sources of microelements in dietary supplements for animals are inorganic salts: sulphates, oxides or chlorides. Drawback of inorganic forms, such as sulfates is high reactivity and destructivity due to high hygroscopicity and solubility in water [191]. Organic forms represent significantly less oxidative properties, thus protecting some vitamins present in feeds, e.g., vitamin E [191]. It has been demonstrated that the inorganic forms are absorbed and retained in low extent [190]. An alternative to inorganic supplements are organic preparations, whereby microelements are bound with amino acids, proteins, organic acids (propionic, lactic) or other organic molecules [192, 193]. Amino acid chelates are produced in the reaction of dissolved salts with amino acids. Microelement cation is bound with both groups: amino (by covalent bond) and carboxyl (by ionic bond) [193].

Many reports of in-vivo experiments showed that the organic forms are characterized by higher bioavailability expressed as higher retention and higher tissue concentrations than in groups of animals fed with inorganic products [192, 194]. The studies included investigations on growth performance, health, reproduction, utilization of nutrients and metabolism of minerals, as well as distribution in vital organs and immune response [190, 192, 194]. Studies on bioavailability of microelements from organic preparations showed increased absorption as compared with inorganic forms. These investigations were carried out on cattle, pigs, rats and ewes [190]. For instance, bioavailability of Zn from zinc-methionine is 228 percent as related to zinc sulfate (100 %), while zinc oxide showed 61 percent [192].

The biological feed supplements where microelement ions are bound with the biomass seem to be a promising perspective. The chemical mechanism of the process is similar as in amino acid chelates, whereby microelement cations are bound with carboxyl group. In the case of biosorption, carboxyl group is provided by the biomass – an insoluble material. Microelements supplemented on biological carrier are characterized with higher bioavailability than from inorganic mineral feed supplements, which was shown in feeding experiments on laying hens and swine. The advantage of an application of biosorption in order to enrich the biomass with microelements is lower price of the biomass as carrier of microelements when compared to amino acids.

REFERENCES

[1] Mandal GP, Dass RS, Isore DP, Garg AK, Ram GC. Effect of zinc supplementation from two sources on growth, nutrienthet utilization and immune response in male crossbred cattle (Bosindicus x Bostaurus) *bulls. Anim. Fed. Scien. Techn.*,2007; 138: 1-12.

[2] MacDonald RS. The role of zinc in growth and cell proliferation. *J. Nutr.,*2000; 130: 1500–1508.

[3] Chesters JK. Trace-element-gene interactions with particular reference to zinc. *Proc.Nutr.Soc.*,1991; 50: 123–129.

[4] Droke EA, Spears JW.In-vitro and in-vivo immunological measurements in growing lambs fed diets deficient, marginal or adequate in zinc.*J. Nutr. Immunol.*,1993; 2:71–76.

[5] Hambidge KM, Casey CE, Krebs NF. *Trace-elements in Human and Animal Nutrition.*W. Mertz, ed., Academic Press, New York, 1986; Vol. 2, p. 1.

[6] Shinde P, Dass RS, Garg AK, Chaturvedi VK, Kumar R. Effect of Zinc supplementation from different sources on growth, nutrient digestibility, blood metabolic profile and immune response of male guinea pig. *Biol. Trace Elem. Res.,*2006; 112: 247-262.

[7] Haan de JB, Tymms MJ, Cristiano F, Kola I. Expression of copper/zinc superoxide dismutase and glutathione peroxidase in organs of developing mouse embryos, fetuses, and neonates. *Pediatric Res.,*1994; 35: 188–196.

[8] Zapf J, Froesch R. Insulin-like growth factors/somatomedins: structure, secretion, biological actions and physiological role. *Hormone Res.,* 1986; 24: 121–130.

[9] Sackett RL, McCusker RH. Multivalent cations depress ligand affinity of insulin-like growth factor-binding proteins-3 and -5 on human GM-10 fibroblast cell surfaces. *J. Cell.Biochem.*,1998; 69: 364–375.

[10] *National Research Council. Nutrient Requirements of Dairy Cattle.* National Academy Press, Washington, 2001.

[11] Hawk SN, Uriu-Hare JY, Daston GP, Jankowski MA, Kwik-Uribe C, Rucker RB, Keen CL. Rat embryos cultured under copper-deficient conditions develop abnormally and are characterized by an impaired oxidant defense system. *Teratology*,1998; 57: 310-320.

[12] Camaschella C, PaganiE.Iron and erythropoiesis: a dual relationship.*Int. J.Hematol.,*2011; 93:21-26.

[13] Legrand D, ElassE, CarpentierM, Mazurier J.Lactoferrin: a modulator of immune and inflammatory responses. *Cell. Mol. Life Sci.,*2005; 62: 2549- 2559.

[14] Ward PP, Paz E, Conneely OM. Multifunctional roles of lactoferrin: a critical overview. *Cell. Mol. Life Sci.*,2005; 62: 2540–2548.

[15] Adlerova L, Bartoskova A, Faldyna M.Lactoferrin: a review. *Vet. Med. (Praha),*2008;53,457–468.

[16] Omata Y, Satake M, Maeda R, Saito A, Shimazaki K, Yamauchi K, Uzuka Y, Tanabe S, Sarashina T, Mikami T. Reduction of the infectivity of Toxoplasma gondiiand Eimeriastiedaisporozoites by treatment with bovine laktoferricin. *J. Vet. Med.Sci.*,2001; 63: 187–190.

[17] Cirioni O, Giacometti A, Barchiesi F, Scalise G. Inhibition of growth of Pneumocystis cariniiby lactoferrins alone and in combination with pyrimethamine, clarithromycin and minocycline. *J. Antimicrobial Chemotherapy*,2000; 46: 577–582.

[18] Cornish J, Callon KE, Naot D, Palmano KP, Banovic T, Bava U, Watson M, Lin JM, Tong PC, Chen Q, Chan VA, Reid HE, Fazzalari N, Baker HM, Baker EN, Haggararty NW, Grey AB, Reid IR.Lactoferrin is a potent regulator of bonecell activity and increases bone formation in vivo. *Endocrinology,* 2004; 145: 4366–437.

[19] Anderson RA, Bryden NA, Evock-Clover CM,Steele NC.Beneficial effects of chromium on glucose and lipid variables in control andsomatotropin-treated pigs are associated with increased tissue chromium and alteredtissue copper, iron, and zinc. *J. Anim. Sci.,* 1997; 75: 657-661.

[20] Borel JS, Majerus TC, Polansky MM, Moser PB, Andersen RA. Chromium intake and urinary chromium excretion of trauma patients. Biol. Trace Elem. Res.,1984; 6: 317.

[21] Mertz W.Chromium: History and nutritional importance. *Biol. Trace Elem. Res.*, 1992; 32: 2.

[22] EastmondDA., MacGregor TJ, Slesinski RS.Trivalent Chromium: Assessing the Genotoxic Risk of an Essential Trace-element and Widely used Human and Animal Nutritional Supplement. *Crit. Rev. Tox.,*2008; 38: 173-190.

[23] Chu FF, Doroshow JH, Esworthy RS. Expression, characterization and tissue distribution of a new cellular seleniumdependentGSHPx. *J. Biol. Chem.*,1993; 268: 2571-2576.

[24] Avissar N, Ornt DB, Yagil Y, Horowitz S, Watkins RH, Kerl EA, Takahashi K, Palmer I. S., Cohen H. J. Human Kidney Proximal Tubules Are the Main Source of Plasma Glutathione Peroxidase. *Am. J. Physiol.*,1994; 266: 367-375.

[25] Ursini F, Maiorino M, Gregolin C. The selenoenzyme phospholipid hydroperoxide glutathione peroxidase. *Biochim. Biophys. Acta,*1985; 839: 62-70.

[26] Sun Q-A, Wu Y, Zappacosta F, Jeang KT, Lee BJ, Hatfield DL, Gladyshev VN. Redox Regulation of Cell Signaling by Selenocysteine in Mammalian ThioredoxinReductases. *J. Biol. Chem.,*1999;274: 24522-24530.

[27] Renko K, Hofmann PJ, Stoedter M, Hollenbach B, Behrends T, Kohrle J, Schweizer U, Schomburg L. Down-regulation of the hepatic selenoprotein biosynthesis machinery impairs selenium metabolism during the acute phase response in mice. *Faseb. J.*,2009; 23(6):1758–1765.

[28] AkessonB, Bellew T, Burk RF. Purification of selenoprotein P from human plasma. *Bioch. Bioph. Acta*,1994; 1204: 243-249.

[29] Seiler A, Schneider M, Forster H, Roth S, Wirth EK, Culmsee C, Plesnila N, Kremmer E, Radmark O, Wurst W, Bornkamm GW, Schweizer U, Conrad M. Glutathione peroxidase 4 senses and translates oxidative stress into 12/15-lipoxygenase dependent- and AIF-mediated cell death. *Cell Metab.*,2008; 8: 237–248.

[30] Yeh JY, Vendeland SC, Gu QP. Dietary selenium increases selenoprotein W levels in rat tissues. *J. Nutr.,*1997; 127: 2165-2172.

[31] Loflin J, Lopez N, Whanger PD, Kioussi C.Selenoprotein W during development and oxidative stress. *J. Inorg. Biochem.,*2006; 100: 1679–1684.

[32] Beilstein MA, Vendeland SC, Barofsky E, Jensen ON, Whanger PD. Selenoprotein W of rat muscle binds glutathione and an unknown small molecular weight moiety. *J. Inorg. Biochem.,*1996; 61: 117–124.

[33] Reeves MA, Hoffmann PR. The human selenoproteome: recent insights into function and regulation. *Cell. Mol. Life Sci.*,2009; 66: 2457-2478.
[34] Takeda A. Manganese action in brain function. *Brain Res. Rev.*,2003; 41: 79–87.
[35] Finley JW, Davis CD. Manganese deficiency and toxicity: are high or low dietary amounts of manganese cause for concern? *Biofactors*,1999; 10: 15–24.
[36] Underwood EJ. *Trace-elements in Human and Animal Nutrition*, 4th ed. Academic Press, New York, 1977; pp. 170–190.
[37] Harrison JH, Hancock DD, Conrad HR. Vitamin E and selenium for reproduction of the dairy cow. *J. Dairy Sci.*,1984; 67: 123–132.
[38] Corah LR, Ives S. The effects of essential trace minerals on reproduction in beef cattle. *Vet. Clin.North Am. Food Anim. Pract.*,1991; 7: 40–57.
[39] McKenzie RL, Rafferty TS, Beckett GJ. Selenium: an essential element for immune function. *Immunol. Today*,1998; 19: 342–345.
[40] Segerson EC, Libby DW. Ova fertilisation and sperm number per fertilised ovum for selenium andvitamin E treated Charolais cattle. *Theriogenology*,1982; 17: 333–341.
[41] Olson JD. The role of selenium and vitamin E in mastitis and reproduction of dairy cattle. *Cattle Pract.*,1995; 3: 47–49.
[42] Liu CH, Chen Y, Chen M, Zhang J, Zhang Z, Huang M, Huang Y, Su Q, Lu ZH, Lu R, Lu X, Feng D, Zheng PL. Preliminary studies on influence of selenium to the developments of genital organs and spermatogenesis of infancy boars. *Acta Vet. Zootech. Sin.*,1982; 13: 73–77.
[43] Marin-Guzman J, Mahan DC, Chung YK, Pate JL, Popo WF. Effects of dietary selenium and vitamin E on boar performance and tissue responses, semen quality and subsequent fertilization rates in mature gilts. *J. Anim. Sci.*,1997; 75: 2994–3003.
[44] Rollin F, *Lebreton P, Guyot H. Trace-element*s deficiencies in Belgian beef and dairy herds in 2000–2001. Proc. 22th World Buiatrics Congress, Hanover, 2002; p. 72.
[45] Buergelt CD, Sisk D, Clhenoweth PJ, Gamboa J, Nagus R. Nutritional Myodegeneration Associated with Dorsal Scapular Displacement in Beef Heifers. *J. Comp. Path.*,1996; 114: 445-450.
[46] Ramirez CE, Mattioli GA, TittarelliCM, Giuliodori MJ, Yano H. Cattle hypocuprosis in Argentina associated with periodicallyflooded soils. *Livestock Prod. Sci.*,1998; 55: 47-52.
[47] Prohaska JR. Biochemical change in copper deficiency. *J. Nutr. Biochem.*,1990; 1: 452-461.
[48] Sukalsky KA, LaBerge TP, Johnson WT. In vivo oxidative modification of erythrocyte membrane proteins in copper deficiency. *Free Radic. Biol. Med.*,1997; 22: 835—842.
[49] Cerone SI, Sansinanea AS, Streitenberg SA, Garcı´a MC, Auza NJ. Cytochrome c-oxidase, Cu,Zn-superoxide dismutase and caeruloplasmin activities in copper-deficient bovines. *Biol. Trace. Element Res.*,2000; 73: 269-278.
[50] Cockel KA, Belonje B. The carbonyl content of specific plasma proteins is decreased by dietary copper deficiency in rats. *J. Nutr.*,2002; 132: 2514-2518.
[51] Hawk SN, Lanoue L, Keen CL, Kwik-Uribe CL, Rucker RB, Uriu-Adams JY. Copper-deficient rat embryos are characterized by low superoxide dismutase activity and elevated superoxide anions. *Biol. Reprod.*,2003; 68: 896-903.
[52] Pan YJ, Loo G. Effect of copper deficiency on oxidative DNA damage in Jurkat T-lymphocytes. *Free Radic. Biol. Med.*,2000; 28: 824-830.

[53] Abba M, De Luca JC, Mattioli G, Zaccardi E. Dulout FN. Clastogenic effect of copper deficiency in cattle. *Mutat. Res.*,2000; 466: 51-55.

[54] Picco SJ, De Luca JC, MattioliG, Dulout FN. DNA damage induced by copper deficiency in cattle assessed by the comet assay. *Mutat.Res.*,2001; 498: 1-6.

[55] Picco SJ, Abba MC, Mattioli GA, Fazzio LE, Rosa D, De Luca JC, Dulout FN. Association between copper deficiency and DNA damage in cattle. *Mutagenesis,*2004; 19: 453- 456.

[56] Lukasewycz OA, Prohaska JR. Lymphocytes from Copper-deficient Mice Exhibit Decreased Mitogen Reactivity." *Nutr. Res.*,1983; 3: 335.

[57] Lukasewycz OA, Prohaska JR, Meyer GS, Schmidt JR, Hatfield SM, Marder P. Alterations in Lymphocyte Subpopulations in Copper- deficient Mice." *Infect. Immun.,*1985; 48: 644.

[58] Flynn A. "Control of in vitro Lymphocyte Proliferation by Copper, Magnesium, and Zinc Deficiency." *J. Nutr.,*1984; 114: 2034.

[59] Vyas D, Chandra RK. "Thymic Factor activity, Lymphocyte Stimulation Response, and Antibody-producing Cell in Copper Deficiency." *Nutr. Res.*,1983; 3: 343.

[60] Suttle NF, Jones DG. Copper and Disease Resistance in Sheep: A Rare Natural Confirmation of Interaction Between a Specific Nutrient and Infection." *Proc. Nutr. Soc.,*1986; 245-317.

[61] Ahmed WM, El Khadrawy HH, Hanafi EM, Amal R, Hameed A, Sabra HA. Effect of Copper Deficiency on Ovarian Activity in Egyptian Buffalo-cows. *World J. Zool.*,2009; 4 (1): 1-8.

[62] Tessman RK, Lakritz J, Tyle, JW, Casteel SW, Williams JE, Dew RK. Sensitivity and specificity of serum copper determination for detection of copper deficiency in feeder calves. *J. Am. Vet. Med. Assoc.*,2001; 218: 756-760.

[63] Sharma MC, Joshi C, Pathak NN, Kaur H. Copper status and enzyme, hormone, vitamin and immune function in heifers. *Res. Vet. Sci.,*2005, 79 (2): 113-123.

[64] Corah L. Trace mineral requirements of grazing cattle. *Anim. Feed Sci.Technol.*,1996; 59: 61-70.

[65] Guyot H, Sargerman C, Lebreton P, Sandersen C, Rollin F. Epidemiology of trace-elements deficiences in Belgian beef and diary cattle herds. *J. Trace Elem. Med.Biol.,*2009; 23: 116-123.

[66] Pitts WJ, Miller WJ, Fosgate OT, Morton JD, Clifton CM. Effect of zinc deficiency and restricted feeding from two to live month of age on reproduction in Holstein bulls *J.Dairy Sci.*,1966; 49: 995.

[67] Kendall NR, McMullen S, Green A, Rodway RG. The effect of a zinc, cobalt and selenium soluble grass bolus on trace-element status and semen quality of ram lambs. *Anim. Product. Sci.*,2000; 62: 277-28.

[68] Beach RS, Gershwin ME, Hurley LS. Growth and development of postnatally zinc-deprived mice. *J. Nutr.,*1980; 110: 201.

[69] Beach RS, Gershwin ME, Hurley LS. Gestational zinc deprivation in mice: Persistence of immunodeficiency for three generations. *Science*,1982a; 218: 469.

[70] Beach RS, Gershwin ME, Hurley LS. The reversibility of developmental retardation following murine fetal zinc deprivation. *J. Nutr.,*1982b; 112 (1): 169.

[71] Clauss M, Lendl C, SchramelP, Streich WJ. Skin lesions in alpacas and llamas with low zinc and copper status – a preliminary report. *Vet. J.,*2004; 167: 302-305.

[72] Machen M, Montgomery T, Holland R, Braselton E, Dunstan R, Brewer G, Yuzbasiyan-Gurkan V. Bovine hereditary zinc deficiency: lethal trait A 46. *J. Vet. Diagn. Invest.*,2006; 8: 219-227.
[73] *National Research Council. Nutrient Requirements of Dairy Cattle*. National Academy Press, 1984, Washington.
[74] Suttle N. Assessing the needs of cattle for trace-elements. *Farm Animal Practice*, 2004.
[75] Soetan KO, Olaiya CO, Oyewole OE. The importance of mineral elements for humans, domestic animals and plants. *A review. Afr. J. Food Sci.,*2010; 4: 200-220.
[76] Puls R. *Mineral levels in animal health*. Sherpa Int. 1988.
[77] Merck VM. The Merck veterinary manual. 6thed. *A handbook of diagnosis, therapy and disease prevention and control for the veterinarian.* Merck and Co. Inc. Rahway, 1986, New Jersey.
[78] Gordon RF. Poultry Diseases. The English language book society and bailliere. *Tindall,* 1977, London.
[79] Carvahlo PR, Goncalves Pita MC, Loureiro JE. Manganese deficiency in bovines: Connection between manganese metalloenzyme dependent in gestation and congenital defects in newborn calves. *Pak. J. Nutr.,*2010; 9: 488-503.
[80] Shrader RE, Erway LC, Hurley LS. Mucopolysaccharide synthesis in the developing inner ear of manganese-deficient and pallid mutant mice. *Teratology*,1973; 8: 257-266.
[81] Georgievskii VI, Annenkov BN, Samokhin VI. *Mineral Nutrition of Animals*, 1982, Butterworth.
[82] Gruden N. Suppression of transduodenal manganese transport by milk diet supplemented with iron. *Nutr. Metab*.,1977; 21: 305-309.
[83] Southern LL, Baker DH.Eimeriaacervulina infection in chicks fed deficient or excess levels of manganese. *J. Nutr*.,1983; 113: 172.
[84] Wilson TG. Bovine functional infertility in Devon and Cornwall: Response to manganese therapy. *Vet. Rec.,*1966; 79: 562.
[85] Maas J. Relationship between nutrition and reproduction in beef cattle. *Vet. Clin. N.Am. Food Anim. Pract.*,1987; 3: 633.
[86] Leach RM. *Biochemical role of manganese.* Trace-element metabolism in animals 2nd ed. University of Baltimore, 1974.
[87] Frieden E. *Biochemistry of the essential ultratrace-elements.* Plenum Press,1984, New York.
[88] Uppala RT, Roy SK, Tousson A, Barnes S, Uppala GR, Eastwood D.Induction of cell proliferation micronuclei and hyperdiploidy/diploidy in the mammary cells of DDT and DMBA-treated pubertal rats.*Environ. Mol. Mutagen*.,2005; 46: 43-52.
[89] Murray RK, Granner DK, Mayes PA, Rodwell VW. *Harper's biochemistry*. 25th ed. McGraw-Hill, Health Profession Division, 2000.
[90] Brown M. Harnessing chromium in the fight against diabetes. *Drug Discovery Today*,2003; 8: 962-963.
[91] Wennberg A. Neurotoxic effect of selected metals. *Scand. J. Work. Environ. Health*.,1994; 20: 65-71.
[92] Juturu V, Komorowski JR. Chromium supplements, glucose and insulin responsem. *Am. J. Clin. Nutr.,*2003; 78: 190.
[93] Frank A, Danielsson R, Jones B. Experimental cupper and chromium deficiency and additional molybdenum supplementation in goats. II Concentration of trace and minor

elements in liver, kidneys and ribs: haematology and clibical chemistry. *Scien. TotalEnviron.,*2000; 249: 143-170.
[94] Langard S. Chromium. In: *Metals in the environment.* Academic Press,1980, London.
[95] Chojnacka K.Biosorption and accumulation- the prospects for practical applications. *Environment Int.,*2010; 36: 299-307.
[96] Durand M, Komisarczuk S. Influence of majors minerals on rumen microbiota. *J. Nutr.,*1988; 118: 249–260.
[97] Owusu-Asiedu A, Nyachoti CM, Marquardt RR. Response of early-weaned pigs to an enterotoxigenic Escherichia coli (K88) challenge when fed diets containing spray-dried porcine plasma or pea protein isolate plus egg yolk antibody, zinc oxide, fumaric acid, or antibiotic *J. Anim. Sci.,*2003; 81: 1790-1798.
[98] Minakowski D, Tywończuk J. Animal feeding vs. requirements of intensive livestock production. *RaportRolny,* 31, 2004.
[99] Kuczewska H. Feed additives for sale and use in animal nutrition. Centr. Zw. Spółdz. Rolniczych, Warszawa, 1979.
[100] Kujawiak R. Effect of premix on the efficiency of poultry production, *Sano-Modern Animal Nutrition*, Sekowo, 2003. <<http://www.ppr.pl>> date of entry 12 IV 2010.
[101] McDowell LR. *Minerals in animal and human nutrition.* Elsevier Science B.V. Holand, 2003.
[102] Wierny A. *Mineral additives in animal nutrition.* PaszaPrzem., 2002; 4: 9.
[103] Notice of Minister of Agriculture and Rural Development about feed additives and feed materials. *List of feed materials*, 7 January 2004.
[104] Kaczmarek A. *Farming cattle*, Ed. Univ. Agricultural University, Poznan, 1994.
[105] Wójcik S. The utility of minerals in animal nutrition. *PaszaPrzem.*, 2002; 4: 6.
[106] Jamroz D. *Animal nutrition and feedstuffs, Physiological and biochemical basis of animal nutrition*, Ed. Science. PWN, Warsaw, 2009; pp 61-90.
[107] Smith OB, Akinbamijo OO. Micronutrients and reproduction in farm animals. *Anim, Reprod. Sci.*, 2000; 60–61: 549-560.
[108] Smulikowska S. *Feeding Standards for Swine.* Omitech, Warsaw, 1996.
[109] Korol W. Mineral feed additives-regulation: Poland-EU. *PaszaPrzem.,* 2002; 4: 18.
[110] Dach J, Starmans D. Heavy metals balance in Poland and Dutch agronomy: Actual state and previsions for the future. *Agric. Ecosyst. Environ.*, 2005; 107: 309-316.
[111] Marchetti M, DeWayne AH, Tossani N, Marchetti S, Ashmead SD. Comparison of the Rates of Vitamin Degradation when Mixed with Metal Sulphates or Metal Amino Acid Chelates. *J Food Comp Anal.,* 2000; 13: 875-884.
[112] Chowdhury SD, Paik IK, Namkung H, Lim HS. Responses of broiler chickens to organic copper fed in the form of copper–methionine chelate. *Anim. Feed Sci. Technol.,* 2004; 115: 281-293.
[113] Ashmed HD. The absorption and metabolism of iron amino acid chelate. *ArchivosLatinoamericanos de Nutricion*, 2001; 51:13-21.
[114] Albion Advanced Nutrition, *The nutrition review. Albion Research Notes*, 2006; 1:4.
[115] http://www.aafco.org/Portals/0/Public/2009_annual_reports-compiled-REVISED.pdf>> date of entry 12 IV 2010, AAFCO 100th ANNUAL MEETING, Washington.
[116] Spears JW. Organic trace minerals in ruminant nutrition. *Anim. Feed Sci. Technol.,* 1996; 58: 151-163.

[117] Albion Human Nutrition, Molecular Weight and the Metal Amino Acis Chelate. *Albion Research Notes*, 1995; 4:1.

[118] Ammerman CB. *Bioavailibility of Nutrients for Animals*, Academic Press. San Diego, 1995.

[119] Pott EB, Henry PR, Ammerman CB, Merritt AM, Madison JB, Miles RD. Relative bioavailibility of copper in a copper-lysine complex for chicks and lambs. *Anim. Feed Sci. Technol.,* 1994; 45: 193-203.

[120] Baker DH, Odle J, Funk MA, Wieland TM. Bioavailibility of copper in cupric oxide, cuprous oxide and in a lysine complex. *Poult Sci.,* 1991; 70: 177-179.

[121] Aoyagi S, Baker DH. Nutritional evaluation of a copper-methionine complex for chicks. *Poult. Sci.,* 1993; 72: 2309-2315.

[122] Aoyagi S, Wedekind KJ, Baker DH. Estimates of copper bioavailability from liver of different animal species and from feed ingredients derived from plants and animal. *Poult. Sci.,* 1993; 72: 1746-1755.

[123] Zanetti MA, Henry PR, Ammerman CB, Miles RD. Estimation if the relative bioavailability of copper sources in chicks fed conventional dietary levels. *Br. Poult. Sci.,* 1991; 32: 583-588.

[124] Ledoux DR, Henry PR, Ammerman CB, Roa PV, Miles RD. Estimation of the relative bioavailability of inorganic copper sources for chicks using tissue uptake of copper. J. *Anim. Sci.*, 1991; 69: 215-222.

[125] Aoyagi S, Baker DH. Bioavailibility of copper in analytical-grade and feed-grade inorganic copper sources when fed to provide copper at levels below the chicks requirement. *Poult. Sci.,* 1993; 72: 1075-1083.

[126] Bunch RJ, McCall JT, Speer VC, Hays VW. Copper supplementation for weanling pigs. *J. Anim. Sci.,* 1965; 24: 995-1000.

[127] Allen MM, Barber RS, Braude R, Mitchell KC. Further studies on various aspects of the use of high-copper supplements for growing pigs. *Be. J. Nutr.,* 1961, 15: 507-522.

[128] Buescher RG, Griffin SA, Bell MC. Copper availability to swine from Cu64 labeled inorganic compounds. *J. Anim. Sci.,* 1961; 20: 529-531.

[129] Poitevinit AL. Determination of the true biological availability of ferrous carbonate. *Feedstuffs,* 1979; 51: 31.

[130] Fritz JC, Pla GW, Robert T, Boehne JW, Hove EL. Biological availability in animals of iron from common dietary sources. *J. Agric. Food Chem.*, 1970; 18: 647-651.

[131] Pla GW, Fritz JC. Availability of iron. J. Assoc. of Anal. Chem., 1970; 53: 791-800.

[132] Brady PS, Ku PK, Ullrey DE, Miller ER. Evaluation of an amino acid-iron chelate hematinic for the baby pig. *J. Anim. Sci.,* 1978; 47: 1135-1140.

[133] Spears JW, Schoenherr WD, Keglez EB, Flowers WL, Alhusen HD. Efficacy of iron methionine as source of iron for nursing pigs. *J. Anim. Sci.*, 1992; 70 (Suppl. 1) : 243. (abstract)

[134] Baker DH, Halpin KM. Efficacy of manganese-protein chelate compound compared with that of manganese sulfate for chicks. *Poult. Sci.*, 1987; 66: 1561-1563.

[135] Henry PR, Ammerman CB, Miles RD. Relative bioavailability of manganese in a manganese Methionine complex for broiler chicks. *Poult. Sci.,* 1989; 68: 107-112.

[136] Southern LL, Baker DH. Excess manganese ingestion in the chick. *Poult. Sci.,* 1983; 62: 642-646.

[137] Watson LT, Ammerman CB, Miller SM, Harms RH. Biological availability to chicks of manganese from different inorganic sources. *Poult. Sci.*, 1971; 50: 1693-1700.

[138] Schaible PJ, Bandemer SL, Davidson JA. The manganese content of feedstuff and its relation to poultry nutrition. Michigan Agric. *Exp. Sta. Tech. Bull.* 1938; 159.

[139] Gallup WD, Norris LC. The amount of manganese requires to preventperosis in the chicks. *Poult. Sci.,* 1939; 18: 76-82.

[140] Wedekind KJ Baker DH. Zinc bioavailability in feed-grade sources of zinc. *J. Anim. Sci.,* 1992; 68: 684-689.

[141] Roberson RH, Schaible PJ. The availability to the zinc as the sulfate, oxide or carbonate. *Poult. Sci.*, 1960; 39: 835-837.

[142] Roberson RH, Schaible PJ. The tolerance of growing chicks for high levels of different forms of zinc. *Poult. Sci.*, 1960; 39: 893-896.

[143] Roberson RH, Schaible PJ. The zinc requirement of the chick. *Poult. Sci.*, 1958; 37:1321-1323.

[144] Hahn JD, Baker DH. Growth and plasma zinc responses of young pigs fed pharmacologic levels of zinc. *J. Anim. Sci.,* 1993; 71:3020-3024.

[145] Walter R. *Therapeutic iron complex*. USP 2'877'253, 1959.

[146] Walter R. Process for raising blood serum iron levels and controlling anemia. *USP* 2'957'806, 1960.

[147] Ashmead HH. Soluble non-ferrous metal proteinates. *USP* 4'216'144, 1980.

[148] Ashmead HH. Soluble iron proteinates. *USP* 4'216'144, 1980.

[149] United State Patent 5'504'055, 1996.

[150] Davidson JA. Unlocking micromineral absorption. *Feed-International*, 2007; 28: 15-20.

[151] Miles RD, Henry PR. Relative trace mineral bioavailability. *Ciência Animal Brasileira* 2000; 1: 73-92.

[152] Cao J, Liu LP, Henry PR, Ammerman CB. Solubility and gel chromatographic assessment of some organic zinc sources. *J. Anim. Sci.*, 1998; 76: 188.

[153] Ofertahandlowafirmy: JHBiotech, Biotechnologies for Safer Agriculture <<http://www.jhbiotech.com/animal_products/proteinates.htm>>, date of entry 1 III 2010.

[154] Mrvcic J, Stanzer D, Stehlik-Tomas V, Škevin D, Grba S. Optimization of bioprocess for production of copper-enriched biomass of industrially important microorganism Saccharomyces cerevisiae. *J. Biosci. Bioeng.*, 2007; 103: 331-337.

[155] Tompkins TA, Renard NE, Kiuchi A. Clinical Evaluation of the Bioavailability of Zinc-enriched Yeast and Zinc Gluconate in Healthy Volunteers. *Biol. Trace. Elem. Res.,* 2007; 120: 28-35.

[156] Jamroz D, Podkówka W, Chachulowa J. Żywieniezwierząt i paszoznawstwo, *Wyd. Nauk.* PWN Warszawa, 2004.

[157] Hossain SM, Barreto SL, Silva CG. Growth performance and carcass composition of broilers fed supplemental chromium from chromium yeast. *Anim. Feed. Sci. Technol.,* 1998; 71: 217-228.

[158] Chojnacka K. Biosorption and bioaccumulation – the prospects for practical applications. *Environ. Int.,* 2010; 36: 299-307.

[159] Chojnacka K. The application of biosorption and bioaccumulation of toxic metals in environmental pollution control. W Lewinsky AA (Red.) *Hazardous Materials and*

Wastewater: Treatment, Removal and Analysis, NOVA Science Publisher, Nowy York, 2007; pp 309-359.

[160] Liu Y, Liu Y-J. Biosorption isotherms, kinetics and thermodynamics. *Sep. Purif. Technol.*, 2008; 61: 229-242.

[161] Volesky B. *Biosorption and me. Wat. Res.*, 2007; 41:4017-4029.

[162] Selatnia A, Boukazoula A, Kechid N, Bakhti MZ, Chergui A, Kerchich Y. Biosorption of lead (II) from aqueous solution by a bacterial dead Streptomyces rimosus biomass. *Biochem. Eng. J.*, 2004; 19:127.

[163] Kao W-C, Huang C-C, Chang J-S Biosorption of nickel, chromium and zinc by MerP-expressing recombinant Escherichia coli. *J. Hazard. Mat.*, 2008; 158:100-106.

[164] Choi J, Lee JY, Yang J-S. Biosorption of heavy metals and uranium by starfish and Pseudomonas putida. *J. Hazard Mat.,* 2009; 161: 157-162.

[165] Kapoor A, Viraraghavan T. Fungi as biosorption. W Wase DAJ, Forster CF (Red.). *Biosorbents for Metal Ions*, London, UK, Taylor and Francis, 1997; pp 67–85.

[166] Aksu Z, Çalik A, Dursun AY, Demircan Z. Biosorption of iron(III)–cyanide complex anions to Rhizopusarrhizus: application of adsorption isotherms. *Process Biochem.,* 1999; 34: 483-491.

[167] Akhtar K, Khalid AM, Akhtar MW, Ghauri MA. Removal and recovery of uranium from aqueous solutions by Ca-alginate immobilized Trichodermaharzianum. *Biores. Technol.,* 2009; 100: 4551-4558.

[168] Sag Y, Kutsal T. Determination of the biosorption activation energies of heavy metal ions on Zoogloearamigera and Rhizopusarrhizus. *Process Biochem.*, 2000; 35: 801-807.

[169] Jalali R, Ghafourian H, Asef Y, Davarpanah SJ, Sepehr S. Removal and recovery of lead using nonliving biomass of marine algae. *J. Hazard Mat.*, 2002; 92: 253-262.

[170] Tsai W-T, Chen H-R. Removal of malachite green from aqueous solution using low-cost chlorella-based biomass. *J. Hazard Mat.,* 2010; 175: 844.

[171] Aksu Z. Determination of the equilibrium, kinetic and thermodynamic parameters of the batch biosorption of nickel(II) ions onto Chlorella vulgaris. Process Biochem., 2002; 38: 89-99.

[172] Gokhale SV, Jyoti KK, Lele SS. Kinetic and equilibrium modeling of chromium (VI) biosorption on fresh and spent Spirulinaplatensis/Chlorella vulgaris biomass. *Bioresour. Technol.,* 2008; 99: 3600-3608.

[173] Gong R, Ding Yi, Liu H, Chen Q, Liu Z. Lead biosorption and desorption by intact and pretreated Spirulina maxima biomass. *Chemosphere,* 2005; 58: 125-130.

[174] Gupta R, Ahuja P, Khan S, Saxena RK. HarapriyaMohapatra, Microbial biosorbents: Meeting challenges of heavy metal pollution in aqueous solutions. *Curr. Sci.*, 2000; 78: 967–973.

[175] Vijayaraghavan K, Yeoung-Sang Y. Bacterial biosorbents and biosorption. *Biotech. Adv.,* 2008; 26: 266-291.

[176] Brady JM, Tobin JM. Binding of hard and soft metal ions to Rhizopusarrhizus biomass. *Enzyme Microb. Technol.*, 1995; 17: 791-796.

[177] Zielinska A, Chojnacka K. The comparison of biosorption of nutritionally significant minerals in single- and multi-mineral systems by the edible microalga Spirulina sp. *J. Sci. Food Agri.,* 2009; 89: 2292–2301.

[178] Michalak I, Chojnacka K. The application of microalga PhitophoravariaWille enriched with microelements by biosorption as biological feed supplement for livestock. *J. Sci. Food Agric.*, 2008; 88: 1178–1186.
[179] Zielińska A, Chojnacka K, Dobrzański Z, Górecki H, Michalak I, Korczyński M, Opaliński S. *Lemna minor enriched with microelements via biosorption a new bio-metallic feed additives, (under review).*
[180] Deng SB, Ting YP. Fungal biomass with grafted poly(acrylic acid) for enhancement of Cu(II) and Cd(II) biosorption. *Langmuir*, 2005; 21: 5940-5948.
[181] Vijver MG, Van Gestel CAM, Lanno RP, Van Straalen NM, Peijnenburg W. Internal metal sequestration and its ecotoxicological relevance: a review. *Environ. Sci. Technol.,* 2004; 38: 4705-4712.
[182] Kuroda K, Ueda M, Shibasaki S, Tanaka A. Cell surface-engineered yeast with ability to bind, and self-aggregate in response to, copper ion. *Appl. Microbiol. Biotechnol.,* 2002; 59: 259-264.
[183] Kadukowa J, Vircikova E. Comparison of differences between copper bioacumulation and biosorption. *Environ. Int.*, 2005; 31: 227-232.
[184] Crist RH. Nature of Bonding between Metallic Ions and Algal Cell Walls. *Environ. Sci. Technol.,* 1981; 5: 1212-1217.
[185] Schiewer S, Volesky B. Modeling of the Proton-Metal Ion Exchange InBiosorption. *Environ. Sci. Techno.*, 1995; 29: 3049-3058.
[186] Davis TA, Volesky B. A review of the biochemistry of heavy metal biosorption by brown alga. *Water Res.,* 2003; 37: 4311-4330.
[187] Veglio F, Beolchini F. Removal of metals by biosorption: a review. *Hydrometallurgy,* 1997; 44: 301-316.
[188] Zielinska A. *Elaboration of production process of new generation of mineral feed additives based on microalage biomass, Ph.D thesis*, Wroclaw University of Technology, 2010.
[189] Senthilkumar P, Nagalakshmi D, Ramana Reddy Y, Sudhakar K. Effect of different level and source of copper supplementation on immune response and copper dependent enzyme activity in lambs. *Trop. Anim. Health Prod.*, 2009; 41: 645–653.
[190] Pal DT, Gowda NKS, Prasad CS, Amarnath R, Bharadwaj U, Suresh Babu G, Sampath KT. Effect of copper-and zinc-methionine supplementation on bioavailability, mineral status and tissueconcentrations of copper and zinc in ewes. *J. Trace Elem. Med. Biol.,* 2010; 24: 89–94.
[191] Lu L, Wang RL, Zhang ZJ, Steward FA, Bin Liu XL. Effect of Dietary Supplementation with Copper Sulfate or Tribasic Copper Chloride on the Growth Performance, Liver Copper Concentrations of Broilers Fed in Floor Pens, and Stabilities of Vitamin E and Phytase in Feeds. *Biol. Trace Elem. Res.*, 2010; 138: 181–189.
[192] Shinde P, Dass RS, Garg AK, Chaturvedi VK, Kumar R. Effect of Zinc Supplementation from Different Sources on Growth, Nutrient Digestibility, Blood Metabolic Profile, and Immune Response of Male Guinea Pigs. *Biol. Trace Elem. Res.,* 2006; 112: 247-262.
[193] Pechova A. Misurova L. Pavlata L. Dvorak R. The Influence of Supplementation of Different Forms of Zinc in Goats on the Zinc Concentration in Blood Plasma and Milk. *Biol. Trace Elem. Res.,* 2009; 132: 112–121.

[194] Garg AK, Mudgal V, Dass RS. Effect of organic zinc supplementation on growth, nutrient utilization and mineral profile in lambs. *Anim. Feed Sci. Technol.*, 2008; 144: 82-9.

In: Livestock: Rearing, Farming Practices and Diseases ISBN 978-1-62100-181-2
Editor: M. Tariq Javed

Chapter 3

SOY PROTEIN PRODUCTS: ANTI-NUTRITIONAL FACTORS, CLASSIFICATION, PROCESSING, QUALITY ASSESSMENT, NUTRITIONAL VALUE AND APPLICATION IN ANIMAL FEED

Yueming Dersjant-Li * ***and Manfred Peisker***
ADM Specialty Ingredients, 1540 AA Kooga/dZaan (The Netherlands)

ABSTRACT

This chapter describes different soy protein products based on production processes, their nutritional value and application in animal feed.

There are several commercial soy protein products available in the market. These local products come from different origins, varieties and are produced differently. This causes variations in their nutritional value. For a systematic classification, appropriate terms need to be defined and used to describe different commercial products. First, the production process leads to different protein levels in the final product that can be used for their classification. Based on production process, soy protein products can be classified into two categories: 1) Basic processed soy protein including full fat soybean meal (36% CP), defatted soybean meal (44% CP) and high protein soybean meal (48% CP); 2) Highly processed soy protein including enzymatically/microbially fermented soy protein (52-56% CP), non-classically produced soy protein concentrate (60% CP) and classically produced soy protein concentrate (65-67% CP). Next to the protein level, other components are intrinsically linked to the products in above defined categories, which may have significant consequences for their usage in different animal species and age groups within a species.

Raw soybeans contain high amount of soy anti-nutritional factors (ANFs) that can be divided into heat stable and heat sensitive components. In animal nutrition, the most important heat labile ANFs are trypsin inhibitor activities (TIA) and lectins. The most important heat stable ANFs are antigens and oligosaccharides. High antigen levels can cause allergic response and intestinal damage in calves, young piglets and in some fish species. Oligosaccharides cannot be digested by young calves and monogastric animals,

* Corresponding author: Yueming.dersjant-Li@adm.com

but can be fermented by bacteria, resulting in intestinal disturbance and diarrhea in young animals (such as weaning piglets).

These ANFs must be removed or inactivated by proper treatments before their applications in animal feed. Basic processed soy proteins are typically produced by heat treatment. The proper heating process reduces heat sensitive ANFs, such as trypsin inhibitor activity and lectins. However, large variation may exists for levels of trypsin inhibitor activity among differently processed soy protein products due to heat treatment conditions. The heat stable ANFs, in general, are not removed from these basic processed soy proteins. During the process of making of highly processed soy protein, both heat labile and heat stable ANFs are removed/eliminated, thus the final product have low levels of ANFs. In the process for soy protein concentrate, indigestible carbohydrates are also removed, leading to higher protein content and digestibility compared with basic processed soy products.The basic processed products are valuable sources of protein for poultry, growing and finishing pigs, ruminants and aquatic animals. However, due to the high amounts of heat stable soy ANFs, they have limited application in young animals` feed.Highly processed soy protein products have low soy ANFs; however, the ANF content and indigestible soy carbohydrates content are related to the processing methods and treatment conditions.

Production processes have direct impact on product quality and amino acids digestibility. Highly processed soy protein products have improved protein and AA digestibility when compared to basic processed soy products. The available literature demonstrates that the indigestible carbohydrates and other ANFs play an important role in digestibility of amino acids in soy protein, whereas the molecular size of the protein seems to have no or lesser importance.Due to the differences in production process, the quality assessment criteria are also different for basic processed soy proteins and highly processed soy proteins. For basic processed soy protein, TIA and PDI (Protein Dispersibility Index) are common methods for evaluating heat treatment efficiency. For highly processed soy protein, the following parameters should be considered: crude protein and essential AA content as percentage of crude protein, AA digestibility at different age of animals, indigestible carbohydrates content, soy ANFs level including TIA, antigens, oligosaccharides and soluble non-starch polysaccharides.

The composition and nutritional value of soybeans as well as soybean meal are not only related to production process, but also varies among origin and year of production. These variations affect feed formulations. Awareness of the variation in soybeans and processed soybean meal composition as well as variation on digestibility is important for nutritionists. In the past, animal feeds have been formulated with excess nutrients to meet/exceed the nutritional requirement of animals. Economic pressure and environmental regulations have made modern feed formulation to meet the nutritional requirements more precisely. This optimizes performance and minimizes waste output. However, this concept is not working without precise knowledge of ingredients used in feed formulation. Since soy products are the overwhelming supply source of dietary crude protein and amino acids, the quality assessment of soy products with respect to amino acid composition and digestibility and ANFs disserves utmost attention. Additionally, many different processing technologies are around and may affect soy quality as well.

For basic produced soy protein, the use of table values to formulate animal feed is not advisable seeing the existing variations in composition, ANF content and amino acid digestibility. It is recommended that nutritionists use producers' data sheets wherever available to formulate feed or make own analysis. The use of table values is only advisable in case there are no other sources of data available. Due to the advanced production process, highly processed soy protein can provide more consistent product quality, as these products are tightly controlled to meet nutritional specifications.

In brief, nutritional composition differs between soy protein products due to different processing methods. In practice, differently processed soy products are applicable in feed for different animal species or physiological stages. For example, standard defatted soybean meal with 44percent crude protein is commonly used in diet for growing and finishing pigs. High-Pro soybean meal with 48percent crude protein is commonly used in poultry feed. Highly processed soy products such as fermented soy protein and non-classical soy protein concentrates are commonly used in weaning piglets feed and aqua-feed. Classical soy protein concentrates are commonly used in calf milk replacer and weaning piglets feed.

1. INTRODUCTION

1.1. World Oil Seed Production

FAO statistics [1] demonstrate that soybean is the most important plant protein source from oil seeds, contributing more than 57percent of the total world oil seed production (Figure 1).

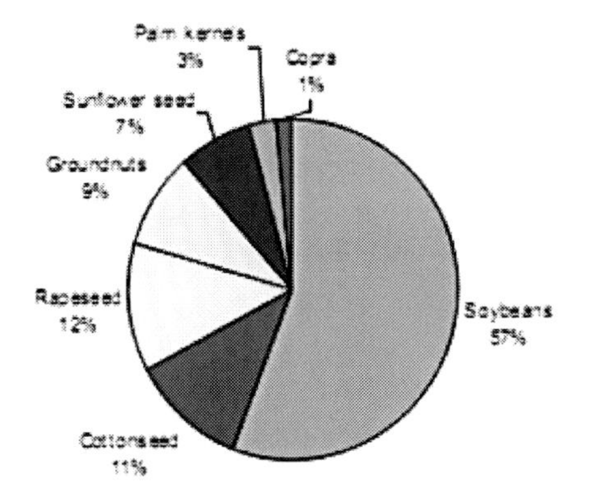

	2007/2008	2008/09 est	2009/10 f'cast
Soybeans	220	211.5	248
Cottonseed	44	40.7	40.2
Rapeseed	48.7	58.4	58.3
Groundnuts	35.4	35.2	33
Sunflower seed	28.9	33.9	31.5
Palm kernels	11.2	11.5	12.1
Copra	5	5.2	5.3
Total	393.2	396.4	428.4

Figure 1. World oil seed production in percentage (data from 2007/2008) and estimated for 2008/2009, forecasted for 2009/2010, in million tons [1].

1.2. Soy Protein in Animal Nutrition

Modern livestock production is unthinkable without soybeans (*Glycine max*). Soybeans represent the most important and abundantly available vegetable protein source. Having balanced amino acids profile and good digestibility, soy protein is by far the main plant protein source used in livestock feed.

It is well known that raw soybeans contain high levels of anti-nutritional factors (ANFs) and thus need to be properly processed before their application in animal feed. Different production processes are used to produce commercial soy protein products. Consequently, the nutritional values differ significantly, not only in protein and amino acids, but also in carbohydrates and the level and activity of soy heat stable ANFs. The quality assessment criteria also differ for different soy products.

Differently processed soy products are applicable in the feed for different animal species or physiological stages. For example, standard defatted soybean meal with 44percent crude protein is commonly used in diets for growing and finishing pigs and high pro-soybean meal with 48percent crude protein in poultry feed. Highly processed soy products such as fermented soy protein are commonly used in feed for weaning piglets and aquatic species. Soy protein concentrates are mainly used in calf milk replacers and weaning piglets feed.

As there are many different types of soy protein products commercially available for animal feed, the diversities of soy protein products requests classification of soy protein products.

2. Soy Anti-Nutritional Factors (ANFs)

Raw soybeans contain ANFs, which can have negative impacts on growth performance of livestock. The ANFs in soybeans can be divided into heat labile and heat stable components. The heat labile ANFs include protease inhibitors, lectins, and goitrogens. The heat stable ANFs include antigens, oligosaccharides, saponins, estrogens, phytate and cyanogens. The most important heat labile ANFs are trypsin inhibitors activities (TIA) and lectins. Heat labile ANFs can be removed by the proper heating of the raw soybeans. Therefore, the levels of these ANFs are closely related to heat treatment conditions. All commercial soy protein products have undergone a heat treatment process. However, the heat stable soy ANFs are not very well recognized. The most important heat stable ANFs are antigens and oligosaccharides. Heat stable ANFs cannot be removed by a standard toasting process. Most of the heat stable ANFs can be removed or inactivated during ethanol extraction, or by fermentation process.

In general, the most important ANFs, which can have a significant negative effect on the growth performance of animals, are antigens, trypsin inhibitors, oligosaccharides, lectins and saponins. The impacts of these ANFs on performance of animals are discussed below.

2.1. Heat Labile ANFs

2.1.1. Trypsin Inhibitor Activity (TIA)

Soybeans contain about the same amounts of Kunitz and Bowman-Birk trypsin inhibitor (protease inhibitors). The latter is a low-molecular-weight protein, which has the ability to inhibit trypsin as well as chymotrypsin. The inhibitors are peptides that can form stable inactive complexes with some of the pancreatic enzymes. Thus, the trypsin inhibitors can reduce the activities of trypsin and chymotrypsin, decrease digestibility of nutrients and consequently depress growth performance of animals.

Soy protein products need a proper thermal processing to reduce trypsin inhibitor activity (TIA) in order to achieve the optimal product quality. The negative effect of TIA has been observed in many animal species, including poultry, calves, pigs and fish. Literature studies [2, 3, 4, 5] show that a TIA level of higher than 3 mg/g had a negative effect on protein digestibility and growth performance of animals. On the other hand, overheating can lower TIA to such low levels that the digestibility of protein may be reduced. It was reported [5] that the optimal growth performance of chicks was obtained at TIA levels between 1-3 mg/g (Figure 2).

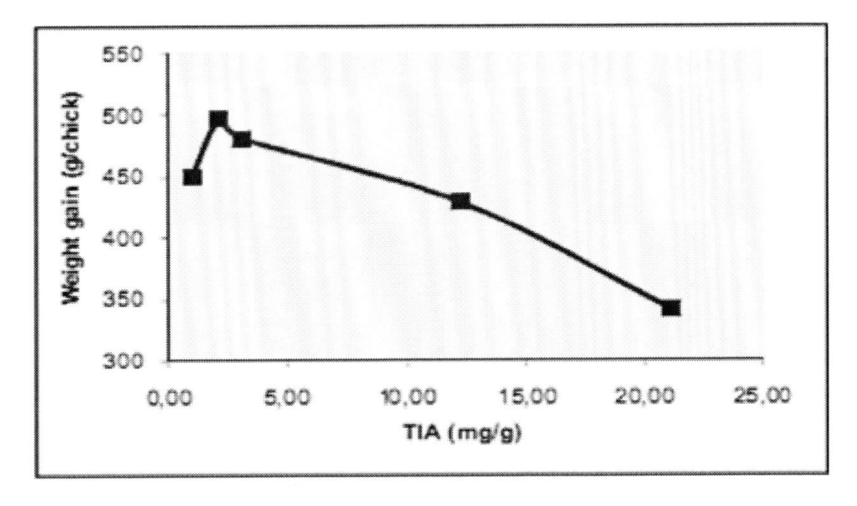

Figure 2. The relationship between TIA levels in soybean meal and growth rate of chickens.

Schulze [6] evaluated the effects of purified soybean trypsin inhibitor on the flow of nitrogen at the distal ileum in young growing pigs. This study demonstrated that trypsin inhibitors substantially increased the ileal flow of both endogenous and exogenous nitrogen. The author suggested that TIA affected exogenous nitrogen and true nitrogen digestibility in a linear manner (Table 1).

Full fat soybean meal and defatted soybean meal (SBM) are processed by heat treatment to reduce TIA. Therefore, TIA is commonly recognized and commonly used as quality assessment criteria for SBM.

Table 1. Effect of added soybean Kunitz soybean trypsin inhibitor and lectins on the flow (percent of N intake) of total, dietary and endogenous N at the ileum of pigs, as determined by using 15N-dilution techniques [6, 7]

	TIA (mg inhibited trypsin/g diet)			Lectin activity (mg/kg diet)		
	0.21	2.49	5.77	0	177	1065
N flow (% of intake)						
Total	14.2^{a}	26.9^{b}	44.7^{c}	13.8^{a}	14.9^{a}	18.7^{b}
Dietary	0.9^{a}	4.1^{a}	16.7^{b}	5.2^{b}	3.8^{a}	6.2^{b}
Endogenous	13.3^{a}	22.8^{b}	28.0^{c}	8.6^{a}	11.1^{b}	12.5^{c}

Within a row, mean values with different superscript letters are significantly different (P<0.05) for a given experiment.

2.1.2. Lectins

Lectins are glycoproteins, which can cause the agglomeration of red blood cells. Lectins bind to the small intestinal mucosa cells and cause their disruption. This damages the intestinal surface and results in reduced nutrient absorption and decreased growth rate of animals.

Supplementation of purified soybean lectins to a diet for young pigs led to a dose-dependent increase of the dietary and endogenous flow of nitrogen [7].

2.2. Heat Stable ANFs

2.2.1. Soy Antigens

Soy antigens are storage globulins, glycinin (called 11S globulin) and β-conglycinin (known as 7S globulin), which represent two-thirds of the proteins in the beans. These proteins have large molecular weight that may escape digestion and are immunogenic in young calves, piglets and in some fish species. They have been suspected to interfere with intestinal function via immunological mechanisms [8].

Antigenic proteins are capable of inducing a humoral immune response when fed to animals [9]. A humoral immune response occurs when specific antibodies are produced and secreted into body fluids such as the blood. Antigenic proteins are known to cause gut wall damage and immunological reaction in the gut linked with disorder in gut function in piglets and veal calves. Antigenic protein stimulates the immune system of young mammals to develop chronic gut hypersensitivity reactions.

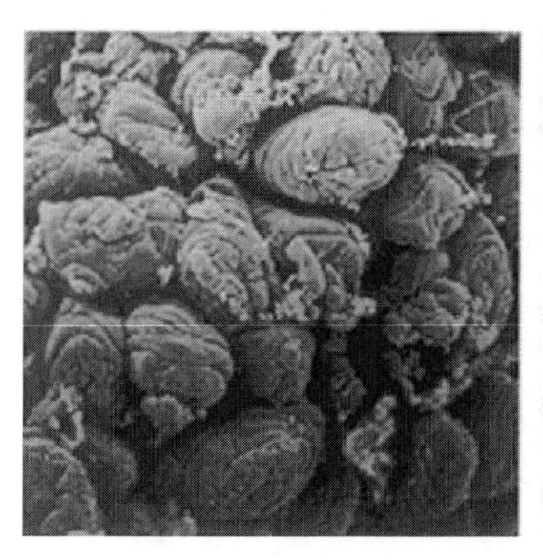

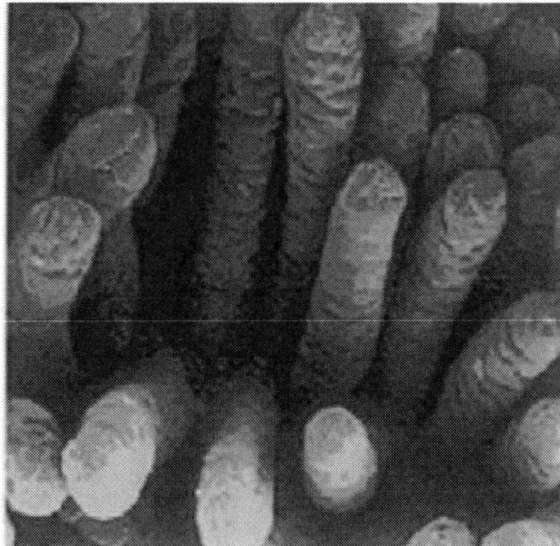

Figure 3. Scanning electron micrographs of mid-jejunal sections of the small intestine of two soy-sensitive animals challenged with a diet containing either a soy flour (left) or an ethanol extracted traditional soy protein concentrate (right) [15].

In the early 1990s, it was recognized that gut hypersensitivity could be transiently observed in piglets soon after weaning. Providing early-weaned piglets with low antigen and highly digestible protein sources is the best way to limit the gut inflammation around weaning [10]. The piglets remain sensitive to antigenic protein until tolerance is fully established by the age of 12 weeks [10]. In contrast, calves consuming high antigenic soy protein develop more severe, long lasting immune mediated gut hypersensitivity reaction to antigenic soy products. This can result in increased endogenous gut protein secretion and change in gastro-intestinal motility and morphology as well as gut permeability. Based on direct skin tests and *in- vitro* lympho-proliferation, Lalles *et al.* [11] reported that β-conglycinin is the most allergenic protein in soybean. The glycinin is also immunogenic. Motility disruption in sensitized calves appears on an average higher than 7 and 6 mg of immuno-reactive glycinin and β-conglycinin/g of crude protein intake, respectively [12]. In pre-ruminant calves fed antigenic heated soy flour, it was demonstrated by ELISA that around 10 and 1percent of undigested glycinin and β-conglycinin, respectively reached the terminal ileum in an immuno-reactive form [13]. Poultry appears to be less sensitive to soy antigens.Young animals are very sensitive to antigens present in processed soybean products. The absorption of soy antigens can generate antibody production in the bloodstream, damage the intestinal surface, reduce the absorption of nutrients, and decrease the performance of animals. Many studies have reported that calves or piglets exposed to low quality, high antigen soybean meal had intestinal disturbances as characterized by reduced villus length and surface area [14, 15]. In fish, intestinal damage was observed after feeding soybean meal containing high levels of ANFs [16, 17, 18].

No allergic reaction has been observed in calves, piglets and salmon when fed diets based on refined soy protein such as soy protein concentrate containing very low levels of antigen activity. Figure 3 demonstrates that feeding a diet containing low antigen soy protein concentrate to calves assured healthy villi whereas feeding a diet containing high antigen soy flour damaged the intestinal surface [15]. It was observed that piglets fed a diet containing a soy protein concentrate had normal serum antibody titers specific to soy protein, whereas the serum antibody titers were high in pigs fed a diet containing antigenic soybean meal [14]. The effect of soy antigens on serum antibody immunoglobulin G titers is illustrated in Figure 4.

Soy antigens cannot be removed by standard toasting process commonly used to produce soybean meal. Extrusion process at high temperature and moisture may reduce antigen activities to some but usually non-sufficient extent. Other methods such as enzymatic or microbial fermentation processes can also reduce antigen activities.

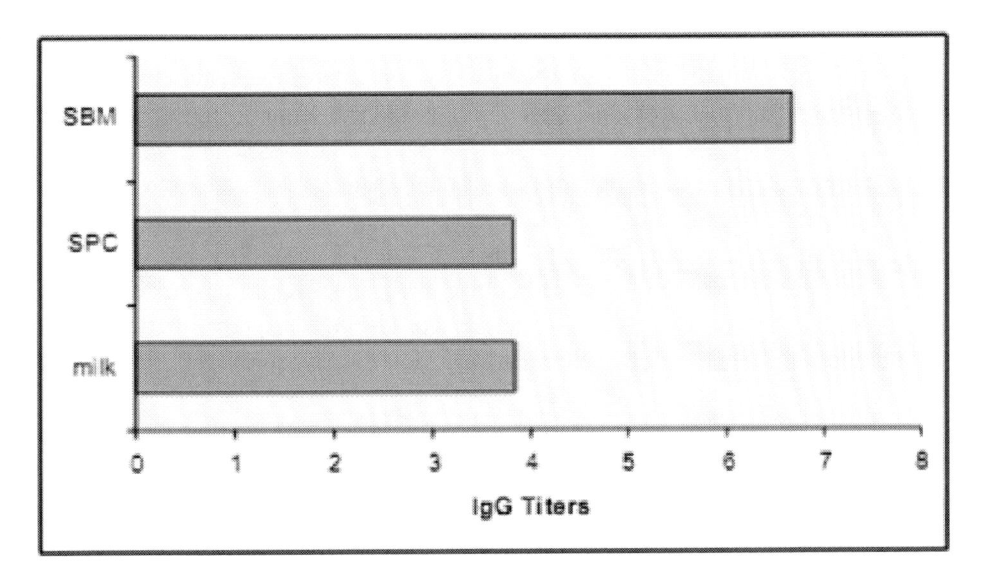

Figure 4. Effect of soy antigen on serum antibody immunoglobulin G titers to soy protein in pigs (SBM: soybean meal; SPC: soy protein concentrate).

Table 2 shows the effect of processing methods on selected anti-nutritional factor levels, confirming that antigens are resistant to toasting process.

Highly processed soy protein such as feed grade soy protein concentrate typically contains less than 50 mg/kg glycinin antigens, whereas soybean meal contains typically 40,000 to 66,000 mg/kg glycinin antigens.

Table 2. Content of selected ANFs and protein dispersibility index (PDI) in various toasted soy products [9]

	PDI, %	TIA, mg/g	Lectins, mg/g	Antigens, mg/kg
Untoasted soy flour	90	23.9	7.3	610,000
Slightly toasted soy flour	70	19.8	4.5	575,000
Normally toasted soy flour	20	3.1	0.05	125,000
Ethanol extracted SPC	6	2.5	<0.0001	<20

2.2.2. Soy Oligosaccharides

Oligosaccharides cannot be broken down in the digestive system of young calves, monogastric animals and aquaculture species owing to the lack of a specific enzyme - alpha-galactosidase. Intestinal bacteria can ferment oligosaccharides and this causes flatulence and diarrhea. Soy oligosaccharides are important heat stable ANFs, these components cannot be removed by heat treatment. In contrast to mannan-oligosaccharides derived from yeast, which are commonly used as prebiotics due to their binding capacity to pathogenic bacteria at low dietary inclusion levels, soy oligosaccharides do not possess prebiotic properties.

Iji and Tivey [19] reported that the effect of natural raffinose series oligosaccharides (raffinose, stachyose and verbascose) is different from synthetic (industrial) products on the productivity of broiler chickens. The use of natural raffinose series oligosaccharides can have negative effects on animal health and productivity, but there are beneficial effects from synthetic oligosaccharides. Due to the absence of α-galactosidase in the upper gastrointestinal tract of monogastric animals, the raffinose series oligosaccharides remain undigested up to the lower gut, where they are fermented by intestinal bacteria with release of gases, which have been associated with flatulence in non-ruminant animals. The digestion of raffinose series oligosaccharides may also alter the osmotic differences between the mucosa and plasma, and this may account for the diarrhea observed in animals on diets with high levels of legume seeds [19]. Iji and Tivey [19] have related the different responses between synthetic and natural series raffinose with the usually very low dietary concentrations of synthetic raffinose.

The major oligosaccharides in soybeans are raffinose and stachyose that are present at approximately 1 and 6percent in SBM dry matter, respectively [20]. It was observed that soy oligosaccharides increased microbial activities, chyme production, total volatile fatty acids and total biogenic amines content; but reduced nutrient digestibility and increased the frequency of occurrence of diarrhea in weaning piglets [21, 22, 23]. Soy oligosaccharides did not stimulate the growth of bifidobacteria selectively, indicating that they do not exhibit prebiotic properties [24].

2.2.3. Saponins

Saponins are glycosides having a bitter taste, form foam in water, and have the ability to hemolysis red blood cells. Soy saponins can create an undesirable taste in animal feed and may alter the intestinal functions of animals. The presence of saponins in soy products is highly dependent on the methods of preparation. Soybean meal and soy flour contain between 0.43-0.67 percent soy saponins [25]. Soy protein concentrates produced by ethanol extraction are devoid of saponins since alcohol is a 'bond-breaker' and helps to remove saponins from the proteins.

2.2.4. Phytic Acid

Phytic acid is the acid form of the anion phytate, which cannot be hydrolyzed by enzymes secreted into the gut by livestock animals. Phosphorous in phytate has low bio-availability. Moreover, phytate can form complexes with a variety of minerals, such as calcium, copper, iron, thus reduce the availability of these nutrients. However, the adverse effect of phytate can be reduced by using feed enzymes, e.g., phytase. There are phytase pre-treated soy protein products commercially available, the phytase pre-treatment reduces phytate content and improves phosphorous digestibility [26].The dietary use of exogenous phytase to enhance phosphorous digestibility is now a common practice.

2.2.5. Other ANFs

Other soy ANFs such as estrogens are not very well investigated. Inconsistent results have been reviewed in literature on anti-nutritional or beneficial effects of soy estrogens, which may be related to dosage levels. As the concentration of estrogen is low, its anti-nutritional effect on animals is less important compared to other ANFs listed above.

2.3. Summary

The proximate composition of ANFs in soy protein products is presented in Table 3. This Table gives only an indication on the levels of ANFs in different soy products. The actual level of ANFs in toasted full fat soybean and soybean meal may vary to a large extent between samples, due to differences in origin and processing conditions of soy.

Table 3. Proximate composition of ANFs in raw soybeans, toasted full fat soybeans, soybean meal and soy protein concentrate (SPC)

	Raw soybeans	Toasted full fat soybean	Soybean meal	SPC
Trypsin inhibitor (mg/g)	45-50	1-8	1-8	1-3
Glycinin antigen (ppm)	180000	>50000	40000-66000	1-10
β-conglycinin antigen (ppm)	69000	16000	16000	1-10
Lectins (ppm)	3500	10-200	10-200	<1
Oligosaccharides (%)	14	14	15	1-3
Saponins (%)	0.5	0.5	0.6	0

The content of above mentioned soy ANFs is closely related to the applied processing technologies, which will be discussed below.

3. Production Process and Classifications

3.1. Production Process

*Full fat soybean meal:*Full fat soybean meal is processed by heat treatment of whole soybeans. The commonly used technologies to treat soybeans are: 1) cooking, 2) roasting, (e.g., rotating drum systems, fluidized bed, cascading chamber, micronizing, jet-sploding and microwave) and 3) dry or moist extrusion [27].

Cooking is a relatively simple process. The raw soybeans are immersed in water and boiled for 30 to 120 minutes. They are then mechanically dried, or dried in air by spreading them out on the ground. These beans are fed whole, ground or crushed. This is a traditional method used normally by local soy producers [27].

Roasting is a process of dry heating soybeans (e.g., 110-170°C), the heating temperature depends on the type of equipment used and the desired nutritive value of the full-fat soybean meal. Micronizing is a process of cooking soybeans by the heat that is generated by vibrating molecules under the influence of infrared rays. With Jet-sploding, the beans are sent through a stream of air that is pre-heated to 140 to 315°C. This drives the intracellular water close to the boiling point (~95°C) and the developing vapor pressure pops up the seeds. After heating, normally the soybeans are placed into a roller mill to facilitate the release of the intercellular fat and make the final product [27].

The principle of an extrusion process is the application of high temperature (140 to 170°C) and pressure (up to 50 bar) for a short time period (<90 seconds) and sudden pressure release at the extruder outlet. In principle, dry and wet extrusion is possible. In dry extrusion, the product is cooked in its own moisture for 20 to 40 seconds, reaching temperatures of 120 to 165°C. Wet extrusion involves pre-conditioning with steam and possibly water and optional injection of saturated steam into extruder barrel. An additional drying step is required [27] in wet extrusion. It was observed that wet extrusion is more effective in denaturing the ANFs than dry extrusion [27]. The level of ANFs reduction is also related to extrusion temperature, moisture, pressure and type of extruders (single or twin screw extruder).

In these heat treatment processes, heat labile ANFs such as trypsin inhibitor activity are reduced. Due to the different processes applied and varying processing conditions, variations in trypsin inhibitor activities may exist between different full-fat soybean products. Proper processing conditions should enable sufficient inactivation of the heat labile soy ANFs, whereas heat stable ANFs are not removed during above described processing of full-fat soybean meal.

Defatted soybean meal 44percent CP: Defatted soybean meal (SBM) is a by-product from soy oil extraction. Several factors influence the concentration of CP and AA in SBM. The method of extraction can result in differences in the CP and AA concentration in SBM. The expeller method is the traditional way of removing oil from the seed. With this method, approximately 6percent oil retains in the meal. Currently the most common commercial procedure of oil extraction is the solvent extraction method, which is an efficient way of removing oil from soybean. Solvent extracted SBM contains less than 1percent oil and has greater concentrations of CP and AA than expelled SBM. The raw soybeans are first cleaned, de-hulled and cracked, and solvent extracted to remove soy oil. The defatted soy flakes are then heat treated to recover the extraction solvent and to reduce the heat labile ANFs. For 44percent CP soybean meal, the soy hulls are added back to soybean meal.

In the standard soybean meal process, different heat treatment steps may be involved. The first step is the conditioning temperature (approximately 71°C, 11% moisture), which is not high enough to denature the protein or inactivate trypsin inhibitors. After oil extraction, the remaining hexane in the meal (about 35%) is removed by directly injecting live steam. The defatted meal contains approximately 16 to 24percent moisture is then heat treated by either toasting or cooking at 105 to 110°C for 15 to 30 minutes to reduce heat labile ANFs. A sufficient amount of heat is needed to reduce the moisture level to about 10 to 12percent. The resulting product normally has a KOH protein solubility ranging between 80 to 85percent. Excessive heat treatment may reduce available lysine content; a reduction as much as 40percent was reported, supposedly due to Maillard reaction [28].

Hi-pro soybean meal 48percent CP: The same process is used as above, but without adding back soy hulls. Thus, the product has higher crude protein and lower fiber content compared to 44percent CP soybean meal. For both 44 and 48percent CP SBM products, variations in processing conditions may lead to large variations in TIA levels. Heat stable ANFs are still present at high levels.

Enzymatically or microbially fermented soy protein: This type of soy protein is produced by fermentation/enzyme treatment of defatted soy flakes or hi-pro soybean meal. In case a microbial fermentation process is applied for manufacturing, microorganisms with known features are added stepwise to trigger a fermentation process. During fermentation, soy antigens are denatured and the oligosaccharides are degraded in the product by the microorganism`s own enzyme activity. The fermentation process is facilitated by the use of a mold or bacteria or both with *Aspergillus oryzae* and *Bacillus subtilis*as the predominant strains, respectively [29, 30].The fermentation process increases the concentration of CP. The AA profile of the fermented product is different from conventional SBM because of the preferential utilization of some AA by the fungi [29].

Soy proteins produced by fermentation/enzyme treatment process have higher protein levels (50 –56%) when compared to soybean meal. However, the protein levels in these products are lower than soy protein concentrate, as the soluble non-starch polysaccharides are not removed.

Classical SPC: Soy protein concentrate (SPC) production process removes soluble carbohydrates and contains approximately 70percent CP on DM basis. SPC can be produced by 3 different methods, e.g., alcohol extraction, acid extraction, or moist heat and water leaching extraction. Currently commercially available SPCs are produced by ethanol extraction process.

The classical soy protein concentrate (SPC) is produced starting from high quality soybeans, using low temperature to separate soy oil and obtains the high quality soy white flakes. To produce white flakes, that is, excess hexane is drained by low heat vacuum drying, with minimum protein denaturation. The white flakes are subjected to an ethanol/water extraction process to remove the soluble carbohydrates. During ethanol/water extraction, the heat stable ANFs, e.g., antigens are de-natured and indigestible soy carbohydrates including oligosaccharides and soluble non-starch polysaccharides (NSP) are removed. After ethanol/water extraction, the product is steam heated in order to remove heat labile ANFs. In this way, both heat stable and heat labile ANFs are removed during SPC processing.

Because the heat treatment is applied after the removal of soluble carbohydrates, this processing method has the advantage of reducing the likelihood of Maillard reactions during the heat treatment and improves the availability of amino acids.

Non-classical SPC: This is a soy protein concentrate (SPC) produced differently from classical soy protein concentrate. It is produced by ethanol/water extraction of hi-pro soybean meal instead of soy white flakes. In contrast to the classical soy protein concentrate, soluble carbohydrates are not removed before heat treatment; this may increase the risk of Maillard reaction during the heating process.

The production process to make different soy products is illustrated in Figure 5.

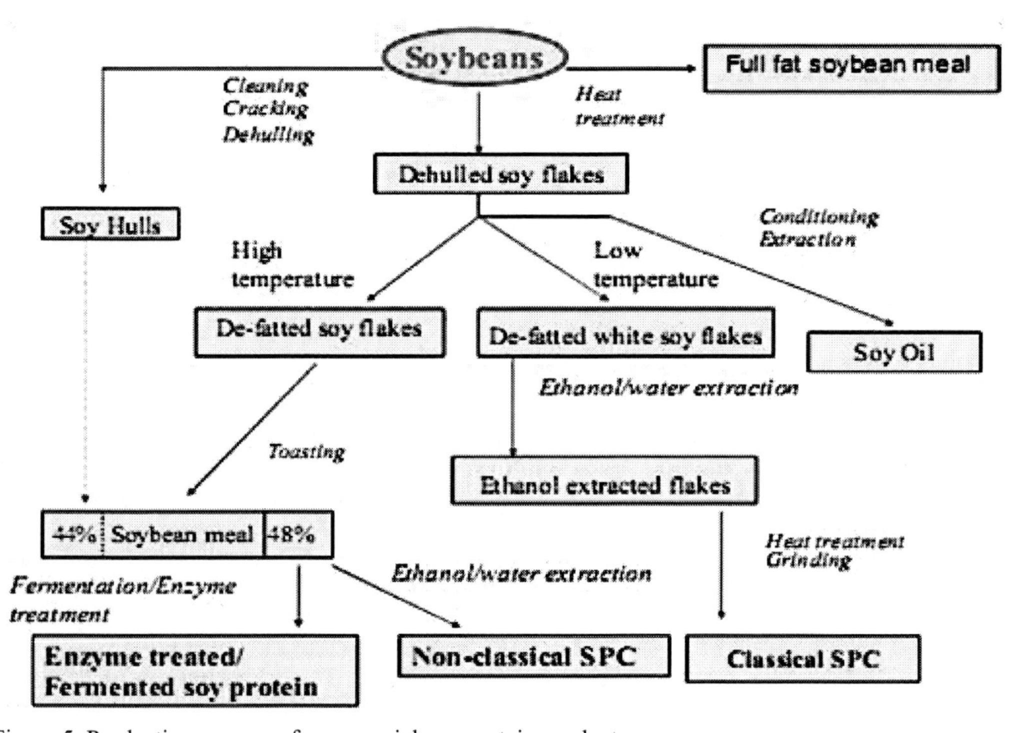

Figure 5. Production process of commercial soy protein products.

3.2. Effect of Production Process on Animal Performance

Hancock *et al.* [3, 4] determined the effect of ethanol extraction of soybean flakes on performance of piglets, function and morphology of pig intestine (Table 4). A 3x3 factorial experimental design was used in this study, including three ethanol extraction treatment methods and three levels of heat treatments. The ethanol extraction methods includes: 1), without ethanol extraction; 2): ethanol extraction after heat treatment (non-classical SPC) and 3): ethanol extraction before heat treatment (classical SPC). The heat treatments were autoclaving at 5, 20 and 60 minutes, respectively.

The villus size (area, height and perimeter length) tended to be greater in pigs fed with soy protein product heated after ethanol extraction and exposed to the intermediate level of heat treatment (Table 4). The improved intestinal morphology corresponded well with improved growth performance. The digestibility of nitrogen and the lysine concentration are higher in soy protein prepared by heat treatment after ethanol extraction, as compared with soybean protein produced by heat treatment before ethanol extraction (especially regarding the over-processing condition). Soy protein produced by ethanol extraction before heating significantly improved the feed intake, growth rate and feed utilization efficiency in piglets (Table 4). The improved feed utilization efficiency may be related to the increased villus size and improved lysine digestibility, as indicated by a high plasma lysine concentration.

This study demonstrated that soy protein quality is closely related to processing technologies.

3.3. Classification of Soy Products

There are many commercial soy protein products available in the market, thus different terms are used to describe these soy protein products. Different production processes for different soy products result in different crude protein (CP), carbohydrates, ANFs levels of the final products. Soy protein products in general can be simply classified based on their crude protein content (Table 5). A highly purified soy protein isolate (90% CP on DM basis) is not included in Table 5, as soy protein isolate is produced mainly for food applications and may not be economical to be used in animal feed.

Table 4. Effect of ethanol extraction and heat treatment of soybean flakes on product composition, average daily weight gain (ADG), average daily feed intake (ADFI) and gain/feed ratio, plasma urea and lysine concentration, digestibility of nitrogen and morphology of the intestine in piglets [3, 4]

	Without ethanol extraction			Heat treatment before ethanol extraction			Heat treatment after ethanol extraction		
Min. of autoclaving	5	20	60	5	20	60	5	20	60
Analysis of the soy product, %:									
Crude protein	50.8	50.9	51.1	65.6	67.1	65.7	66	67.2	66.9
Lysine	3.6	3.6	2.8	4.5	4.5	3.6	4.7	4.6	4.3
Arginine	3.9	4.1	3.5	4.8	5.0	4.5	5.1	5.2	5.1
Histidine	1.3	1.2	1.2	1.6	1.6	1.6	1.7	1.7	1.7
Isoleusine	2.2	2.1	2.2	3	3.1	3.1	3.1	3.2	3.2
Met+cys	2.1	2.0	1.9	2.6	2.7	2.6	2.6	2.6	2.5
Trypsin inhibitor, mg/g	27.8	4.4	0.3	14.5	2.1	0.6	11.2	3.1	0.4
Results:									
ADG, g/d	65	318	176	188	319	202	227	378	341
ADFI, g/d	321	638	461	448	645	511	478	713	711
Gain/feed	0.167	0.499	0.385	0.410	0.493	0.388	0.476	0.533	0.480
Plasma urea, mg/dl	44.3	39.0	42.6	31.8	36.3	45.8	34.5	32.1	36.6
Plasma lysine, %	1.24	1.13	0.69	1.11	1.09	0.92	1.17	1.16	1.32
N digest %	72.6	80.9	80	79.6	83.1	80.8	80.7	84.7	82.3
N retention, g/d	5	6	5.1	5.9	6.8	5.3	6.1	6.8	5.9
Duodenum villus height, μm	370	493	461	437	474	379	454	539	520

Table 5. Classification of soy protein products based on crude protein (CP) content

Classification		Typical CP content % (as is)	Term of soy products	Production process
Basic Processed Soy Products (< 50% CP)		36	Full fat soybean meal	Heat treatment
		44	De-fatted soybean meal Or soybean meal	De-hulled, defatted soy flakes, heat treatment, add back soy hulls
		48	Hi-pro soybean meal	De-hulled, defatted soy flakes, heat treatment
Highly Processed Soy Products (> 50% CP)	Enzyme treated/Fermented Products (50 > 60% CP)	52-56	Enzyme treated/ Fermented soy protein	Enzymatically or microbially fermentation of hi-pro SBM
	SPC (> 60% CP)	60-62	Non-Classical SPC	Produced by ethanol/water extraction of hi-pro SBM
		63-67	Classical SPC	Produced by ethanol/water extraction of white flakes

4. NUTRITIONAL VALUE

4.1. Nutritional Value of Different Soy Protein Products

The typical nutritional values of soy protein products are summarized in Table 6. Highly processed soy protein products have low levels of both heat labile and heat stable soy ANFs. This makes the product more suitable for application in young animals` feed. The ANF levels, however, may differ between different products produced at different fermentation or ethanol extraction conditions.

Table 6. Typical composition of commercial soy protein products (based on NRC [31], 1998 or producer's data sheet)

	Full fat SBM	SBM 44% CP	Hi-pro SBM	Fermented soy	Non-Classical SPC	Classical SPC
Composition, %, as is						
CP	36	44	48	52-56	60-62	63-67
Fat	18	1-3	1-3	1-2.5	2	0.5-1
Moisture	10	10-11	10-11	8-9	10	5-7
Crude fiber	5.5	7	3-3.5	3-4.5	<5	3-4
NFE[1]	30.5	30-35	30-35	24-28	18	16-18
Ash	4.5	6.3	5.8	6.2-6.8	<5	5.6-7
AA, % CP						
Lys	6.31	6.46	6.43	6.1-6.2	5.67-7	6.3-6.5
Thr	4.01	3.95	3.94	3.93-4.0	3.67-4.5	4.0-4.2
M+C	3.07	2.99	3.00	2.86-3.02	2.67-3.34	2.91-3.0

[1] NFE: Nitrogen free extract = 100-CP-Fat-Fiber-water-ash, e.g., carbohydrates fraction.

Table 7. Typical nutritional values of soy protein products [31]

	Soybean meal	De-hulled SBM	Full fat SBM	Soy protein concentrate*
Crude protein, % as is	43.8	47.5	35.2	65
Crude fat, %	1.5	3.0	18.0	1
ME swine (kcal/kg)	3180	3380	3690	3931**
Lysine, %	2.83	3.02	2.22	4.23
Methionine, %	0.61	0.67	0.53	0.91
True ileal dig Lys swine, % (NRC)	89	90	86	95
True ileal dig Met swine, % (NRC)	91	91	85	94
SID*** Lys piglets		84	78	91
SID Met piglets		87	76	92
True ME poultry (kcal/kg)	2486	2590	3454	2870
True dig Lys poultry	92.2	91.2	86.9	92.2
True dig Met poultry	91.2	91.9	86.0	91.2

* A classical feed grade soy protein concentrate; ** ME for weaning piglets. *** SID – standardized ileal digestibility of piglets determined by Urbaityte [33].

Different production processes result in differences in indigestible carbohydrates content, especially soluble NSP content. Fermented soy protein contains higher amounts of soluble NSPs compared to SPC, as these soluble NSPs are not removed during the fermentation process (see Table 6).

It is generally accepted that increasing dietary NSPs level can have negative effects on the digestibility and rate of absorption of nutrients. The fermentation of NSP in the intestine may cause flatulence and diarrhea. Pluske *et al.* [32] reported that an elevated level of dietary soluble NSP increased pathogenesis of swine dysentery in weaned pigs. Reducing the quantity of fermentable substrate entering the large intestine reduces the incidence of swine dysentery. The authors suggested that soluble NSPs in weaner diets cause proliferation of *E. coli* in the small intestine and are detrimental for piglet growth. A level of soluble NSPs less than 1percent in the diet is associated with lesser disease incidence.

The nutritional values for different soy protein products commonly used in feed formulation are compared and presented in Table 7.

4.2. Variation in SBM Composition

The composition of soybeans is influenced by its growing conditions (temperature, precipitation, soil type, latitude etc.). Soybeans harvested from different countries, even from the same country but from different locations, can vary in their composition. Consequently, soy protein products produced from soybeans collected from different locations can differ in their nutritional composition. This will impact on feed formulations.

4.2.1. Variation in Origin

Soybeans produced under varying environmental conditions vary in their composition and quality. Soybean meal composition varies among samples collected from different origins.

Table 8. Nutritional composition of SBM collected from different origins [34]

% DM	n	DM	CP	CF	NDF	EE
Argent.	97	88.6	51.8^{c}	5.3^{b}	10.6^{b}	2.0^{a}
Brazil	81	88.8	53.0^{b}	6.3^{a}	12.0^{a}	2.0^{a}
USA	134	88.6	54.4^{a}	4.2^{c}	8.7^{c}	1.8^{b}
SEM		0.04	0.09	0.06	0.09	0.03
% CP	n	Lys	Met	Cys	Thr	Trp
Argent.	62	6.14^{b}	1.38^{a}	1.52^{a}	3.96	1.38^{a}
Brazil	68	6.10^{b}	1.32^{b}	1.45^{b}	3.93	1.36^{b}
USA	108	6.21^{a}	1.38^{a}	1.52^{a}	3.94	1.38^{a}
SEM		0.011	0.002	0.003	0.004	0.002
% DM	n	Saccharose	Stachyose	Raffinose		
Argent.	50	7.5^{a}	5.5^{b}	1.3^{b}		
Brazil	41	7.0^{b}	5.3^{b}	1.4^{a}		
USA	81	7.8^{a}	6.4^{a}	1.1^{c}		
SEM		0.11	0.08	0.03		

$^{a,b)}$ Values in the same column followed by different letters are significantly different (P<0.05).

Table 9. Comparison of the composition of soybeans from Brazil, China and the United States [35]

	Brazil	China	United States
Samples analyzed	48	49	36
States/Provinces involved	5	6	15
Dry matter	86.98^{a}	92.33^{b}	93.86^{c}
Ash	5.10^{a}	5.42^{b}	5.4^{b}
Crude protein	40.86^{a}	42.14^{b}	41.58^{ab}
Lipid	18.66^{a}	17.25^{b}	18.70^{a}
Neutral detergent fiber	13.36	13.79	13.85
KOH protein solubility %	27.68^{a}	29.18^{b}	30.80^{c}
Amino acids			
Lysine	2.48^{a}	2.56^{ab}	2.60^{b}
Methionine	0.28^{a}	0.41^{b}	0.42^{b}
Phenylalanine	2.04^{a}	2.05^{a}	2.19^{b}
Threonine	1.56^{a}	1.60^{a}	1.75^{b}
Cystine	0.75^{a}	0.88^{b}	0.61^{c}

$^{a,b)}$ Values in the same row followed by different letters are significantly different (P<0.05).

In a study to investigate soybean meal composition from different origins, a total of 312 samples were collected during 2007 to 2009 and were analyzed [34]. Seventy five percent of the samples were collected from the country of origin and 25percent of the samples were collected from European ports (Spain, Portugal, France, Germany and Poland) with control of port of origin. The origin of the 312 samples was 43percent from USA, 31percent from Argentina and 26percent from Brazil. The samples were analyzed for amino acids composition and *in vitro* digestibility, quality assessment parameters, mineral and sugar content.

Table 10. Within country variation in soybeans composition(%, as is) [35]

Locations							
Soybeans collected from Brazil:							
	B1	B2	B3	B4	B5		
Crude Protein	42.32	40.3	41.23	39.39	41.14		
Lipid	19.75	18.63	18.6	18.02	18.29		
Lysine	2.54	2.3	2.59	2.5	2.49		
Methionine	0.28	0.32	0.28	0.22	0.3		
Cystine	0.78	0.87	0.72	0.59	0.78		
CP from NPN*	3.18	5.24	1.97	1.53	1.4		
Soybeans collected from US, Maturity zone: North to South US							
	1	2	3	4	5	6	7
Crude Protein	40.21	40.96	40.69	43.19	40.72	39.97	44.54
Lipid	18.46	17.89	18.05	18.89	19.65	19.02	19.54
Lysine	2.52	2.5	2.55	2.51	2.64	2.62	2.78
Methionine	0.26	0.4	0.48	0.57	0.34	0.39	0.39
Cystine	0.62	0.62	0.61	0.62	0.61	0.62	0.61
CP from NPN*	0.79	0.38	-0.3	3.04	-1.98	-1.85	-0.26
Soybeans collected from China							
	C1	C2	C3	C4	C5	C6	
Crude Protein	42.57	40.83	43.73	41.05	40.69	43.61	
Lipid	14.51	17.12	17.48	17.96	17.74	16.38	
Lysine	2.6	2.46	2.71	2.47	2.33	2.77	
Methionine	0.44	0.44	0.46	0.42	0.36	0.37	
Cystine	0.91	0.91	1.03	0.87	0.75	0.78	
CP from NPN*	3.08	2.89	1.6	3.09	4.89	0.93	

* NPN- Non Protein Nitrogen: Crude protein from NPN is expressed as: Crude protein – TAA (total amino acids).

As shown in Table 8, SBM from Argentina has low CP and lysine content, while SBM from USA have higher CP and lysine content as percentage of protein. Grieshop and Fahey [35] reported that soybean samples collected from different countries differ in their composition (Table 9). Soybeans from China had higher crude protein content and lower fat content than those from Brazil (Table 9).Differences also occurred in the amino acids content between U.S. and Brazilian beans. For most assessed amino acids, the content was significantly higher in U.S. beans. Within country variations in soybean composition also exists (Table 10). These differences are largely due to different climatic conditions (maturity zones) within the soy planting area. The data also show that the level of Non-Protein-Nitrogen (NPN) is much lower for U.S. beans compared with Brazilian or Chinese beans.

4.2.2. Variation in Years

The crude protein and amino acids content in soybean meal may vary between production years [36] (Figure 6).

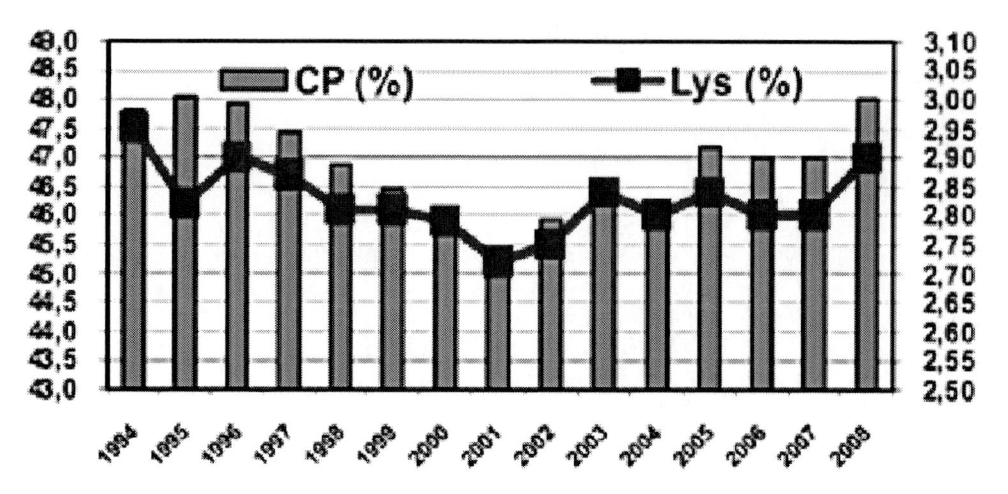

Figure 6. CP and Lysine content (%) in SBM in different years [36].

The variation in soy products composition requests proper analysis to get optimal efficiency in animal feeding.

4.3. Digestibility

4.3.1. Soy Processing and Digestibility

Growing conditions of soybeans and processing technology have effects on amino acids digestibility. Karr-Lilienthal *et al.* [37] evaluated amino acids digestibility of soybean meal produced from five major soybean producing countries, i.e., Argentina, Brazil, China, India and the United States. Within a country, they differentiated according to quality classes. The quality of soybean (i.e., low, intermediate and high) was determined by criteria such as color, protein content, and/or processor history. They found that there is a large variation in AA digestibility between samples collected from different countries. Table 11 compares digestibility of crude protein, total amino acids (TAA), and the essential amino acids lysine, methionine and threonine

Table 11. True ileal crude protein and amino acids digestibility (%) in pigs fed semi purified diets containing soybean meals prepared in original countries or prepared from soybeans grown in original countries and processed under uniform conditions in the United States [37]

	SBM sourced in different countries					
	SBM[1)]	Low	Intermediate	High	Standard[2)]	SEM
From Argentina						
CP	73.3^{a}	82.1^{b}	84.4^{bc}	84.1^{bc}	90.1^{c}	2.8
TAA	73.7^{a}	86.2^{b}	87.9^{bc}	86.4^{b}	91.3^{c}	2.42
Lysine	75.4^{a}	86.0^{b}	86.8^{bc}	86.2^{b}	90.2^{c}	1.91
Met	81.9^{a}	91.3^{b}	91.1^{b}	90.9^{b}	92.1^{b}	0.97
Thr	79.4	92.6	82.9	83.0	85.8	1.72
From Brazil						
CP	51.4^{a}	72.2^{b}	78.9^{bc}	77.4^{bc}	87.9^{c}	4.2

	SBM sourced in different countries					
	SBM[1)]	Low	Intermediate	High	Standard[2)]	SEM
TAA	53.5[a]	78.4[b]	83.1[b]	83.5[b]	90.8[c]	3.18
Lysine	55.5[a]	76.7[b]	81.7[b]	81.6[b]	90.4[c]	3.29
Met	62.6[a]	84.5[b]	87.4[b]	88.0[b]	92.6[c]	2.72
Thr	50.1[a]	72.5[b]	76.7[b]	77.7[b]	86.8[c]	4.2
From China						
CP	54.0[a]	78.1[b]	87.7[c]	86.2[c]	86.1[c]	2.42
TAA	60.4[a]	83.5[b]	89.6[c]	89.0[c]	87.7[bc]	2.73
Lysine	60.3[a]	81.9[b]	87.4[c]	88.1[c]	87.7[c]	2.11
Met	62.3[a]	87.4[b]	91.4[bc]	92.8[c]	90.9[bc]	3.00
Thr	55.4[a]	83.8[b]	89.9[c]	85.3[bc]	83.4[b]	4.90
From India						
CP	53.7[a]	81.9[b]	82.2[b]	87.1[b]	86.5[b]	2.48
TAA	58.9[a]	86.4[b]	86.1[b]	88.7[b]	89.3[b]	1.85
Lysine	65.2[a]	88.1[b]	87.4[b]	89.0[b]	83.9[b]	3.45
Met	66.0[a]	88.6[b]	88.1[b]	90.9[b]	90.9[b]	2.61
Thr	57.1[a]	80.7[b]	80.2[b]	83.7[b]	84.0[b]	1.95
From United States						
CP	84.7[a]	82.5[a]	92.9[b]	94.0[b]	95.7[b]	4.57
TAA	81.9[a]	88.7[b]	88.7[b]	91.0[b]	94.3[b]	4.24
Lysine	82.8[a]	87.0[ab]	90.9[b]	90.6[b]	93.1[c]	2.71
Met	86.4[a]	89.9[ab]	92.7[bc]	93.2[bc]	93.7[c]	1.61
Thr	76.0[a]	81.9[ab]	86.8[bc]	86.7[bc]	89.6[c]	3.66

[1)] SBM prepared in the United States under uniform conditions from soybeans collected in different countries.

[2)] Standard: SBM purchased on the market in the United States, as a control for each trial.

[a,b)] Values within same row followed by different letters are significantly different ($P < 0.05$).

Based on the results in Table 11, the authors suggested that the soybean meal produced in Argentina and Brazil were less digestible compared to soybean meal produced in USA. Whereas the soybean meals produced in China, India and the United States have similar digestibility. Remarkable differences occurred for beans collected in different countries and processed under the same uniform conditions in the U.S.

4.3.2. Soy Composition vs Digestibility

In a recent study, 22 SBM samples from three origins (USA [8], BRA [7], and ARG [7]) were tested in six replications (6 birds) per treatment. Apparent ileal digestibility (AID) of CP and AA was measured at 21 to 24 days of age. It was observed that in general, the digestibility of CP and lysine (Table 12) is significantly co-related to CP, sucrose, total and reactive lysine and TIA content [34].

In another study, AID of CP and AA was determined in broilers with six replications (16 broilers) per treatment at 21 to 24 days of age. Six SBM treatments from different origins (USA [2], ARG [1], BRA [3]) were used as a sole protein source [38]. In this study, it was observed that the digestibility of CP and AA (Table 13) is positively related to CP, Sucrose (for AMEn) and negatively related to TIA and NDF.

Table 12. Digestibility of CP and AA in SBM determined in broilers [34]

	Avg	Range		Avg	Range		Avg	Range
TIA, mg/g	2.72	1.8-4.2	CP	47.7	46-49.9	AID		
KOH sol., %	83.2	67.9-90.2	Lys	2.94	2.69-3.09	CP piglets[1]	76.8	69.9-79.8
PDI, %	13.5	8.7-16.6	Met	0.65	0.62-0.70	CP broilers	84.5	82.1-88.0
Raf + stach., %[2]	6.20	5.5-7.0	Cys	0.68	0.63-0.72	Lys	87.6	85.0-90.5
Reactive lys, %	86.2	84.0-87.8	NDF	8.9	6.8-14.3	Met	87.9	85.9-90.7
In vitro Boisen[3], %	87.4	84.6-90.5	Sucrose	6.3	5.4-7.5	Cys	70.0	66.4-75.0

[1] Hohenheim University (standardized) (n = 18); [2] Raffinose and stachyose; [3] 2 steps.

Table 13. Ileal digestibility of SBM in broilers [38]

Country	CP %	TIA mg/g	AID, % DM	N	GE	Lys	Met	Cys
ARG	46.1	6.5	75.6[b]	77.9[b]	80.3[b]	80.9[b]	84.1[c]	55.1[b]
BRA-1	45.5	5.1	75.2[b]	79.0[bc]	79.1[b]	83.5[a]	85.7[ab]	55.5[b]
BRA-2	47.2	4.1	76.7[b]	79.2[bc]	80.3[b]	84.4[a]	86.5[b]	56.4[b]
BRA-3	45.2	5.1	76.8[b]	77.3[c]	80.4[b]	77.8[c]	81.9[d]	56.9[b]
USA-1	50.6	2.4	81.8[a]	82.1[ab]	85.8[a]	84.0[a]	86.3[b]	62.9[a]
USA-2	48.6	1.8	82.3[a]	85.5[a]	85.3[a]	85.1[a]	88.8[a]	65.8[a]
SEM (n = 6)			1.02	1.46	0.89	0.78	0.66	1.94

Table 14. Apparent Ileal digestibility of CP and AA of SBM from different origins in broilers [39]

	BRA	ARG	USA-1	USA-2
CP	47.6	46.3	46.6	48.1
Total lys[1]	2.69	2.91	2.93	3.02
React. Lys	84.0	86.6	86.0	85.8
AID				
CP	82.0	84.0	85.0	84.1
Lys	85.0	87.9	87.4	87.7
Met	87.0	88.0	88.7	87.8
Cys	67.2	68.1	72.0	70.7

[1] Lys profile: 5.65, 6.29, 6.34 and 6.28% of CP, respectively.

Corchero *et al.*[39 cited by 34] determined the digestibility of CP and AA in soybean meal from different origins and observed that digestibility of CP and AA is related to total and reactive lysine in broilers (Table 14).

4.3.3. Digestibility of Highly Processed Soy Protein Product

In a recent study, standardized ileal AA digestibility in different soy protein products was determined in weaning piglets [33]. In this study, piglets were implanted with T-cannula at 21 to 23 days of age, the test diet was fed at 28 days of age and ileal digestibility was measured at 34 to 55 days of age (7 to 10 kg body weight). The tested protein sources included feed grade soy protein concentrate, high-pro soybean meal, full fat soybean meal, enzymatically and microbially fermented soy protein (Table 15). Clearly, amino acids digestibility in different commercial available soy protein products is related to the production process. The processing technologies used for SPC and enzymatically fermented soy protein improved amino acids digestibility when compared to soybean meal. This study also demonstrated that further processing from soy protein concentrate to soy protein isolate had no benefit on amino acids digestibility in piglets (data not shown), therefore using soy protein isolate in piglet's feed is not cost effective.

Table 15. Content (%) and standardized ileal digestibility (SID) of crude protein and amino acids in different soy products

	SPC*	SP_{A1}	SP_{A2}	SP_{A3}	SP_b	SB_e	SBM_{hp}
Content, % DM							
CP	69-78	57	61.4	62	58	43	55.5
EE	0.4-1.9	2	2.4	2.1	3.95	22.2	2.16
TIA, mg/g	0.86-1.73	1.43	1.6	1.8	1.1	7.36	2.22
SID, %							
CP	86^{efg}	82^{cde}	84^{cdf}	86^{dfg}	74^{ab}	73^{a}	80^{bd}
Lys	90^{ghj}	80^{def}	81^{eg}	88^{ej}	71^{bd}	78^{cde}	84^{eh}
Met	90^{cd}	88^{c}	87^{bc}	91^{cf}	86^{bc}	76^{a}	87^{bc}
Cys	72^{fh}	58^{bde}	67^{cdgf}	74^{efhjk}	55^{bc}	58^{bd}	72^{efhj}
Thr	81^{gj}	74^{bcd}	75^{cdf}	82^{fgk}	66^{a}	67^{ab}	76^{cg}
Trp	84^{gj}	77^{bfh}	78^{cfg}	84^{jgh}	71^{b}	71^{bc}	82^{efj}
Val	87^{efgh}	82^{cde}	82^{cdf}	88^{dgh}	74^{ab}	71^{a}	81^{bd}
Ile	88^{fg}	83^{cdh}	85^{cf}	88^{dfg}	77^{ab}	72^{a}	81^{bc}

* SID SPC-- mean values of 4 classical SPC products, no significant difference among these 4 products. $SP_{A1\text{-}3}$ – enzymatically fermented soy protein; SP_b -- microbially fermented soy protein; SB_e – extruded soybeans; SBM_{hp} – high protein soybean meal.

[a,b,c] - values with a common superscript are not significantly different [33].

As shown in Table 15, the extruded full fat soybean meal had high TIA level that might partially explain the low AA digestibility. The high protein soybean meal was properly heat treated based on its TIA value, the low AA digestibility value could be due to the high heat stable soy ANFs in this product, as the production process for this product did not remove the heat stable ANFs, such as antigens and soy oligosaccharides.

4.3.4. AA Digestibility of Intact Soy Protein vs Hydrolyzed Peptides

It has been suggested that for fermented soy protein, the enzyme treatment has broken down the large soy protein molecules to smaller peptides, which are more easily digestible. Fermentation and enzyme treatment process reduced molecular size of protein. It is reported that 53.8percent of protein is in a range of 19 to 48 kD in soybean meal, while 86.6percent of protein is in a range of 9 to 18 kD in fermented soybean meal [29].

Yang *et al.* [30] determined ileal AA digestibility of soybean meal (SBM), soy protein concentrate (SPC) and one enzymatically and two microbially fermented soy proteins in weaning piglets. In this study, it was observed that soy protein concentrate has significantly or numerically higher ileal amino acids digestibility when compared to microbially fermented soy protein that contains smaller peptides in weaning piglets at 28 days of age (Table 16).

Table 16. True ileal digestibility of amino acids in weaning piglets

Protein source*	SBM	SPC	SP[1]	FSP-A	FSP-B
Arg	85.61^{b}	88.91^{ab}	90.48^{a}	90.88^{a}	87.91^{ab}
His	81.14^{b}	88.59^{a}	82.33^{b}	83.84^{b}	84.57^{ab}
Ile	75.61^{c}	84.76^{a}	82.81^{ab}	81.36^{ab}	80.23^{b}
Leu	81.40^{b}	85.31^{a}	84.15^{ab}	85.65^{a}	84.18^{ab}
Lys	83.42^{b}	88.48^{a}	84.96^{ab}	86.13^{ab}	85.27^{ab}
Met	83.48	88.37	86.52	84.04	86.81
Phe	80.11^{b}	82.57^{ab}	86.62^{a}	79.61^{b}	81.41^{b}
Thr	80.30^{c}	86.43^{a}	84.28^{ab}	80.15^{c}	82.03^{bc}
Val	79.79^{b}	82.62^{ab}	86.20^{a}	84.27^{ab}	84.55^{ab}
Sub mean	81.21^{b}	86.23^{a}	85.37^{a}	83.99^{ab}	84.11^{ab}

* SBM: soybean meal; SPC: soy protein concentrate; SP[1]: enzymatically fermented soy protein; FSP-A: microbially fermented soy protein by *A. oryzae*; FSP-B: fermented soy protein by *A. oryzae+B. subtilis.*

Table 17. Standardized ileal digestibility (%) of amino acids in experimental diets in weaning piglets* [40]

Protein sources	SBM	SP[1]	FSP[2]
Arg	90.9^{a}	98.2^{b}	93.5^{ab}
His	84.0^{a}	88.9^{ab}	84.4^{a}
Ile	82.9^{a}	89.8^{b}	85.8^{ab}
Leu	82^{a}	89.3^{b}	85.4^{ab}
Lys	79.2^{ab}	88.3^{b}	77.2^{a}
Met	85.5^{a}	92.2^{b}	88.3^{ab}
Phe	84.1^{a}	91.9^{b}	87.2^{ab}
Thr	77.4^{a}	85.8^{ab}	78.5^{ab}
Trp	84.8^{a}	87.5^{ab}	83.5^{ab}
Val	81.9^{a}	89.5^{b}	84.3^{ab}

1: SP is enzymatically fermented soy protein, 2: FSP is a using fermentation technology hydrolyzed soy product, described as a new generation functional peptide made from solid-state fermentation of soybean meal.

$^{a, b, c}$ Means within a row lacking a common superscript letter differ ($P < 0.05$).

* Data are least square means of 7 observations per treatment.

It was observed that a soy product containing high levels of hydrolyzed soy peptides (<20 kDa) did not have improved AA digestibility when compared to traditional soybean meal in weaning piglets [40] (Table 17).

Similarly, it was observed that intact wheat gluten had higher AA digestibility than hydrolyzed wheat gluten in weaning piglets. These data indicate that hydrolyzed protein does not necessarily have improved amino acids digestibility compared to intact soy protein.

The literature data indicated that soy ANFs and indigestible carbohydrates content are main factors, which can have negative effects on AA digestibility.

5. Quality Assessment

5.1. Quality Assessment Methods

Soy protein products need a proper thermal processing to reduce the heat labile ANFs, i.e., trypsin inhibitor activity and lectins and to achieve the optimal product quality. Figure 7 shows the relationship between heating time and quality of soybean meal. Insufficient thermal treatment does not properly reduce the ANFs level thus the potential nutritional value of soy is not attained and the target animal cannot exploit the full potential of the soy. Overheating on the other hand can lead to Maillard reactions or protein/amino acid cross-linking reactions that reduce amino acids availability and thus present a detrimental effect on nutritive value. Assessment of different technological processes for soy treatment can be based on this approach, i.e., how close comes the result to the maximal nutritional potential (e.g., true ileal amino acid digestibility) and how big is the "window" between under- and over-processing.

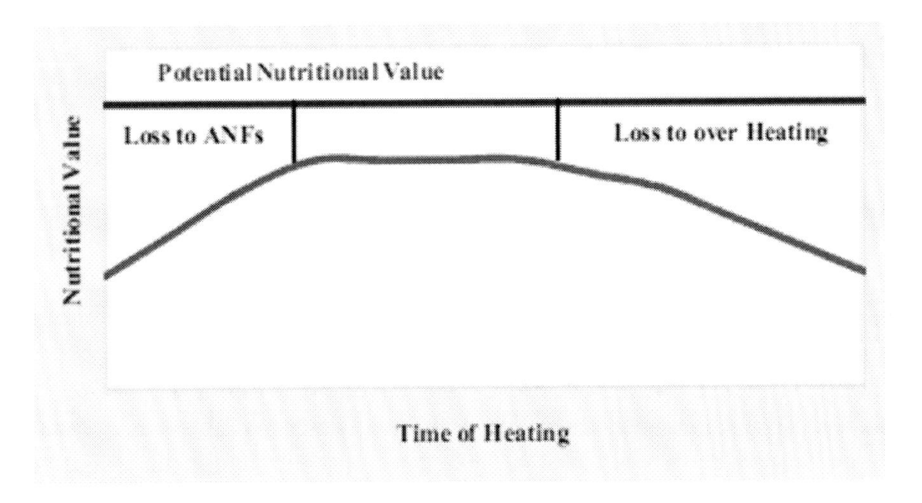

Figure 7. Relationship between heat treatment and product quality (41].

5.1.1. Methods Commonly Used for Soybean Meal Quality Assessment

Different methods are used to assess soybean meal quality, including direct measurement of trypsin inhibitor activity and indirect methods such as urease index, KOH protein solubility and protein dispersibility index (PDI). Direct measurement of TIA gives the most accurate indication on the degree of heat treatment and soy protein quality, however, it requires high skills and it is relatively difficult to measure. Among the indirect methods, it has been demonstrated that PDI is the simplest, most consistent and most sensitive method [42]. While urease index and KOH protein solubility showed a poor consistency and sensitivity to heat treatment.

Protein Dispersibility Index (PDI): PDI is a method to measure the amount of soy protein that dispersed in water after blending a sample with water in a high-speed blender.

Procedure is indicated below [43]:

- First to determine nitrogen content of soy sample by using official methods.
- Then to place a 20g sample of a soybean product in a blender.

- Add 300 ml of deionized water (30°C).
- Stir at 8500 rpm for 10 min. [44].
- Filter and centrifuge for 10 min. at 1000g.
- To analyze nitrogen content of the supernatant.
- To calculate the results as percentage of the original nitrogen content of the sample.

KOH protein solubility: This method determines the percentage of protein (nitrogen) that is solubilized in a potassium hydroxide (KOH) solution.

Urease Index: Urease is an inherent enzyme in soy that is denatured by heat treatment. Urease index can be measured and expressed as either 1), pH increase due to ammonia release (AOCS Ba 9-58), or, 2), direct measurement of nitrogen release (EU-method 71/250/EEC).

The recommended values of TIA, PDI, KOH protein solubility and Urease Index are presented in Table 18.

Table 18. Summary of quality control tests for heat treated SBM [42, 43]

Test	Duration	Skill level	Accuracy	Recommended values
Urease Index, pH change	20 min	Low	Average	0.02-0.30
Urease Index, N release	40 min	Average	Average	Max 0.4 mgN/g/min 30°C
KOH solubility	20 min	Low/average	Average	70-85%
PDI	10 min	Low/average	Good	15-30%
TIA	>24h	High	Good	<5 mg/g*

* In practice, for young animals, the recommended TIA level is below 2-3 mg/g.

5.1.2. Comparison of TIA with PDI, KOH Protein Solubility and Urease Index

Batal *et al.* [42] determined the effect of heating time on TIA, Urease Index, KOH protein solubility and PDI values in soy flakes, and tested the impact on weight gain and gain/feed ratio in chicks. Based on the data from this study, the relationship between weight gain and KOH protein solubility, urease Index, PDI and TIA, respectively is analyzed by using a linear regression model (Figure 8).

As shown in Figure 8, the highest R^2 is observed from the linear regression between TIA and growth performance in chicks, confirming that TIA gives an excellent indication on SBM quality. Among the indirect measurements, PDI gives the best correlation (R^2 = 0.90) to weight gain of chicks, it is the best indicator of heat treatment conditions and quality of SBM. KOH protein solubility and Urease index showed a poorer correlation compared with the PDI, to weight gain of the animals. This indicates the limitations of these methods to reflect the degree of proper heat treatment to achieve the best SBM quality. Repeated experiments by Batal *et al.* [42] showed similar results, that is, PDI is more sensitive to heat treatment compared to KOH protein solubility. For example, with increasing heat treatment from 0 to 30 min., the PDI decreased from 71 to 14percent, while KOH protein solubility decreased from 89 to 67percent.

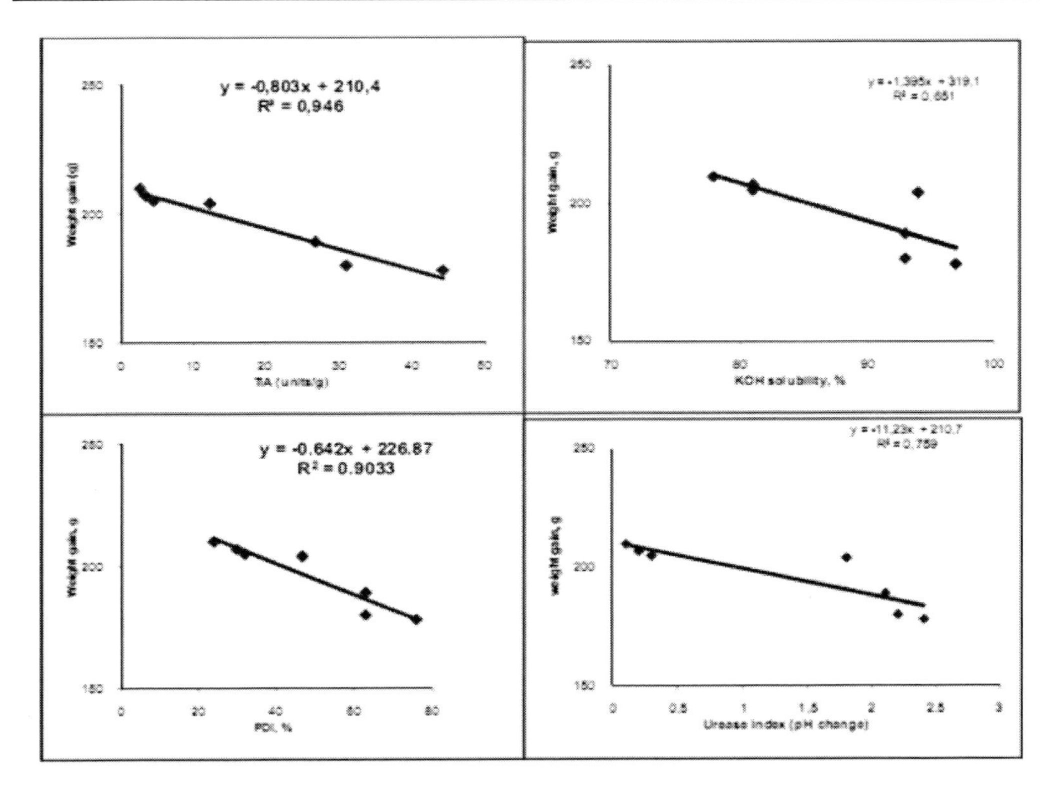

Figure 8. The relationship between weight gain of the chicks and TIA, KOH protein solubility, PDI values and Urease Index in SBM.

Araba and Dale [5] tested the influence of heating a raw hexane-extracted SBM on soy protein quality and growth performance of chicken. It was reported that weight gain of chicken is clearly related to TIA levels in SBM (Table 19).

Table 19. The effect of autoclaving time on SBM quality and growth performance in 0 to 21 day chick [5]

Heating time (min)	Weight gain	Feed/gain	KOH, %	Urease activity	TIA, mg/g
EXPERIMENT 1	g/chick			(pH change)	
0	342[c]	2.55[ab]	99.2	2.4	21.1
15	461[a]	2.10[c]	83.2	0.12	2.20
30	407[b]	2.33[bc]	58.5	0.00	0.50
60	296[d]	2.76[a]	38.0	0.00	0.10
EXPERIMENT 2					
0	342[d]	2.44[a]	99.2	2.4	21.1
5	429[c]	2.29[ab]	87.7	2.04	12.2
10	481[b]	2[c]	79.1	0.23	3.07
15	496[a]	2.09[bc]	74.9	0.00	2.1
20	450[bc]	2.03[c]	71.8	0.00	0.97

Means in a column with no common superscript are significantly different ($P<0.05$).

Overheating, however, reduced TIA level below 0.5 mg/g and resulted in reduced growth performance, due to the heat damage of protein and amino acids. A TIA level of 0.5 mg/g indicates over-heating. The determination of the point of 'over processed' is more difficult to assess than 'under processed'. When using indirect methods, KOH solubility is considered to be useful for estimation of over-heating and PDI is good method for determine under-heating for basic processed soy products [43].

5.1.3. The Sensitivity of KOH Protein Solubility

The optimal KOH protein solubility values differ from different trials. As shown in Table 19, Araba and Dale [5] observed an optimal heating time at 15min, with a TIA level of about 2 mg/g for both experiment 1 and 2. The KOH protein solubility values at the optimal heating time (i.e., 15 min), however, was 83percent in experiment 1 and 75percent in experiment 2, although the same lot of raw SBM were used for both experiment 1 and 2. On the other hand, Batal *et al.* [42] reported that, autoclaving soy flakes from 0 to 36 min changed KOH protein solubility from 97 to 78percent in experiment 1; autoclaving soy flakes from 0 to 30 minutes changed KOH protein solubility from 89 to 67percent in experiment 2; and autoclaving soy flakes from 0 to 12 minutes changed KOH protein solubility from 90 to 79percent, with a lowest KOH protein solubility value of 72percent at 9 min autoclaving in experiment 3. These data indicate that KOH protein solubility has a poor consistency and sensitivity to heat treatment of soy flakes.

The variation in optimal KOH protein solubility with reference to animal performance observed in different trials may be explained by the poor consistency of KOH protein solubility measurements. It was observed that KOH protein solubility of the same SBM samples measured by different laboratories gave different values (Table 20). This clearly indicates that KOH protein solubility method may be too imprecise to be used reliably to predict feeding value of SBM [45].

Table 20. KOH protein solubility value of SBM samples analyzed by different labs [45]

	US	Arg.	Brazil	Malaysia
Lab 1. KOH-PS, %	79	78	67	83
Lab 1. urease (pH change)	0.02	0.02	0.03	0.03
Lab 2. KOH-PS, %	92	84	87	93
Lab 2. urease (pH change)	0.09	0.09	0.09	0.09

5.1.4 In Vitro Digestible AA and Heat Damage Indicator (HDI)

Literature studies showed that heat treatment not only reduces amino acids content but also reduces their digestibility [46]. Lysine is well recognized as the most heat labile amino acid. Parsons *et al.* [46] observed that excessive heat treatment decreased lysine content by 16percent and its digestibility by 25percent (Table 21). Thus, the total digestible lysine content was reduced by 36percent by over-heating treatment. Moderate heat treatment reduced digestible lysine content by 23percent.

Formulating diets using table values for standardized ileal amino acid digestibility would overestimate the nutritional value of such heat treated ingredients, which may result in reduced animal performance. Whilst amino acid analysis of raw materials will identify amino acid losses in heat damaged material, which will be considered in least cost formulation, adjustment of amino acid digestibility is not commonly used. A new technique was recently developed [47] that allows identification and ranking of heat damage of soybean meal, which is called Heat Damage Indicator (HDI). With a system called AminoRED [47], based on NIR (near-infrared analysis) techniques, it predicts the degree of heat damage and its impact on amino acids digestibility. Depending on the degree of heat damage, amino acid digestibility value of SBM can be adjusted.

Table 21. Effects of heat treatment on amino acid content and digestibility of soybean meal in poultry

Autoclaving time (minutes)	Lysine	Cystine	Methionine	Threonine
Content, %				
0	3.27	0.70	0.71	1.89
20	2.95	0.66	0.71	1.92
40	2.76	0.63	0.70	1.87
Digestibility, %				
0	91	82	86	84
20	78	69	86	86
40	69	62	83	80
Digestible amino acid content, %				
0	2.98	0.57	0.61	1.59
20	2.3	0.46	0.61	1.65
40	1.9	0.39	0.59	1.50

By analysis of over 48000 SBM samples, the most frequently measured value for soybean meal was an HDI of 12. Extreme heat damage as indicated by HDI values over 50 were observed only infrequently. The histogram of HDI over 48000 SBM samples is shown in Figure 9 [48].

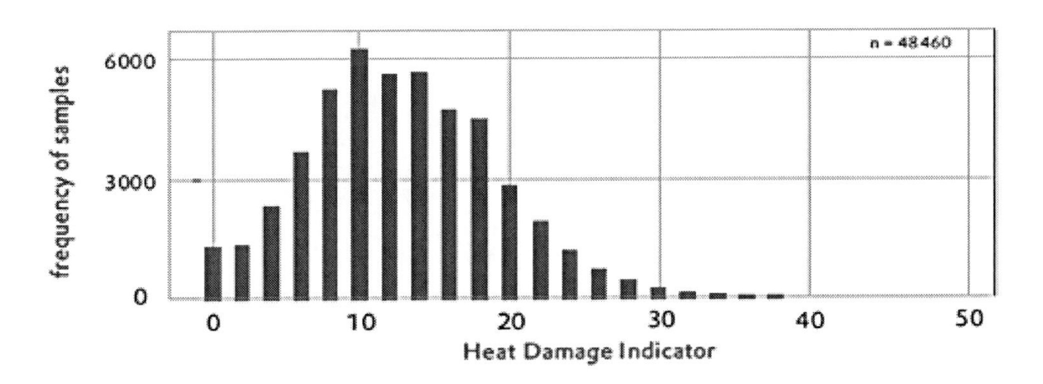

Figure 9. Distribution of HDI measured in SBM samples [48].

NIR technology can be used as a tool to predict the digestible amino acid of feedstuffs. It is useful for ranking of different feed materials, or the same feed material (such as DDGs) from different origins and produced with different processing technologies [49]. The principle of NIR technology is: a feedstuff is illuminated with light of a specific and known frequency (or wavelength) in the near-infrared region. The machine measures the absorption of light by the feedstuff expressed as the difference between the amount of light emitted by the NIR machine and the amount of light reflected by the sample. The NIR measurements are calibrated with large amount of samples and validated by *in-vivo* studies in order to predict the composition and digestibility of feedstuffs. The advantage of this method is the rapidity of the test.

Another *in-vitro* amino acids digestibly measurement is called Immobilized Digestive Enzyme Assay (IDEA), this method measures the α–amino groups liberated during protein peptide bond hydrolysis. Two systems are available for IDEA analysis. The original system needs a total assay time of 2.5 days, which used pepsin in a low pH digester followed by neutralization and digestion with chymotrypsin, trypsin, and intestinal peptidase in a second digester. The other system is designed for rapid specific testing of soybean meal that takes about 24 hours. Correlation (R^2) with *in vivo* rooster protein digestibility assays (lysine) for soybean meal was 0.91 and 0.88 for original IDEA values and IDEA kit values, respectively [50, 51].

As lysine is very sensitive to heat treatment, measurement of reactive lysine by homoarginine method will provide more precise insight on available lysine content in heat treated soybean meal [52]. However, this method is laborious and not commonly used in practice.

5.2. SBM Quality

Mateos [34] reported results on soybean meal quality assessment. In this study, a total of 312 samples were collected during 2007 to 2009 and were analyzed. The samples were analyzed for TIA, PDI, KOH solubility and Urease activities. Table 22 shows that composition and protein quality of soybean meal vary among origins. Significant differences in Trypsin inhibitor activity (TIA) and KOH solubility were also observed, indicating differences in processing conditions.

Table 22. Nutritional composition and quality parameters of SBM of different origins [34]

	N	TIA, mg/g	PDI, %	KOH, %	Urease, g N/g
Argent.	90	3.0^b	16.8^b	82.3^c	0.02^b
Brazil	78	3.0^b	15.6^c	84.7^b	0.03^a
USA	124	3.9^a	19.9^a	87.5^a	0.02^b
SEM		0.05	0.24	0.20	0.008

a,b,c: means in a column with no common superscript are significantly different ($P<0.05$).

An interesting observation is the effect of storage time on PDI and KOH solubility (Table 23). PDI significantly and KOH solubility numerically decreased during storage, while TIA maintained constant during storage. This confirmed that TIA is the most reliable quality assessment parameter for soybean meal quality, in terms of protein and amino acids digestibility.

Table 23. Effect of storage time on SBM quality parameters

Item	Storage period, wks.				SEM (n = 7)
	0	24	48	80	
UA[1]	0.03	0.02	0.02	0.02	0.010
PDI, %	20.2[a]	16.5[ab]	14.9[b]	12.7[b]	1.18
KOH sol., %	85.1	84.2	81.4	82.9	1.15
TIA[2]	4.6	4.6	4.4	4.4	0.40

$P > 0.10$ except for PDI ($P < 0.01$).

[1] Urease: mg N_2/g x min. [2] mg/g SBM

5.3. Quality Assessment of Soy Protein Concentrate

In the processing of soy protein concentrate (SPC), soluble carbohydrates and heat stable ANFs are removed/denatured by ethanol/water extraction. The difference in processing of SBM and SPC is illustrated in Figure 5.

Soy protein concentrate is processed differently from SBM, consequently, the quality assessment parameters and their optimal levels for SPC are different from SBM. For alcohol extracted SPC, the ethanol extraction process denatures protein and reduces the heat-stable ANFs such as antigens. At the same time, denaturation of protein by ethanol extraction reduces PDI and KOH protein solubility. Thus, the low PDI and KOH protein solubility in ethanol extracted SPC is due to protein denaturation by ethanol/water extraction and not due to over-heating. For this reason, the indirect methods, such as KOH protein solubility and PDI, are not recommended for assessment of protein quality of ethanol extracted SPC. Trypsin inhibitor activity, however, is the best method in assessment of protein quality for ethanol extracted SPC.

Feed grade SPC is designed for the feed of young animals. Young animals are very sensitive to ANFs, especially antigens. High antigens in soy products can cause allergic response, damage in intestinal tissues and increase the stress of weaning animals. Soy protein concentrate has very low antigen levels compared to SBM. Furthermore, oligosaccharides are removed during the processing of SPC; this consequently reduces the incidence of diarrhea and assures better intestinal health in weaning animals.In brief, for soybean meal, TIA and PDI are preferred methods to assess soy protein quality. KOH protein solubility and urease index are applicable but are not sensitive enough and show higher variability in the results.

For ethanol extracted SPC, KOH protein solubility is not applicable for quality assessment since ethanol extraction process denatures protein and reduces protein solubility. The quality control criteria for feed grade SPC should be focused on the level of ANFs, including antigens, trypsin inhibitor activity and indigestible carbohydrates.

The accuracy and suitability of some quality control tests on assessment of soy protein products quality is illustrated in Table 24.

Table 24. Accuracy and suitability of different methods for quality assessment of different soy protein products

Test	Toasted full fat soybean	SBM	SPC
TIA	++++	++++	++++
PDI	+++	+++	n.s.*
KOH solubility	++	++	n.s.
Urease Index	+	+	n.s.

* not suitable.

6. Application

6.1. Basic Processed Soy Protein

The basic processed soy protein products, including full fat soybean meal, defatted soybean meal and hi-pro soybean meal are commonly used protein sources for poultry, grower and finisher pigs. Currently, almost all commercial feed products contain soybean meal. A plethora of research has been conducted to search for alternative protein sources for SBM, such as using other oil seed meals or biofuel by-products. However, alternative protein sources can only partially replace soybean meal.

Compared to other plant protein sources, soybean meal has the advantage of balanced amino acids profile, excellent amino acid digestibility and less variation in nutrient composition and digestibility. Other plant protein products, such as rapeseed meal and DDGS products can have large variation in nutritional value and amino acid digestibility compared to soybean meal.However, as these basic processed soy products are produced by heat treatment, they can contain high levels of heat stable ANFs. This limits their application in young animals' feed, such as for weaning piglets, calf milk replacer and for fish and shrimps.

Full-fat soybean meal can be used in swine, poultry and ruminants diets. For swine diets, it is recommended that the maximum inclusion level of full-fat soybean is 15 to 20percent in swine grower-finisher diets and 25percent in sow diets. High level of full-fat soybeans may result in a softer carcass fat.

High pro soybean meal is mainly used in piglets, growing pigs and in poultry diets.

De-fatted soybean meal (44% CP) is mainly used in growing/finishing swine diets. Both Lo- and Hi-Pro soybean meals may be used in aqua-feeds and pet food, though with limited inclusion levels.

6.2. Highly Processed Soy Protein

Highly processed soy products have high digestible protein content, low indigestible carbohydrate and soy ANFs level. However, due to the high cost of production process, these products are more expensive compared to basic processed products. These products are mainly used in the feed of young animals such as weaning piglets feed, calf milk replacer and feed for fish and shrimps. These products have added value for young animals feed and the price of these products cannot be compared with soybean meal on per unit of protein basis.

Enzymatically or microbially fermented soy protein is used mainly in piglets` feed, aqua-feeds and pet food.

Classical SPC is used in the feed for young animals. Feed grade classical SPC is used mainly for calf milk replacer, for piglets' and broiler pre-starter feed, pet food and aqua-feeds.

Non-Classical SPC is mainly used in aqua-feed and piglets feed.

Literature studies and practical experiences have demonstrated that, due to the low ANFs and indigestible carbohydrate content, the use of highly processed soy products in weaning piglets feed or calf milk replacers can improve intestinal health and improve growth performance.

Yang *et al.* [30] determined the effect of different soy products on growth performance of weaning piglets during 0 to 14 days and 0 to 35 days post weaning. The authors observed that growth performance of weaning piglets is related to dietary soy protein sources. Highly processed soy protein improved growth performance of weaning piglets when compared to basic processed soybean meal (Table 25).

Table 25. Effect of different soy protein sources on growth performance in weaning piglets

Protein source *	SBM	SPC	SP[1]	FSP-A	FSP-B
Day 0-14					
ADG (g)	267[c]	291[a]	282[b]	280[b]	281[b]
ADFI (g)	430[c]	446[ab]	449[a]	431[bc]	443[abc]
F/G	1.60[ab]	1.53[c]	1.57[b]	1.60[a]	1.61[a]
Day 15-35					
ADG (g)	430[c]	477[a]	461[ab]	453[b]	469[ab]
ADFI (g)	684[b]	711[a]	700[ab]	685[b]	704[a]
F/G	1.56[a]	1.49[b]	1.52[ab]	1.51[ab]	1.53[ab]
Day 0-35					
ADG (g)	369[c]	403[a]	391[b]	389[b]	390[b]
ADFI (g)	582[b]	605[a]	599[a]	584[b]	599[a]
F/G	1.58[a]	1.50[c]	1.53[bc]	1.54[bc]	1.55[ab]

a.b.c—Values with different superscripts in the same row significantly differ ($P<0.05$).

* SBM: soybean meal; SPC: soy protein concentrate; SP[1]: enzymatically fermented soy protein; FSP-A: fermented soy protein by *A. oryzae*; FSP-B: fermented soy protein by *A. oryzae*+*B. subtilis*.

Table 26 indicates general considerations about possible applications of commercially available soy protein products in animal feed.

Table 26. Application of different soy protein products in animal feed

	Full fat SBM	SBM 44%CP	Hi-pro SBM	Fermen ted soy	Non-Classical SPC	Classical SPC
CP, % as is	36-38	43-44	46-49	52-56	60-62	63-67
Grower/finisher pigs	++	+++	++			
Poultry	++		+++			
Weaning piglets	-	-	L*	+++	++	+++
Calf milk replacer	-	-	-	+		+++
Poultry pre-starter	L	-	L	+++	++	+++
Aqua-feed	L	-	L	+++	+++	+++

* Limited inclusion levels.

REFERENCES

[1] FAO statistics. *Food and Agriculture Organization of the United Nations*.http://faostat.fao.org/. 2009.

[2] Wilson RP, Poe WE. Effects of feeding soybean meal with varying trypsin inhibitor activities on growth of fingerling channel catfish. *Aquaculture,*1985; 46: 19-25.

[3] Hancock JD, Peo ER, Jr. Lewis AJ, Moxley RA. Effects of ethanol extraction and heat treatment of soybean flakes on function and morphology of pig intestine. *J. Anim. Sci.*, 1990a; 68: 3244-3251,.

[4] Hancock JD, Peo ER, Jr. Lewis AJ, Grenshaw JD. Effects of ethanol extraction and duration of heat treatment of soybean flakes on the utilization of soybean protein by growing rats and pigs. *J. Anim. Sci.*,1990b;68: 3233-3243.

[5] Araba M, Dale NM. Evaluation of protein solubility as an indicator of under processing soya bean meal. *Poultry Sci.*,1990; 69: 1749-1752.

[6] Schulze H. *Endogenous ileal nitrogen losses in pigs. PhD thesis*. Department of animal nutrition, Wageningen Agriculture University, Wageningen, the Netherlands. 1994.

[7] Schulze H, Saini HS, Huisman J, Hessing M, van den Berg W, Verstegen MWA. Increased nitrogen secretion by inclusion of soy lectin in the diets of pigs. *J. Sci. FoodAgri.,*1995; 69: 501-510.

[8] Tukur HM, Lalles JP, Plumb GW, Mills ENC, Morgan MRA, Toullec R. Investigation of the relationship between in vitro ELISA measures of immunoreactive soy globulins and in vivo effects of soy products. *J. Agric. Food Chem.*,1996; 44: 2155-2161.

[9] Huisman J, Tolman GH. Antinutritional factors in the plant proteins of diets for non-ruminants. In: P.C. Garnsworthy, W. Haresign and D.J.A. Cole (eds), *Recent advances in animal nutrition.* Butterworth-Heinemann ltd, oxford, UK. 1992; pp3-31.

[10] Lallès JP, Jansman AJM. Recent progress in the understanding of the mode of action and effects of antinutritional factors from legume seeds in non ruminant animals. In: A.J.M. Jansman, G.D. Hill, J. Huisman and A.F.B. van der poel (Eds), *Recent advances of research in antinutritional factors in legume seeds and rapeseed.* Wageningenpers, Wageningen, the Netherlands. 1998; pp219-232.

[11] Lallès JP, Dréau D, Salmon H, Toullec R. Identification of soyabean allergens and immune mechanisms of dietary sensitivities in preruminant calves. *Res. Vet. Sci.*,1996; 60: 111-116.

[12] Lallès JP, Benkredda D, Toullec R. Influence of soy antigen levels in milk replacers on the disruption of intestinal motility patterns in calves sensitive to soya. *J. Vet. Med.*,1995; 42: 467-478.

[13] Tukur HM, Lalles JP, Mathis C, Caugant I, Toullec R. Digestion of soybean globulin, glycinin, α-conglycinin and β-conglycinin in the preruminant and the ruminant calf. *Can. J. Anim. Sci.*,1993; 73: 891-905.

[14] Li DF, Nelssen JL, Reddy PG, Blecha F, Klemm RD, Giesting DW, Hancock JD, Allee GL, Goodband RD. Measuring suitability of soya bean products for early-weaned pigs with immunological criteria. *J. Anim. Sci.*,1991; 69: 3299-3307.

[15] Pedersen HCE. Studies of soya bean protein intolerance in the preruminant calf. *PhD dissertation*. University of Reading. UK. 1986.

[16] Van den Ingh TSGAM, Krogdahl A, Olli JJ, Hendriks HGCJM, Koninkx JGJF. Effects of soybean-containing diets on the proximal and distal intestine in Atlantic salmon (Salmosalar): a morphological study. *Aquaculture,*1991; 94: 297-305.

[17] Baeverfjord G, Krogdahl A. Development and regression of soybean meal induced enteritis in Atlantic salmon, Salmosalar L., distal intestine: a comparison with the intestines of fasted fish. *J. fish Dis.*,1996; 19: 375-387.

[18] Krogdahl A, Bakke-McKellep AM, Røed KH, Baeverfjord G. Feeding Atlantic salmon SalmosalarL. soybean products: effects on disease resistance (Furunculosis), and lysozyme and IgM levels in the intestinal mucosa. *Aquaculture Nutrition,*2000; 6: 77-84.

[19] Iji PA, Tivey DR. Natural and synthetic oligosaccharides in broiler chick diets. *World's poultry Sci. J.*,1998; 54: 129-143.

[20] Grieshop CM, Kadzere CT, Clapper GM, Flickinger EA, Bauer LL, Frazier RL, Fahey Jr GC. Chemical and nutritional characteristics of united states soybeans and soybean meals. *J. Agric. Food Chem.*,2003; 51: 7684-7691.

[21] Veldman A, Veen WAG, Barug D, van Paridon PA. Effect of α-galactosides and α-galactosidase in feed on ileal piglet digestive physiology. *J. Anim. Physio. Anim. Nutr.*,1993; 69: 57-65.

[22] Smiricky MR, Grieshop CM, Albin DM, Wubben JE, Gabert VM, Fahey Jr GC. The influence of soy oligosaccharides on apparent and true ileal amino acid digestibilities and fecal consistency in growing pigs. *J. Anim. Sci.*,2002; 80: 2433-2441.

[23] Zhang LY, Li DF, Qiao SY, Wang IT, Bai L, Wang ZY, Han IK. The effect of soybean galactooligosaccharides on nutrient and energy digestibility and digesta transit time in weanling piglets. *Asian-Aust. J. Anim. Sci.*,2001; 14: 1598-1604.

[24] Fledderus J. Possibilities of soy concentrate in piglets feeds without AGPs. *Schothorst Feed Research Report No.* 728. 2005.

[25] Ireland PA, Dziedzic SZ, Kearsley MW. Saponin content of soya and some commercial soya products by means of high performance liquid chromatography of the sapogenins. *J. Sci. Food Agric.*,1986; 34: 694-698.

[26] Vielma J, Mäkinen T, Ekholm P, Koskela J. Influence of dietary soy and phytase levels on performance and body composition of large rainbow trout (Oncorhynchus mykiss) and algal availability of phosphorus load. *Aquaculture,*2000; 183: 349-362.

[27] Mateos GG, Latorre MA, Lázaro R. *Processing soybeans*. American Soybean Association. ftp://74.127.44.23/Backup/pdf/processsb.pdf.
[28] K. C. Rhee, *Effects Of Processing On Nutrient Content of Soybean meal*. http://feedmanufacturing.com/articles/feed-nutrition/soy_processing/.
[29] Hong KJ, Lee CH, Kim SW. Aspergillusoryzae GB-107 fermentation improves nutritional quality of food soybeans and feed soybean meal. *J. Med. Food.*,2004; 7:430-436.
[30] Yang YX, Kim YG, Lohakare JD, Yun JH, Lee JK, Kwon MS, Park JI, Choi JY, Chae BJ. Comparative efficacy of different soy protein sources on growth performance, nutrient digestibility and intestinal morphology in weaned pigs. *Asian-Australasian J.Anim Sci.*,2007; 20: 775-783.
[31] NRC, National Research Council. *Nutrient requirements of swine*. 1998.
[32] Pluske JR, Siba PM, Pethick DW, Durmic Z, Mullan BP, Hampson DJ. The incidence of swine dysentery in pigs can be reduced by feeding diets that limit the amount of fermentable substrate entering the large intestine. *J. Nutr.*,1996; 126: 2920-2933.
[33] Urbaityte R. *Assessment of standardised ileal crude protein and amino acid digestibility in protein supplements for piglets. PhD thesis*. Institute of Animal Nutrition, University of Hohenheim. Cuvillierverlag, Göttingen, Germany. 2009.
[34] Mateos GG. *Influcence of origin on nutritional value of soybean meal*. Soybean meal quality workshops Germany, Netherlands, Denmark and Poland, September 21-24. American soybean association international marketing. 2009.
[35] Grieshop CM, Fahey Jr GC. Comparison of quality characteristics of soybeans from Brazil, China and the United States. *J. Agric. Food Chem.*,2001; 49: 2669-2673.
[36] Fickler J. *Latest Amino Acid Data for Soybean Meal and other protein-rich Feed Ingredients*. Soybean meal quality workshops Germany, Netherlands, Denmark and Poland, September 21-24. American soybean association international marketing. 2009.
[37] Karr-Lilienthal LK, Merchen NR, Grieshop CM, Flahaven MA, Mahan DC, Fastinger ND, Watts M, Fahey Jr GC. Ileal amino acids digestibility by pigs fed soybean meals from five major soybean-producing countries. *J. Anim. Sci.*,2004; 82: 3198-3209.
[38] de Coca-Sinova A, Valencia DG, Jimenez-Moreno E, Lazaro R, Mateos GG. Apparent Ileal Digestibility of Energy, Nitrogen, and Amino Acids of Soybean Meals of Different Origin in Broilers. *Poultry Sci.*,2008; 87: 2613–2623.
[39] Corcheroet al (2009), *cited by Mateos GG*. 2009.
[40] Cervantes-Pahm SK. *Apparent and standardized ileal amino acid digestibility of soybean products fed to pigs*. Masters thesis, 2008, University of Illinois, Urbana.2008; pp 142.
[41] Dersjant-Li Y, Cai YJ, Peisker M. Quality assessment of soy protein concentrate. *Asian Pork Magazine*, Jun/July, 2006.
[42] Batal AB, Douglas MW, Engram AE, Parsons CM. Protein dispersibility index as an indicator of adequately processes soybean meal. *Poultry Sci.*,2000; 79: 1592-1596.
[43] van Eys JE, Offner A, Bach A. *Manual of quality analyses for soybean products in the feed industry*. American Soybean Association (ASA), 2004.
[44] AOCS 1993, *cited by van Eys et al.*, 2004.
[45] Swick RA. Assessing the value and quality of soybean meal. *Int. AQUAFEED*. July-August, 41-46, 2004.

[46] Parsons CM, Hashimoto K, Wedekind KJ, Han Y, Baker DH. Effect of over-processing on availability of amino acids and energy in soybean meal. *Poultry Sci.,*1992; 71: 133-140.

[47] Fickler J. Identifying heat damaged soybean products by Near-infrared analysis. *Afma Matrix.* December 2010: 10-13.

[48] Evonik Degussa GmbH. AMINONews® Information for the Feed Industry. Special Edition. June 2010. *Evonik Degussa Health and Nutrition, feed additives*, Hanau-Wolfgang, Germany.

[49] van Kempen T. NIR technology: *Can we measure amino acid digestibility and energy values?* 12th Annual Carolina Swine Nutrition Conference, 1996. http://www.ncsu.edu/project/swine_extension/nutrition/miscellaneous/theo96csnc.htm.

[50] Schasteen C, Wu J, Schulz M, Parsons C. An enzyme-based protein digestibility assay for poultry diets. *Multi-State Poultry Meeting* May 14-16, 2002.

[51] Schasteen CS, Wu J, Schulz MG, Parsons CM. Correlation of an immobilized digestive enzyme assay with poultry - true amino acid digestibility for soybean meal. *Poultry Sci.*,2007; 86: 343–348.

[52] Fontaine J, Zimmer U, Moughan PJ, Rutherfurd SM. Effect of heat damage in an autoclave on the reactive lysine contents of soy products and corn distillers dried grains with solubles. Use of the results to check on lysine damage in common qualities of these ingredients. *J. Agric. Food Chem.*, 2007; 55:10737–10743.

In: Livestock: Rearing, Farming Practices and Diseases
Editor: M. Tariq Javed
ISBN 978-1-62100-181-2

Chapter 4

BANGLADESH POULTRY SECTOR: GROWTH, COMPETITIVENESS AND FUTURE POTENTIAL

IsmatAra Begum[a,b,*], Sanzidur Rahman[c], Mohammad Jahangir Alam[a,d], JeroenBuysse[a], and Guido Van Huylenbroeck[a]

[a]Department of Agricultural Economics, Ghent University, 653 Coupure Links, 9000 Ghent, Belgium
[b]Department of Agricultural Economics, Bangladesh Agricultural University, Mymensingh-2202, Bangladesh
[c]School of Geography, Earth and Environmental Sciences, the University of Plymouth, Plymouth, PL4 8AA, UK
[d]Department of Agribusiness and Marketing, Bangladesh Agricultural University, Mymensingh-2202, Bangladesh

ABSTRACT

Since the 1960s, the world poultry meat production has been growing faster than any other meat, indicating its rising performance. Bangladesh poultry sector has emerged with great potential during the past two decades. Similar to other developing countries worldwide, Bangladesh also has a long historical record of poultry rearing under the scavengingsystem. The poultry sector in Bangladesh has not yet been fully industrialized and/or transformed from scavenging to a commercial system, although the sectorhas potential to generate income, employment as well as to fill meat consumption deficiency. The present chapter examines thetrends in production and consumption ofpoultry meat over a 35 year time period (1971–2005) in Bangladesh. Next, it examines the responses

* Corresponding authors at: Department of Agricultural Economics, Bangladesh Agricultural University, Mymensingh-2202, Bangladesh and/or School of Geography, Earth and Environmental Sciences, the University of Plymouth, Plymouth, PL4 8AA, UK.
E-mail addresses: ishameen@yahoo.com (IsmatAra Begum), srahman@plymouth.ac.uk, (Sanzidur Rahman), alambau2003@yahoo.com, Jahangir.Mohammad@UGent.be (Mohammad Jahangir Alam), J.Buysse@UGent.be (Jeroen Buysse), Guido.VanHuylenbroeck@UGent.be (Guido Van Huylenbroeck).

ofthe country's poultry and related input marketsto the poultry development policies ofthe 1990s. The chapter also examines the comparative cost competitiveness of poultry meat in relation to world's three major poultry meat suppliers (USA, Brazil and Thailand). Results reveal that transformation of the sector is slow although it has great potential. Demand for poultry meat is strong and the income elasticity of demand is the highest (0.65) as compared to other substitutes (meats and fish). Policy recommendation includes speeding the pace of the transformation process of the sector towards commercialization through various means (e.g., by favoring contract farming) and increase production in order to meet the growing demand for poultry meat.

Keywords: Poultry sector development, meat demand, growth, competitiveness, Bangladesh

1. INTRODUCTION

Poultry meat is the fastest growing meat of global total meat production. World production of broiler meat has been growing faster than any other meats since the 1960s. Worldwide, poultry meat is recognized as the fastest growing sector in meat production. Poultry meat growth rate is comparatively higher (5%) than that of pork (3%) and beef (2%) [1]. It was estimated that by 2015 poultry meat will account for 40percent of all animal protein consumed worldwide [2]. Although demand for poultry meat was high in the developed countries during the 1960s, the demand is also increasing in the developing countries from the 1990s. From 1994 to 2004, theper capita poultry meat consumption has increased by 50percent in the developing countries and 29percent in the developed countries [3]. Poultry meat share in the total meat intake has also increased by 22and 14percent, respectively in the developed and developing countries [3]. Thus, it is clear that future development in the poultry sector has the potential to significantly change the consumption of poultry meat in the developing as well as the developed countries.

Like other developing countries, Bangladesh also has a long historical record of raising poultry under traditional backyard farming or scavenging farming system. About 89percent of the rural households rear poultry and the average number of chicken per household is 6.2 [4]. The history of commercial poultry farming is very recent. Since 1980, the commercial poultry farms started to gain popularity among farmers as a main or subsidiary income source. Gradually people became interested in raising hybrid day-old chicks after knowing about their high meat productivity.

In developing countries, small poultry farming plays a central role in economic development. In Bangladesh, poultry sub-sector is an important avenue to reduce malnutrition problems because poultry meat contributes a substantial 37percent of the total meat production in Bangladesh. This contribution is much higher than what it was before. Ahmedand Haquenoted that poultry meat has contributed about 22 to 27percent of the total animal protein supply in the country during the 1980s [5, 6]. Livestock sub-sector provides full time employment to about 20percent of the total population and about 50percentof the population is engaged in this sub-sector as part timers. Thus, poultry sector also provide self-employment opportunities. Poultry farming also has considerable potential to provide income-earningopportunities, especially for those who have limited land. Small farmers can start poultry farm at their homestead area. Commercial poultry farming may also provide opportunities to create other industries like feed mills, hatcheries, etc. Poultry farming could also play a key role in poverty alleviation in Bangladesh. Thus, the poultry sub-sector is an important avenue to reduce protein deficiency, poverty and unemployment problems of Bangladesh. However, in most of the Asian countries including Bangladesh, the poultry industry is more diverse and less developed on several aspects as compared to the West. A major part of poultry meat still comes from scavenging system where the native chicken has relatively low productivity, low laying rate, high mortality and inconsistent supply because of less organized production system and management practices. However, the structure of any poultry market is affected by three factors. These are: (a) the availability of resources (favorable agro-climatic condition and inputs such as land, labor, capital, feed and technology), (b) consumer preference; and (c) thegovernment policy.

Given this backdrop, the objective of this chapter is to analyze the present situation of the poultry sector in Bangladesh. This chapter traces the process of transformation in the poultry economy of Bangladesh through changes in poultry production, marketing, government policies and consumption. At the end set some recommendations to overcome the problems faced by thissector.

The chapter is organized as follows. Section 2 briefly reviews the poultry farming system of Bangladesh. Section 3 provides an overview of the Bangladesh poultry sector, including the growth rate and trend of poultry population, meat production, per capita consumption of meat, wholesale/retail price of poultry and income elasticity of poultry meat. Section 4 discusses poultry development policies and examines the responses of the country's poultry and related input and output markets to these policies and cost competitiveness of poultry meat. Section 5 narrates discussions, conclusions and policy implications.

2. The Poultry Farming System in Bangladesh

Rural poultry sector has an important role in developing economies, generating income and employment of small and landless farmers and for improving nutritional status of the population [7, 8, and 9]. In Bangladesh, rural farmers consistently keep/raise poultry birds at a small scale largely because there are very little requirements in terms of feeding, housing, water and other production input. Anything that the farmers get from this farming system is their benefit, because they rarely invest on feed, housing, equipment or there are no other operational costs. The chickens of multi-age flocks are allowed to roam freely during the day to scavenge for feed in and outside the house. These include worms, insects, weeds, grasses and anything that can be obtained from rubbish dumps. Some owners provide supplementary grains or household leftovers in the evenings including the leftover bread. No formal house is provided to the birds except that they are allowed to live on space under the paddy storage, locally called *Gola* or the birds are confined in the corners of farmer's dwelling house at night and are set free in the morning or they live in earthen houses made of mud with only one hole for entrance and exit. Thus, in building a small poultry house, farmers use locally available materials like mud or clay, waste bamboo pieces and pieces of wood and straw. That is why scavenging poultry farming system is also recognized as the low input/low output systems.

In contrast, the history of commercial poultry farmingin Bangladesh is very recent. Commercial poultry chicks were not available in the countryuntil a few decades ago.

Commercial poultry farming started in the country as early as 1930 following the introduction of improved variety of birds from the west those are distributed to rural poultry farmers. In the 1930s, government farms were established to supply purebred exotic chicks and pullets of White Leghorn (WL), Rhode Island Red (RIR), and New Hampshire to the farmers. Local birds (*deshi*) were also crossed with WL, RIR, New Hampshire, and Black Minorca to improve the production potential. This continued up to the 1970s. During the war of liberation in 1970s, the government installations were badly affected and the breed stock was lost. During the 1980s, through the Directorate of Livestock Services (DLS), the government of Bangladesh made a special effort to rebuild the government farms and side-by-side developed a program for distribution of exotic cockerels in exchange for local chicken. Around the same time, Bangladesh Airlines (known as Biman) started a commercial poultry farm under the name of *Biman* Poultry Complex at Savar, Dhaka. At first, it was established only for the flight catering purposes, but afterwards the farm started to sell day-old chicks to the private farms.

DLS and *Biman* Poultry complex had a great impact on the growth of commercial poultry farming system in the country. Since 1980, commercial poultry farm started to gain popularity among farmers as a main or subsidiary income source. Day by day, people became interested in raising hybrid day old chicks after recognizing their high productivity. Although the commercial poultry farming system was started in 1980, the country could not get rid of an acute shortage of meat.

3. Overview of Poultry Sector

3.1. Trends in Poultry Population and Meat Production in Bangladesh

For increasing poultry meat production in the face of rapid population growth, government policies related to poultry sector development changed over the years. To analyze the trends in poultry population and production over the 35 year period (1971-2005), the period is broken down into three sub-phases in order to determine the trends during these phases which signify start of the commercial poultry sector and policy development in sequence. The three phases are:

- Phase I (1971-1980): represents the status of poultry population and production before commercial poultry farming system started in Bangladesh.
- Phase II (1981-1990): represents the status of poultry population and production after commercial poultry farming system started.
- Phase III (1991-2005): represents the status of poultry population and production after poultry development policies imposed by the government in order to expand the commercial poultry farming system as well as poultry meat production.

Table 1 shows the estimated annual compound growth rates of poultry population. The results show that the poultry population has increased rapidly over time. Poultry population has increased at the rate of 0.7, 5 and 4percent per annum during the phasesI, II and III, respectively. Poultry production of Bangladesh now is in the process of transforming from scavenging into commercial farming. Although we see that the growth has improved due to the poultry development policies (1990-95) and the pace of the growth continued more or less same during 1995-2005, but this growth is still not sufficient to meet the growing demand of poultry product (meat, egg).

Table 1. Growth rate of poultry population over time (1971-2005)

Year	Fitted semi-log function	Growth rate (% per annum)	t-value	R^2
Period I (1971 to 1980)	LnY= 10.87+0.007t	0.7	0.4	0.02
Period II (1981 to 1990)	LnY = 10.85+0.05t	5	22.85*	0.98
Period III (1991 to 2005)	LnY = 11.40+0.04t	4	10.08*	0.88

Notes: (a) Estimation used FAOStat data, (b) * indicate significant at 1percent level.

Table 2. Growth rate of poultry production over time (1971-2005)

Year	Semi-log function	Growth rate (% per annum)	t-value	R^2
Period I (1971 to 1980)	lnY= 10.79+0.015t	1.5	0.88	0.09
Period II (1981 to 1990)	lnY = 10.87+0.04t	4	5.58*	0.80
Period III (1991 to 2005)	lnY = 11.26+0.04t	4	9.61*	0.88

Notes: (a) Estimation used FAOStat data, (b) * means significant at 1percent level.

Table 2 presents the estimated growth rate of poultry meat production. It can be seen from Table 2 that the production growth rate is low in phase I (1.5%) but the production has increased at a higher rate (4%) during phases II and III.

3.2. Per Capita Consumption of Meat

There is evidence that during the period from 1980 to 2007, protein availability from animal sources in Bangladesh has fluctuated. Per capita consumption of poultry meat over this period (1980-2007) is shown in Figure 1. Per capita consumption of poultry meat has increased over time, with a stable increase from 1998 to 2007, in comparison with the situation in 1980. It is also evident from Figure 1 that the per capita goat and mutton consumption has increased after 1982; in 1996 it reached the top position and then it decreased.

3.3. Wholesale and Retail Price of Poultry

Prices of poultry are mainly determined by the demand and supply situation of poultry in the market. The usual practice for pricing of poultry is bargaining based on visual estimation of live whole bird in the local market. No enforcement exists for maintaining quality or a standard for weighing in the market. However, now-a-days, selected large commercial farms opened poultry meat sales centers in major cities, where consumers can buy branded meat and meat products (e.g., only leg portions, thighs, or livers, etc.) at fixed prices. Usually, farmers sell poultry to wholesaler at the farm gate. Then the wholesalers sell the product to the institutional buyers (such as hospital, school, university dormitory etc.), sales center, or the retailers. However, prices vary substantially depending on the scale and level of operation of the intermediaries and the structure of the marketing channel in various parts of the country.

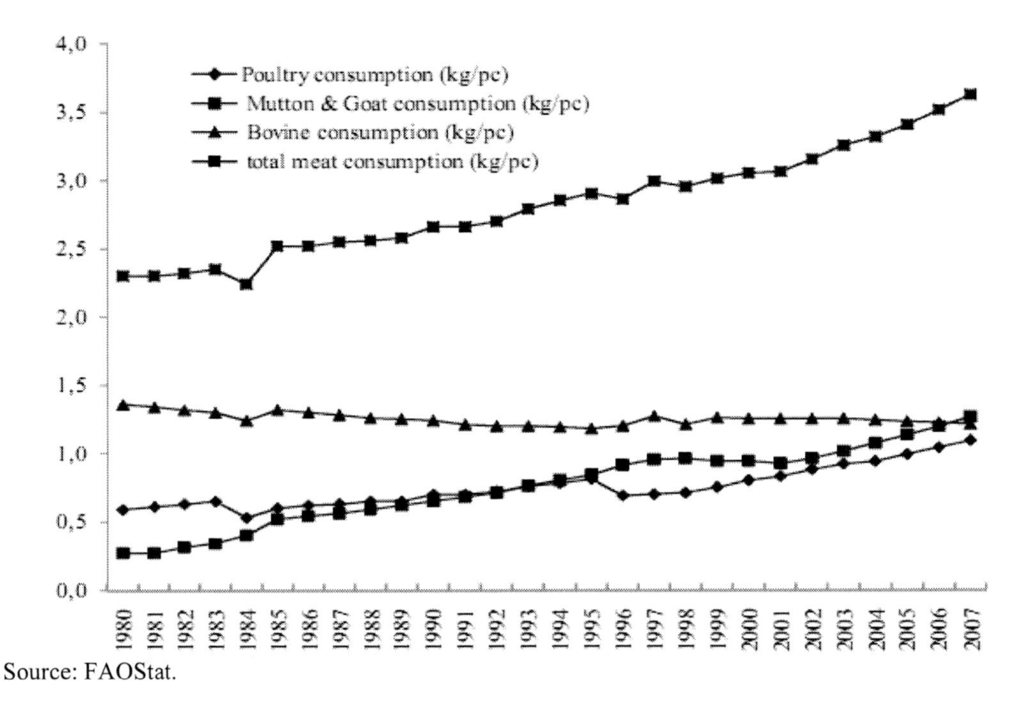

Source: FAOStat.

Figure 1.Trend in annual per capita consumption of meats (1980 to 2007).

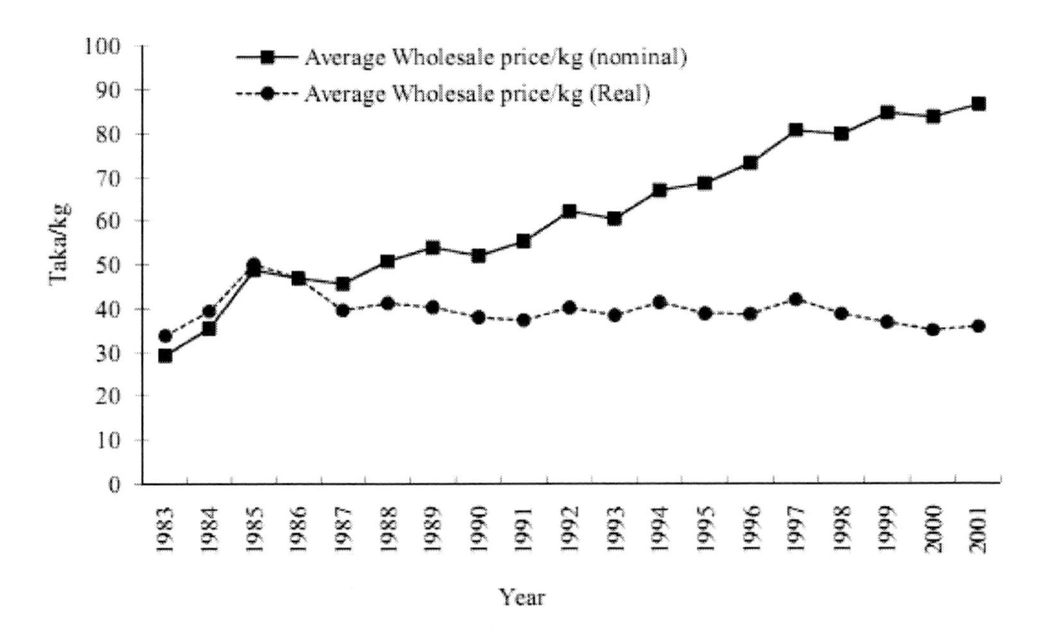

Figure 2. Wholesale nominal and real price of chicken from 1983-2001.

Figure 3. Retail nominal and real price of chicken from 1983-2001.

Figure 2shows the evolution of nominal and real poultry meat wholesale prices during 1983 and 2001. The figure indicates that although the nominal price is increasing but the real price (determined by using Consumer Price Index as deflator) is relatively stable. The pattern is more or less the same for retail prices too (Figure 3).

3.4. The Demand: Income Elasticity of Poultry Meat and Eggs

Income remains the factor with the greatest influence over dietary changes. Research results from USDA's Economic Research Service (ERS) showed that consumption patterns have changed for certain countries in response to rise in income [10]. Income-initiated dietary changes in high-income nations are relatively small as compared with income-initiated dietary changes in lower-income nations. In world development report 2000, the World Bank defines high-income countries as those with 2000 per capita Gross National Product (GNP) above US$9266, middle-income countries as those with 2000 per capita GNP between US$756 and US$9266, and low-income countries as those with 2000 per capita GNP below US$756. Countries in the low- and middle-income groups are generally considered to be developing countries. In countries at the low-income levels, such as Bangladesh, consumer demand for food is driven by the need for individuals to meet basic caloric requirements, leading to diets mainly comprising of carbohydrate-rich products, such as cereals (Table 3). In countries at higher income levels, consumers can readily meet their caloric needs. Income growth among consumers in these countries may lead them to substitute staple foods with more expensive sources of calories, such as meat. Thus, as the economy growsthe structure of meat consumption shifts. In the Middle East and East Asia, increased incomes from rapid industrialization or oil revenues led to greater consumption of meat. In East Asia, per capita meat consumption has risen more than 95percent since the mid 1970's.

Table 3. Per capita meat and cereal consumption in different income countries

Food items	kg per capita			
	1961	1970	1980	1990
Meat				
Low income countries	5.3	7.6	10.0	14.7
	(100.0)	(143.6)	(188.0)	(276.9)
Medium income countries	22.7	26.9	33.6	37.7
	(100.0)	(118.6)	(148.2)	(166.2)
High income countries	54.2	64.8	76.1	80.7
	(100.0)	(119.2)	(140.4)	(148.9)
Cereal				
Low income countries	128.5	148.2	157.1	173.1
	(100.0)	(115.3)	(122.2)	(134.7)
Medium income countries	125.0	131.9	139.9	142.2
	(100.0)	(115.3)	(122.2)	(134.7)
High income countries	122.3	111.7	107.3	108.1
	(100.0)	(91.3)	(87.8)	(88.4)

Source: FAO [11].

Notes: (a) Figures in parentheses shows index 1961=100; (b) Countries are grouped according to the World Bank definition.

Table 4. Consumption of protein from animal sources in differentincome groups

Income group	Monthly household income (Taka1)	No. of households (thousand)	% of house holds	Total No. of members (thousand)	Consumption (kg/capita)					
					Mutton	Beef	Poultry	Others meat	Total meat	Fish
I	(<750-2499)	6958.89	28.54	27379.99	0.08	0.69	0.29	0.06	1.72	54.00
II	(2500-5999)	11234.50	46.13	58743.59	0.05	0.78	0.33	0.04	1.80	48.42
III	(6000-9999)	3615.46	14.83	22191.66	0.05	1.53	0.78	0.18	3.22	52.92
IV	(10000-14999)	1391.35	5.70	9202.09	0.06	1.10	0.53	0.02	2.16	28.26
V	Above 15000	1146.96	4.70	8589.72	0.09	2.02	1.11	0.04	4.14	45.26

Source: Household income and Expenditure Survey [12].

Table 4 shows that group II, which comprises 46 percent of the total number of households and whose monthly income ranges from Taka 2500 to 5999, consumes a very small amount of meat. On the other hand, group V, which comprises less than 5percent of the

[1] 1 US$ =68.25 Taka (July, 2008) [13].

total households and is the highest income group, consumes two times more meat than income group II. This implies that with the increase in income, the meat consumption also increases. Now, let us see the present consumption pattern of Bangladeshi people according to their income level.

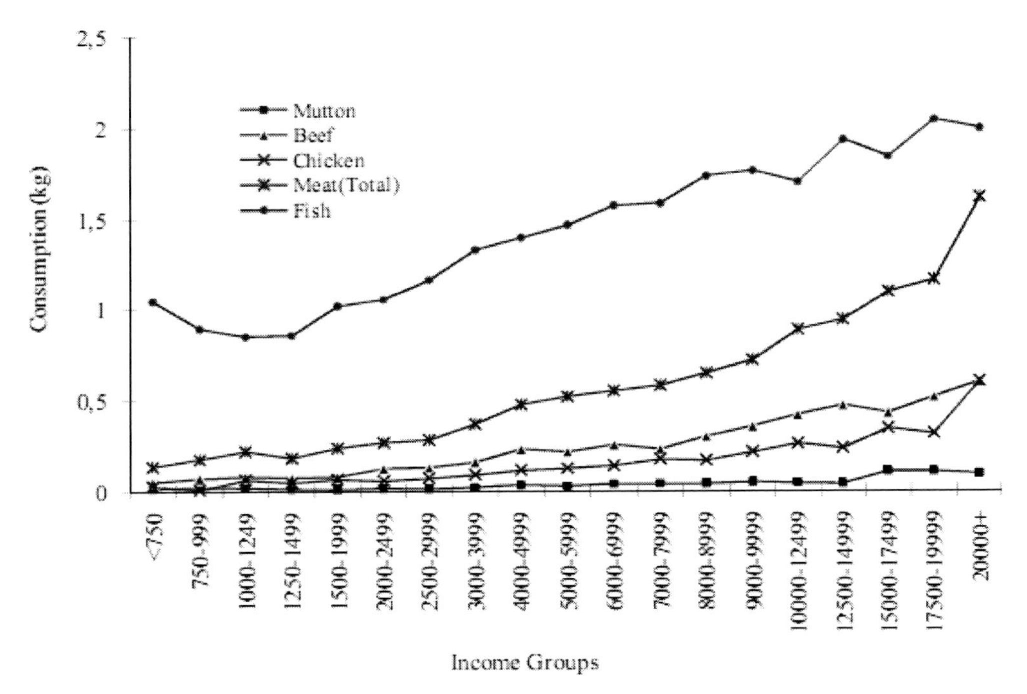

Source: Household Income and Expenditure Survey (HIES) [12].

Figure 4.Per capita annual consumption of meat of different income groups.

Although at present the average level of consumption in Bangladesh is very low, it does not mean that people of all income groups behave similarly. Income elasticity of demand for poultry meat and eggs was estimated from data (7440 households covering the whole country, whereas 5040 household from rural and 2400 household has taken from urban area) on consumption patterns of 19 different income groups as reported in the HIES (Figure 4).

A large number of alternative functional forms are possible for estimating income elasticity. The simple model is a double log function, which is as follows:

$$LnY = LnA + bLnX + U$$

where, Y= Quantity consumed, and X= Income level, U= A random error term which is identically and independently distributed with mean 0 and variance σ_u^2, i.e., N (0 σ_u^2).

Income elasticity of poultry and livestock meat, egg and fish can be seen from Table 6. The results indicate good fit and high R^2. All the coefficients are statistically significant at one percent level. The income elasticity of poultry meat is the highest (0.65) compared to meat from animal source and fish, which means that for a given increase in income, per capita consumption of poultry meat increases more than the consumption of meat from other sources (mutton,beef and fish).

From Table 5, we can see that the dietary habits of Bangladeshi people change when the average income increases. Specifically, as the average income increases, the consumption of animal products also increases. This notion is reconfirmed by the global data in Table 5. The average income of group IV is US$1938, which corresponds to the highest income group of Bangladesh whose monthly income is above Taka 15000. If the average income of this group moves up to group V (average income is US$ 8350) due to economic development, their consumption of animal products will rise almost 4 times from 390 kcal to 1251 kcal and their consumption of vegetable products will fall from 2407 kcal to 2107 kcal.

The evidence, therefore, implies that there is a strong possibility that Bangladeshi people will consume more of animal products as their income increases in the future. Although poultry population and production has increased over time but still per capita availability of meat is much lower than the minimum requirement level [14]. Meat production in Bangladesh has failed to keep pace with the growing meat requirements of a rapidly growing population. To meet the protein gap within a shortest possible time, there was a shift of policy emphasis on commercial poultry rearing in the beginning of 1990's.

Table 5. Dietary pattern of different income classes

Income class	No. of nations	Average income(US$)	DES (kcal)	DES from vegetable product (kcal/day)	DES from animal product (kcal/day)
I	7	187	2125	2014	111
II	3	340	2294	1991	302
III	5	644	2426	2259	167
IV	4	1938	2797	2407	390
V	6	8350	3358	2107	1251
VI	6	13440	3396	2107	1292

Source: FAO [15].

Table 6. Results of estimation of income elasticity

Particulars	Constant	Inx	R^2	n
Beef	-6.07*** (-10.42)	0.59*** (7.14)	0.78	19
Poultry	-7.65*** (-12.60)	0.65*** (9.14)	0.82	19
Fish	1.04*** (9.86)	0.17*** (13.58)	0.91	19

Notes:(a) Income elasticity were estimated by fitting double log; (b) *** indicates significant at 1percent level.

4. GOVERNMENTPOLICIESAND STRATEGIES FOR POULTRYSECTORDEVELOPMENT

In Bangladesh, as elsewhere in the developing countries, in response to meet the demand of protein from animal sources, a policy was set up in the 1990s to stimulate commercial poultry farming. In the Fourth Five Year Plan (FFYP) 1990-95, the government of Bangladesh used an incentive approach to increase the participation of smaller producers through technology transfer and credit schemes. FFYP emphasized the importance of strengthening the organizational and institutional framework related to the poultry sub-sector and improving support services. The main objectives of FFYP were focused on increasing the supply of livestock products through increase in productivity. To increase the supply of poultry products, the strategies and policies in the livestock sector during the FFYP focused on the following major areas:

a. To increase supply of meat with adequate focus is placed on medicine, feed, marketing and other problems along with the development of poultry breeds;
b. Review and necessary changes in the import policy for poultry products;
c. Restrict the issuance of licenses to the large-scale poultry meat production enterprises. However, the establishment farms for the supply of commercial breed chicks will be encouraged;
d. Strengthening of the organizational and institutional framework of the poultry sub-sector for undertaking effective research, work force training and development activities.

The detail strategies of poultry development policy are discussed below:

(I) Input Supply and Pricing

For the supply of day-old chicks, the strategy focused on breeding. Apart from the expansion of the existing dual-purposecrossbreeds, single purpose high yielding breeds were introduced on a large scale. In support of adequate feed supply to poultry farms, a campaign for feed grain was undertaken. Furthermore, both production of maize as well as preparation of balanced feed in the commercial sector was encouraged. In order to increase the availability of feed protein supplements, pilot programs for the collection of slaughter wastes and the processing into bone and blood meal was undertaken. Besides raising production, the government strategies also aim to ensure quality of feed. Feed analytical laboratories were setup in central locations of the most important poultry production areas such as Gazipur, Kishorganj, Dhaka and Chittagong districts. To facilitate increased availability of poultry feed, the government continued restrictions on the export of feed ingredients like wheat bran, rice polish and oilcake.

Strategies also emphasized input price control. Price level for inputs was initially set at a level that allowed for private sector participation. Prices for concentrate feed was fixed at a level since the product was introduced as an integral part of intensive poultry development and supply was foreseen to come from the public sector. The public sector supply was to be gradually withdrawn once the market grows and the private sector assumes its role.For output pricing, the price for meat and poultry products was determined by the forces of supply and demand.

(II) Emphasis on Small Producers

As if in any other developing country, in Bangladesh, development policies of the government have emphasized maintenance and enlargement of the market share of the small producers and for the poultry development policy, the strategies remained the same. As a developing country with a high percentage of people living under poverty, the government cannot freely establish large-scale poultry farms, despite high deficiency in meat supply. For this reason the Poultry Development Policy also emphasized on participation by small farmers.For the welfare of small farmers, licenses for large-scale production of eggs and poultry meat enterprises were restrictively issued. However, the establishment of multiplication farms for the supply of commercial breed chicks was encouraged. Obviously these policies aimed to promote adequate employment and income generation for the poor and small farmers and other groups. Since 1990, the Poultry Development Policy has aimed to maintain and enlarge the market share of small producers. The government issued licenses restrictively to the large-scale producers to protect small-scale producers. Poultry meat production has continued on an upward trend since 1990 when the Poultry Development Policy was initiated, but it has not outpaced the growth of population.

(III) Import Policy

In 1990, the government made proper review and necessary changes in the import policy for poultry products and provided many incentives for this sector, such as the following:

a. This sector gets a tax holiday. No tax is payable for importing the ingredients of poultry feed;
b. The government imposed no embargo or prohibition for importing anything that is directly or indirectly related with this industry;
c. No VAT (value added tax) is payable for marketing the day-old chicks, feeds, eggs or meat;
d. To encourage local investors to produce day-old chicks, the government banned the import of day-old chicks.

(IV)Credit Scheme

Historically, formal institutional credit in the agricultural sector has been targeted principally to the crop sector, but since the 1990s due to the Poultry Development Policy,the government emphasized on credit supply to the poultry sector. Government patronization and efforts of banks helped farmers to easily get bank loans.

The next section examines the responses of the country's poultry and related input markets to the poultry development policies of the 1990s.

4.1. Present Status of the Inputs Market Structure

Market structure of any commodity consists of the characteristics of market organization, which influence the nature of competition and as a result influence formation ofprice within the given market. The setup of the market usually consists of information on the degree of concentration of buyers and sellers, integration, product differentiation and the degree of competition between buyers and sellers [16]. In a competitive business, farms are expected to earn normal profit, otherwise existence of abnormal or super normal profit allows new entrants in the market pushing down the profit rate to the normal level. In a competitive market a trading farm's temporal performance depends on its physical, financial and human resource assets as well as its ability to minimize costs.

4.1.1. Progress of Hatcheries

Rapid demand for day-old chicks justify establishing hatcheries. Available information indicates that there are a total of 120 hatcheries in the country. Among these, 50 are fully functional at present, others are either partially operating or are temporarily closed. Fifty six percentof the active hatcheries are owned by people who have parent stock. Out of 120 hatcheries, 50percent are located in Gazipur, Chittagong and Dhaka districts, where concentration of poultry farms is the highest. Long distance transportation of day-old chicks is costly and that is why hatcheries tend to be established closer to the poultry farms or production areas. Out of the 50 hatcheries, approximately 28 hatcheries are involved in the production of day-old chicks from parent stock. Other hatcheries procure hatching eggs from different sources for selling of chicks [17, 18]. Eleven hatcheries those procure day-old chicks are government owned [19]. AftabBahumukhi Farm Ltd., Paragon Poultry Ltd., Biman Poultry Complex, Quazi Farms Ltd. etc. are some examples of hatchery farms in Bangladesh but their production systems totally depend on import of parent stock from abroad. Bangladesh totally depends on exotic strains of chicken for commercial poultry farming. The productive performance of foreign strains in Bangladesh varies widely. Despite good genetic potentiality, the performance of these strains is satisfactory only when the environment particularly the temperature, nutrition and management are favorable under conditions of Bangladesh. Figure 5 shows the present operational flow of hatcheries in Bangladesh. In the hatchery industry, strain of breeding stock is the main criteria used by buyers and sellers to differentiate products. Hatcheries use different brand names for broiler day-old chicks (DOCs) and some of them have established goodwill among buyers by providing quality DOCs. Currently several strains of broilersare available in the country. In Bangladesh, there are now about 10 international poultry breeding companies those supply day-old chicks all over Bangladesh, either from imported parent stock or from imported hatching eggs. A list of international poultry breeding companies with their broiler strain is given in Table 7. Table 8 shows the commercial broiler production and the number of broiler farms in Bangladesh. Local hatcheries produced about 61 million day-old broiler chicks in the year 2000, which is about half of the present day-old chick demand for commercial farming [20].The hatchery owners set the price of DOCs independently but consider the reaction of competitors in the market. They usually sell DOCs on cash at fixed price to farm owners and agents but provide a commission to the agents. It is important to note that there is no bargaining between buyer and seller of DOCs at any point in the supply chain, it is basically a supply driven market.

Table 9 presents the utilization of production capacity of hatchery in Bangladesh poultry. It is evident from Table 9 that only 35percent of the hatcheries utilize80percentor more of their total production capacity. The hatchery owners sell the DOCs at the hatchery or through their sale center's directly or through sale agent's to poultry farmers. Generally, DOCs are packed in either paper boxes or bamboo baskets. A few hatcheries use their own or hired pick-up vans to transport DOCs from the hatchery to the sale centers and/or agent stores. Generally, most of the time poultry farmers do not transport DOCs in specialized vehicles;

usually they carry on passenger buses, rickshaws or rickshaw-vans. Such transportation is hazardous and increases the likelihood of mortality during movement.

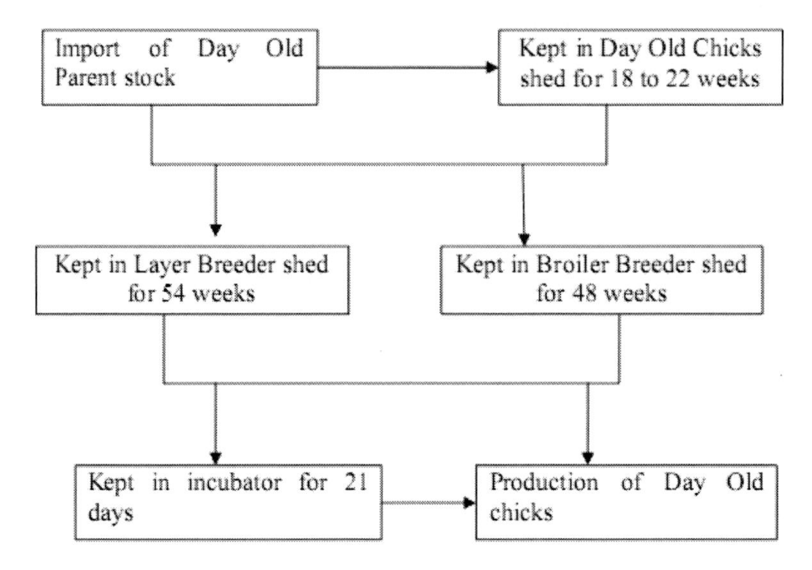

Figure 5. Production processes of day old chicks in Bangladesh.

Table 7. Broiler strains in Bangladesh poultry sector

Breeding Company	Country of Origin	Broiler Strains
Euribrid	Netherlands	Hybro
Arbor Acress Breeder	USA	Arbor Acres
ISA	France	ISA-I757, ISA Redbro
Shaver Poultry Breeding Farm Ltd	Canada	Starbro
Cobb Breeder	England	Cobb
Hubbard	The Netherlands	Hubbard Golden Commet
Ross Breeder Ltd.	UK	Ross 1, Ross 2
Lohman	Germany	Lohman
Indian River	USA	Indian River
Babolna Tetra SL	Korea	Babolna Tetra SL

Source: Poultry KhamarBichitra [21].

Table 8. Commercial broiler production in Bangladesh

Year	Hatchery (Breeding Farm)	Number of Parent Stock	Day Old Chick Production	Broiler Production
1985	2	5,000	600,000	5
1992	6	30,000	3,540,000	4,540
1994	11	65,500	7,750,000	9,920
1995	25	155,000	1,875,000	24,370
2000	50	500,000	6,100,000	82,350

Source: Siddique[17].

Table 9. Utilization of hatchery's production capacity

Utilization of production capacity	Number of Hatchery	% of Total Hatchery
10 - 39%	5	19
40 - 59%	4	15
60 - 79%	8	31
80 - 100%	9	35

Source: Poultry KhamarBichitra [21]; Note: Sample size 26.

4.1.2. Progress of Feed Mills

One of the major problems in the development of the poultry sub-sector in Bangladesh relates to lack of sufficient and appropriate feeds [22, 23]. Two types of feeds are used in poultry farming. One is ready to use manufactured feeds and another is supply of separate feed ingredients for mixing. The ready feeds of different feed mills available in the market are not homogeneous in nature. The manufacturers differentiate poultry feeds on the basis of quality, brand name, sales promotion, and packaging. Different feed mills distribute their products in different ways. Some feed manufacturers distribute feeds through agents, some use wholesalers and retailers; some have their own sales centers. The feed millers set the price of feeds independently but consider the reaction of competitors in the market. Feed millers usually set the prices for wholesalers and *Aratdars*, giving little scope for bargaining except that rates of commission may vary depending on the volume of purchase. Feed millers usually promote their products through advertising and providing quality assurance and incentives such as differential commissions to wholesalers and some millers also provide incentives to farmers. Generally, feed manufacturer fixes the prices for wholesaler. The wholesalers sell feed in both cash and credit to retailers and farmers.In setting selling price, some of the wholesalers charge a fixed margin on the total cost of feed marketing and others add a certain percentage of total cost as profit. The price of feed varies from brand to brand. Feed prices also vary on the basis of mode of payment (cash or credit) and volume of purchase (small or large).

Wheat and maize together constitute over half of the total poultry feed, of which a little less than four-fifths is maize and about one-fifth is wheat. Although the use of wheat and maize for livestock and poultry feed is growing rapidly in developing countries [24], this has not yet reached a significant proportion in Bangladesh. The demand for both livestock and poultry feeds appear to be currently met only from imports, and at seemingly relatively higher costs. In the beginning of the commercial poultry farm, farmers used locally available feed ingredients in their feed formulation. At present, the status of the feed situation has changed.

Most of the ingredients used in the feed formulation are presented in Table 10. Most of the feed ingredients are imported and therefore are subject to the change in the international prices. Maize is imported from India, Brazil etc., as a supplement of domestic supply. Figure 6 presents the year-wise cultivated area of maize from 1979 to 2000 in Bangladesh. From Figure 6, it is clear that although government policy is aimed to increase the production of maize, neither the acreage under maize nor the production of maize increased significantly over time. Figure 6also presents the maize production in Bangladesh over time. This figure shows the main reason why farmers depend on imported maize even now. After the poultry development policy of the government was implemented in 1990, import of maize increased significantly. Figure 7 presents the import value of maize from 1993 to 2002. Aside from maize, commercial poultry feed producers importvarious poultry ingredients from different countries. The country of origin of imported feed is given in Table 11. Poultry feeds are mainly imported from Germany, China, Thailand, India and Taiwan. The exact number of feed mills in operation at present is not known but there are estimated around 35 feed mills with around 850 dealers at the private sector who are producing and distributing poultry feed in the country. These feed mills are owned and operated by the private sector, but their distributional system in rural areas is inadequate and their production does not meet the demand.

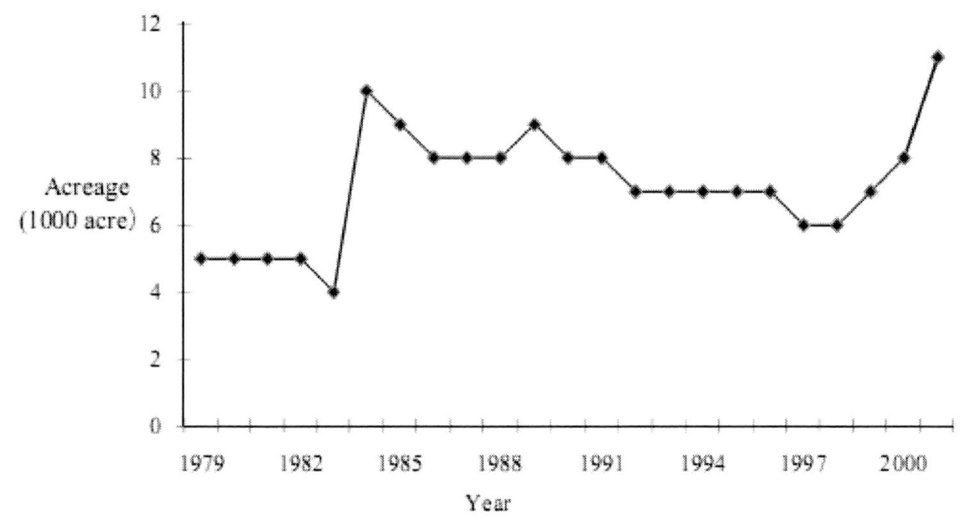

Source: Various issues of BBS [25, 26].

Figure 6.Production area of maize 1979 – 2000.

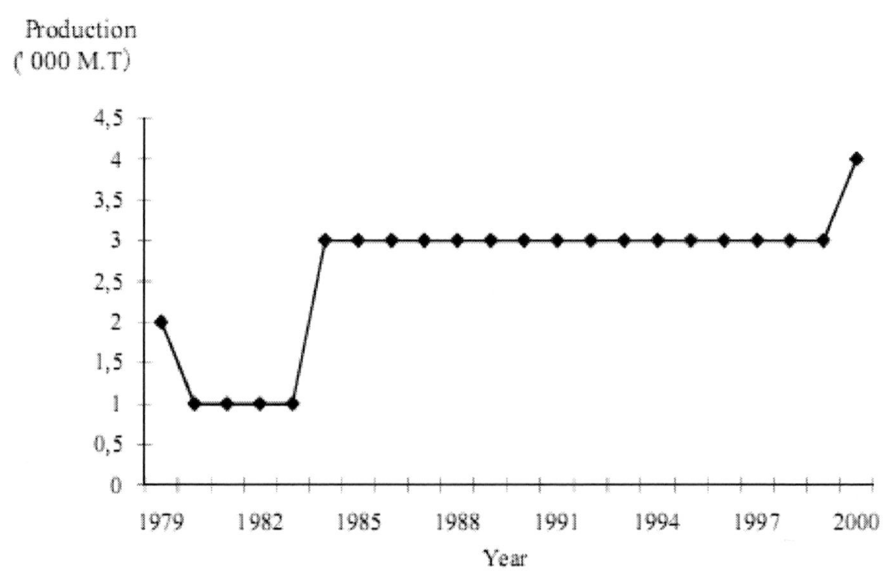

Source: Various issues of BBS [25, 26].

Figure 7.Trend of maize production 1979–2000.

Table 10.Available poultry feed ingredients used in Bangladesh

Ingredients	Locally Produced	Imported
Wheat	x	
Maize	x	x
Rice Polish	x	
Fish Meal	x	
Broken Rice	x	
Wheat Bran	x	
Meat and Bone Meal		x
Soyabean Meal		x
Full Fat Soyabean	x	
Oil Cake	x	
Protein Concentrate		x
Dicalcium Phosphate		x
Mono Calcium Phosphate		x
Limestone	x	
Oyester Shell	x	
Molasses (Raw sugar Juice)	x	
Enzyme		x
Salt	x	
Toxin Binder		
Natural Growth Promoter		x
Vegetabe Fat		x
Vitamin Mineral Premix		x
Choline Chloride		x
Methionine		x
Lysine		x

Source: Field survey of 2003.

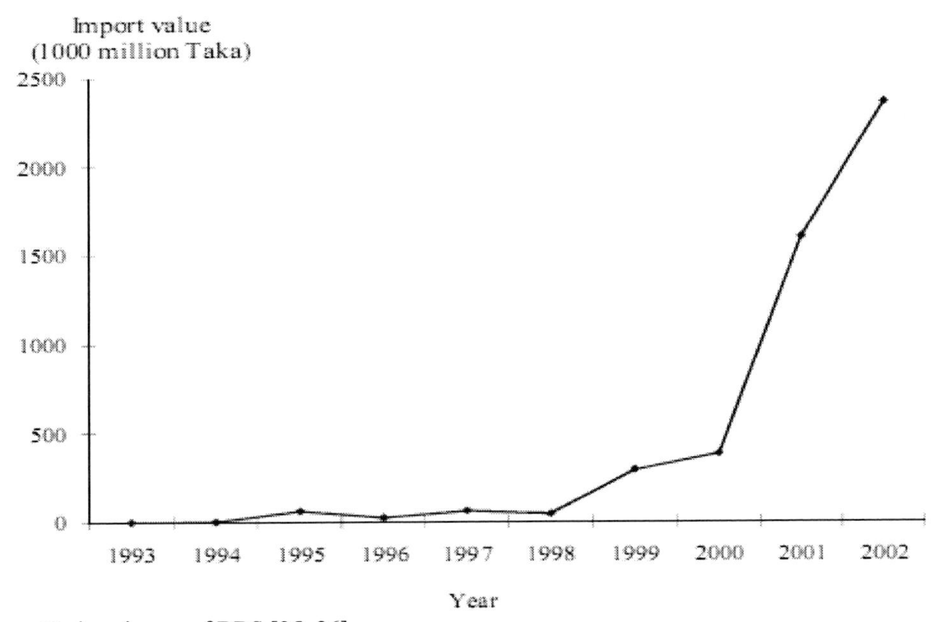

Source: Various issues of BBS [25, 26].

Figure 8. Import value of maize 1993 – 2002.

Table 11. Origin of feed in Bangladesh poultry sector

Brand of feed mill	Importation from	Type and class of feed
Awila	Germany	Pellet (brand Ly)
Jianju	China	Mash and Pellet
KPI	Thailand	Pellet
Spectrams engineering (pvt) ltd	India	Mash and Pellet
Nav Intel (pvt) ltd	India	Mash and Pellet
Chia Tung	Taiwan	Mash and Pellet

Source: Poultry KhamarBichitra[21].

4.1.3. VeterinaryDrugs

The endemic poultry diseases in Bangladesh are Newcastle, Fowl Pox, Fowl Cholera, Fowl Typhoid, Coccidiosis, Gumboro, deficiency diseases and worm infestations, etc. The mortality rate of poultry is high (35 to 40%) due to disease and predators. Available information indicates that more than 100 pharmaceutical companies have been reportedly involved in business of veterinary drugs in the country. An estimated Taka3500 million worth of health and nutritional products are marketed annually in the country [27]. The marketing chain for drugs is simple and composed of 3 actors: the pharmaceutical companies, the wholesalers and the retailers. The pharmaceutical companies distribute drugs to wholesalers. The retailers purchase drugs from wholesalers and sell to the poultry farmers. Although the government gives some necessary vaccines atprice of production cost to help poultry farmers, sometimes farmers rush to buy imported vaccines at a higher price. The Directorate of Livestock Services (DLS) has four field staff and one Livestock Officer in each sub-district level, and they provide services to about 2,00,000 poultry, 50,000 cattle, and 20,000 goats and sheep [19]. The major four types of vaccines including Newcastle disease, Gamboro, Fowl pox, Cholera etc., are produced in Bangladesh. However, vaccines are not available equally in all parts of the country, especially in the remote part of the rural areas, where finding vaccines on time is very rare. Drug prices sometimes increases the production cost of small poultry farmers. For example, the cost of Gumboro vaccine (1000 doses), is between Taka 700 to 720 (retail price). If the flock size is smaller than 1000 birds then farmers have no way to buy below 1000 doses. As a result, since flock size is small like 500 birds or less than 500 birds at most of the farms (85%), use of this particular vaccine will cost Taka 2 or more per bird per vaccination. If such a vaccine is used three times, it will cost Taka 6 or more per bird. If vaccination cost to prevent Gumboro disease alone becomes so high, one can imagine the total cost of executing a vaccination program for a flock.

Vaccination failure is a common report by farmers. Although a number of causes can be held responsible for it, loss of quality during storage, transportation and handling of vaccine appears to be the major contributing factors. Most of the vaccines used in Bangladesh are of foreign origin. There is considerable doubt that these imported vaccines are antigenically competent to deal with viruses prevalent in Bangladesh and that these are received at cold chain facilities for storage. The quality of locally produced vaccines is also questionable. Most commercial farmers use vaccines without knowing the maternal antibody status of the flocks they are raising.

4.1.4. PoultryEquipments

Poultry equipments includebrooder, watering utensils, chick guard, feeder, egg trays, feed trays, and mugs. The marketing chains for equipments are simple and that include the manufacturers, the wholesalers and the retailers. The manufacturers sell equipments to wholesalers. The retailers purchase equipments from wholesalers and sell to poultry farmers.

Manufacturers usually set prices for wholesalers who then set price by adding a fixed margin to the purchase price and marketing cost of equipments. The marketing margins of wholesalers per equipment varyfrom equipment to equipment.

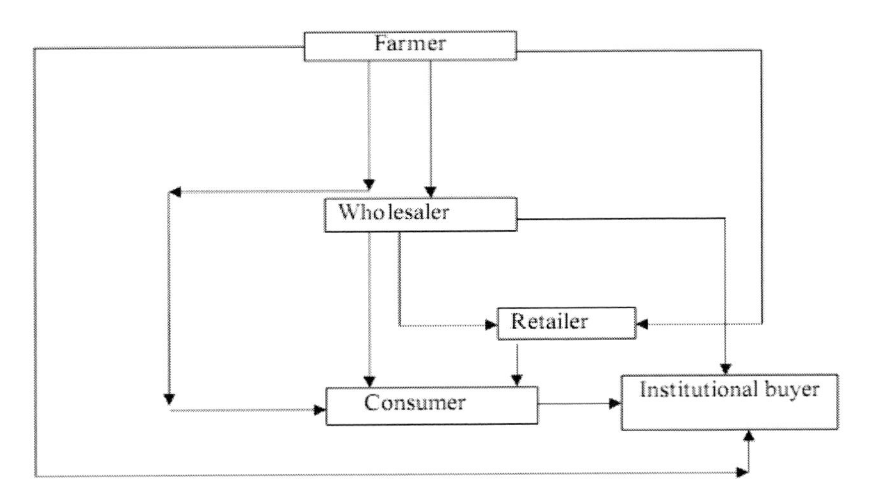

Figure9. Different marketing channels of independent commercial poultry farm.

The commercial poultry farms in Bangladesh face a lotof problems. For example, inadequate knowledge of poultry rearing, a lack of capital, inadequate or unavailability of inputs, outbreak of poultry diseases and lack of institutional credit, etc. Apart from production-oriented problems, farmers also face problems related to marketing of their products. Figure 9 presents different marketing channels of poultry meat in Bangladesh.

Farmers often sell output at lower prices in the local market because of inadequate transport and marketing facilities. Sometimes, they face the problem of selling broiler on time. Hence, they do not often get appropriate return. Another marketing problem mentioned by the producer is the high price of day-old chicks. One reason for this is that chick price varies from strain to strain of the chicken. Another reason is that farm owners fail to get chicks as per their demand. To buy chicks, they need to make advance booking and payments and to wait for a long time for the delivery of chicks. This has given intermediaries a chance to enter into trade channels; theseintermediaries bookchicks on behalf of their customers and sell them to poultry farmers at higher prices. Other marketing constraint faced by the poultry farmers is less availability of feed throughout the year.

4.2. Farmers' Knowledge, Training on Technical Knowledge

Poultry farmers need technical knowledge to run a poultry farm. However, in practice, most of the poultry farmers have started farming without any technical knowledge or institutional training. Consequently, farmers face problems of using modern technology in the production system which limits their potential to maximize poultry output. One of the highly reported problems is inadequate knowledge about poultry rations. Previous research studies have pointed out and recognized similar set of problems about poultry farmers and farming over the past three decades [5, 6, 28, 29, 30, 31, 32, 33, 34 and 35].

4.3. Cost Competitiveness

Commercial poultry started in 1980 and government also took poultry development policy in 1990 to increase commercial poultry production, which resulted in a spectacular increase in the number of poultry farms. Poultry farming has considerable potential for providing income opportunities, especially for those who have limited lands. Small farmers can start poultry farming at their homestead area by using family labor. Poultry farming has been expanding rapidly all over the country (in 1980 it was 787 whereas in 2000 it was 80000) but it has not yet been included in the export basket yet.

Competitiveness relates to poultry farms or poultry industry and not to the nation. Competitiveness is defined as 'the ability of businesses or industries to compete internationally' or 'the ability of businesses or industries in one country to compete withthose of other countries' [36]. Although the Bangladesh poultry industry is very small as compared to neighboring India or the world, there is a strong feeling that Bangladesh can earn foreign exchange by exporting broiler meat. Production cost of poultry products is considered to be low in the country due to low cost of labor and housing. Table 12 presents relative cost competitiveness among the world's three major suppliers with Bangladesh. As can be seen, although Brazil has the lowest production costs of live broiler, but Bangladesh's production cost is close to that of USA. If we consider only cost, then Bangladesh has a potential to export poultry to other countries as it has comparative advantage due to low cost of labor and housing in the country.

Table 12. Broiler production costs (US$/kg)

Item	USA	Brazil	Thailand	Bangladesh
Live broiler	0.56	0.42	0.61	0.53

Source: USDA, Agricultural outlook forum and authors own calculation.

However, before to go to export market, Bangladesh poultry sector should ensure the quality of the products, as well as ensure regular and required amount of products for the export market.

5. Discussions, Conclusions and Policy Implications

Production and consumption of poultry meat have grown faster than any other kind of meat because of its short life cycle, low cost and high consumer preference. Higher poultry demand in the world is the result of shifting taste patterns driven by concerns about the health risks of consuming large quantities of red meat. Empirical observations of changing meat consumption patterns in both the developing and developed worlds bear out these results. The demand for poultry is highly elastic as compared with other meats, thus indicating that marginal expenditures on poultry are likely to increase in the future not only in developed countries but also in developing countries as well as in Bangladesh. High level of vertically integrated contract farming is the key feature of the world's most efficient poultry producers for their improved production and marketing efficiencies and increasing number of value added products at lower cost.

The world poultry market is highly competitive, with an increasing number of efficient producers fighting for market share and the poultry sector is industrializing rapidly, not only in developed countries but in developing countries too. Bangladesh poultry sector is emerging with a great potential. Like other developed countries, Bangladesh needs to emphasize on competitive pricing, aggressive marketing, new products and market developments, safety, quality assurance, biosecurity and the environment. The poultry industry in major exporting countries are characterized by large scale production and high level of vertical integration and by following this system these countries gained from lower production cost and higher incremental return from further processing and value-addition.

From the foregoing discussion it is also clear that for expanding the commercial poultry farming in a short time, the poultry input market in Bangladesh has not been sufficient and well organized yet. The results of the growth rate of poultry population and production indicate that although it is expected due to the poultry development policies (1990-95), there has been an improvement in the poultry sector, but this growth rate is not sufficiently high to meet the ever-growing demand. The consumption behavior of poultry meat also reveals a relatively strong demand for poultry meat. The positive income elasticity of demand for meat implies that the per capita demand for meat would increase in response to an increase in per capita income. The greater availability of poultry meat will depend in large part on the ongoing structural changes in poultry meat production.

Bangladesh poultry sector has not been fully industrialized yet, only hatcheries and feed mills can be defined as industries. Poultry drugs and equipments are partly manufactured locally and partly imported and a full picture of those markets could not support the industry yet. Marketing is one of the weakest links in the poultry development programs. Presently, the market for poultry products is concentrated in peri-urban and urban areas. The overall impression of the market is that of a not well-organizedinstitution with exploitation by intermediariesreducing the margin of profit for farmers who are under threat by vagaries of weather and price fluctuations. There is no legal barrier for any farmer to enter in the hatchery or feed business. Inputs for the hatchery industry, like parent stock, equipments, etc. can be easily imported from abroad with government support. Still the numbers of hatcheries are low. The main problems for the potential farming are the lack of qualified personnels and capital. However, although there are incentive programs for the feed manufacturers but scarcity of raw material along with inadequate and irregular supply of ingredients are the major barriers for the potential investors to enter into the feed industry. The major feed ingredients like maize, soya and oil cakes are presently produced inadequately to meet the demand of poultry sector, so it is necessary to increase their domestic growth rates to fulfill the requirements rather than to import them. Further, availability of these ingredients varies seasonally and due to lack of storage facilities the price fluctuates. The transportation costs also add to the increase in price of feed. Feed alone constitutes about 60 to 70percent of the cost of production of broiler meat and any hike in feed price leads to lower profit to the farmers.

To increase the growth of poultry sector, it needs government support in terms of policies and private entrepreneurs must utilize the support in good faith to increase the growth of feed mills, hatcheries as well as veterinary services and equipments. Since the commercial poultry farming has an enormous potentialto expand through the support of the Poultry Development Policy, but unfortunately there is relatively little evidence that such attempts have been fruitful. This is largely because the rural farmers lack both reliable and cost efficient inputs such as extension advice, mechanization services, day-old chicks, feed and credit in one hand, and guaranteed and profitable markets for their output on the other. Therefore, this is the time to findout the appropriate farming system, which has positive and direct impact on the development of poultry sector in Bangladesh.

Bangladesh poultry sector is now transforming from scavenging to commercial system and contract farming system has started recently. Contract farming is one such institutional organization that is considered to strengthening the supply chain in various economic activities with varying levels of success [14, 37, 38, 39, 40]. However, vertically integrated contract farming is a very new concept in Bangladesh poultry industry. To develop the poultry industry, the policymakers could provide policies to extend the vertically integrated contract farming system all over the country. Contracting system can involve integrating backward to produce own breeding stocks and integrating forward into further processing and retailing and in turns will change consumer preferences. However, this requires transfer of technological information, facilitating grower's access to credit which in turn could expand commercial poultry farming.

The future outlook is positive for the Bangladesh poultry industry because the demand of poultry meat can be expected to grow given its current low level of per capita consumption and anticipated growth in population and household incomes.

REFERENCES

[1] TahaFA. *The Poultry Sector in Middle-Income Countries and Its Feed Requirements: The Case of Eg*ypt. Outlook Report No.WRS03-02.Economic Research Service, USDA (at http://www.ers.usda.gov/publications/WRS03/dec03/wrs0302), 2003.

[2] IFPRI.*International Food Policy Research Institute*, www.cgiar.org/IFPRI, 2000.

[3] FAO. FAOSTAT data.http:// faostat.external.fao.org/default.jsp (accessed April, 2005).

[4] BBS.*Statistical Yearbook of Bangladesh*, 2005. Bangladesh Bureau of Statistics, Statistics Division, Ministry of Planning, Government of the People's Republic of Bangladesh, Dhaka, Bangladesh.

[5] Ahmed R. Prospect and Problems of Broiler Production in Bangladesh. *Proceeding of the First National Conference of Bangladesh Animal Husbandry Association,* February 23-24, 1985, BRAC, Dhaka, Bangladesh.

[6] HaqueQME. Commercial Poultry Farming in Bangladesh. *Proceeding of the First National Conference of Bangladesh Animal Husbandry Association*, February 23-24, 1985, BRAC, Dhaka, Bangladesh.

[7] BembridgeTJ. Impact of maize extension programme in Traskei.*South African Journal of Agricultural Extension*, 1988;17, pp. 22–28.

[8] MokotjoJL. Supply, demand and marketing of principal food grains in Lesotho. Food Security Policies in SADCC regions. In Rukuni, M., Mudimu, G. and Jayne, T.S., (eds). *Proceedings. 5th Annual Conference on Food Security Research in Southern Africa*, 16-18 October 1989, 1990; pp. 204–211.

[9] CreeveyLE. Supporting small-scale enterprise for women farmers in the Sahel.*Journal of International Development,*1991; 3(4): 355–386,.

[10] Regmi A,PompelliG. US food sector linked to global consumers. *Food Review: The Magazine of Food Economics,* US. Department of Agriculture, Economic Research Service, 2002; 25 (1): 39-44.

[11] FAOStat food supply data. Food and Agriculture Organization of United Nations, Rome.http://faostat.external.fao.org/default.jsp (accessed December, 2001).

[12] HIES.*Statistical Yearbook of Bangladesh*, 2003. Planning Division, Ministry of Planning, Government of the People's Republic of Bangladesh.

[13] Bangladesh Bank, Economic Trends, 2008. Statistics Division, Bangladesh Bank.

[14] Begum IA. An assessment of vertically integrated contract poultry farming: a case study in Bangladesh. *International Journal of Poultry Science*, 2005a; 4(3):167-176.

[15] FAO.*The State of Food and Agriculture*.Food and Agriculture Organization, Rome,1991.

[16] HarissB. There is method in my Madness or is it vice versa? *Measuring Agricultural Performance, in Agricultural and Food Marketing in Developing Countries.* John, A. (ed), Redwood Book Ltd., Trowbridge, Wiltshire, UK, 1993.

[17] SiddiqueSA. Desher Protein GhattiMetateBeboharkary Poultry ShilperUddogAbongJatioOrthonititeErObodan (Book written in Bengali language), *The Monthly and Daily Barta*, August, 2000.

[18] ChowdhurySJ. *An Economic Analysis of Broiler Rearing Farms UnderAftabBohumukhi Farm Limited in BajitpurUpozila of Kishorganj District." M.S. thesis*, Department of Agricultural Economics, Bangladesh Agricultural University, Mymensingh, Bangladesh, 2001.

[19] SalequeMA. Scaling Up: *Critical Factors in Leadership, Management, Human Resource Development and Institution Building in Going from Pilot Project to Large Scale Implementation*: the BRAC Poultry Model in Bangladesh, 1999.

[20] Huque QME, Paul DC. Strategies for Family Poultry Production with Special Reference to Women Participation.*Proceeding of 1st SAARC poultry conference*, held on September 24-26, 2001 at Pune, India.

[21] Poultry KhamarBichitra*Year Book of Poultry Business Directory*, 2003. Dhaka.

[22] Mitchell D. *The Livestock and Poultry Sub-sector in Bangladesh*, Mission Report, World Bank, 1997.

[23] AlamJ. *Livestock Resource in Bangladesh–Present Status and Future Potential,* University Press Limited, Dhaka, Bangladesh, 1995.

[24] SarmaJS. Cereal Feed Use in the Third World: *Past Trends and Projections to 2000, Research Report 57*, International Food Policy Research Institute, December, 1986.

[25] BBS.*Statistical Yearbook of Bangladesh*.Bangladesh Bureau of Statistics, Planning Division, Ministry of Planning, Government of the People's Republic of Bangladesh,2000.

[26] BBS.*Statistical Yearbook of Bangladesh*.Bangladesh Bureau of Statistics, Planning Division, Ministry of Planning, Government of the People's Republic of Bangladesh,2003.

[27] AhammadK. *Poultry Business Directory.Bangladesh Poultry Association*, Dhaka, Bangladesh, 2006.

[28] Karim AMA, MainuddinG. *Evaluation on the Field Activities of Backyard Poultry Raiser in Bangladesh,* UNICEF, Dhaka, Bangladesh, 1983.

[29] Islam MM, ShahidullahM. Poultry knowledge of the farmers of a union in Mymensinghdistrict. *Bangladesh Journal of Training and Development*, 1989; 2(1):12-18.

[30] Ukil MA, Paul DC. *Problems and Prospects of Broiler Industry*.Conference of Bangladesh Animal Husbandry Association, Dhaka, Bangladesh, 1992.

[31] BhuiyanAU. *An Economic Analysis of Small Scale Poultry Farming of Kotwali Thana in Mymensingh District.MS Thesis,* Department of Agricultural Economics, Bangladesh Agricultural University, Bangladesh, 1999.

[32] UddinH. *A Comparative Economic Analysis of Broiler and Layer Production in Some Selected Areas of Sadar Thana in Mymensingh District.MS thesis,* Department of Agricultural Economics, Bangladesh Agricultural University, Mymensingh, Bangladesh, 1999.

[33] Begum IA. Vertically integrated contract and independent poultry farming system in Bangladesh: a profitability analysis.*Livestock Research for Rural Development*, 2005b;17(8): 89.

[34] Begum IA,AlamMJ. Contract farming and small farmer: acase study of the Bangladesh poultry sector. *Journal of Bangladesh Studies*, 2005; Vol. 7, No. 2, Penn State Erie, USA.

[35] Begum IA, Osanami F,Takumi K. Performance of vertically integrated contract and independent poultry farms in Bangladesh: a comparative study. *The Review of Agricultural Economics*, Japan, 2005; Vol. 61.

[36] EPAC.*Improving Australia's Competitiveness*.Canberra: Australian Government Publishing Service, 1991.

[37] Glover DJ. Increasing the benefits to smallholders from contract farming: problems for farmers' organizations and policy makers.*World Development,*1987; 15(4):441–448.

[38] Runsten D, Key N. Contract *Farming in Developing Countries: Theoretical Issues and Analysis of Some Mexican Cases*. Report LC/L 989. United Nations Economic Commission for Latin America and the Caribbean, Santiago, Chile, 1996.

[39] FarrellyLL. Transforming Poultry Production and Marketing in Developing Countries: Lessons Learned with Implications for Sub-Saharan Africa. *International Development Working Paper 63*, Department of Agricultural Economics, Michigan State University, USA, 1996.

[40] Eaton C, Shepherd A. Contract Farming: Partnerships for Growth. *AGS Bulletin* 145.FAO, Rome, Italy, 2001.

In: Livestock: Rearing, Farming Practices and Diseases
Editor: M. Tariq Javed
ISBN 978-1-62100-181-2

Chapter 5

PARASITIC DISEASES IN LIVESTOCK UNDER DIFFERENT FARMING PRACTICES: POSSIBILITIES FOR THEIR CONTROL

M. Arias, R. Sánchez-Andrade, J. L. Suárez, P. Piñeiro, R. Francisco, C. Cazapal-Monteiro, F. J. Cortiñas, I. Francisco, A. Romasanta, and A. Paz-Silva*

Equine Diseases Study Group (Epidemiology, Parasitology and Zoonoses), Animal Pathology Department, Veterinary Faculty, Santiago de Compostela University, 27002-Lugo, Spain

ABSTRACT

Three are the main rearing practices classically developed worldwide, intensive, semi-extensive and extensive, on the basis of time the animals graze in the fields. Infection by parasites can be influenced by their management. Most of helminths (cestodes, trematodes, nematodes) present a life-cycle with an external phase in the environment, where resistant forms passed by feces (ova, larvae, cysts) develop to reach the infective stages (larvae, metacercariae) and infection occurs in livestock when they feed on contaminated pastures. Protozoa affecting the digestive apparatus are released to the environment as oocysts and infection is improved in indoor management systems, as occurred with certain ectoparasites (mange and lice). This seems to point that animals under an intensive regime could be mainly exposed to protozoan. By opposite, the livestock under a semi-extensive or extensive management should be infected by helminths and ectoparasites living in the environment. In spite of the trueness of this statement, several aspects related to the rearing of the animals must be taken into account. The administration of herbage to animals maintained indoors might increase the risk of infection by different helminth parasites (*Moniezia*, *Fasciola*, *Paramphistomum*, gastrointestinal nematodes). The supplementation of animals reared on extensive systems by using feeders placed in the grounds, the exposition to protozoan parasites could be enhanced. The main internal parasitism affecting livestock in respect to the type of

---

* E-mail address: adolfo.paz@usc.es

farming have been analyzed. Fecal samples belonging to livestock (cattle, goats, sheep, horses and pigs) under intensive, semi-extensive and extensive regimes were collected and analyzed. The possibilities for controlling the parasitic diseases were discussed.

1. INTRODUCTION

The evolution of humans has run in parallel to the domestication of animals. Firstly, people had to hunt for achieving food products. By the capture of wild animals, meat, skins and different products were obtained, thus allowing the people to get nutrition and protection (clothes) [1, 2]. The requirement for animal-derived food probably served as a stimulus to try the domestication of different animal species [3]. The main objective of rearing animals near to humans was for providing them nutrients and skins. Most important objective was to improve the animals breed, because this will become the guarantee for the supply of food and to meet the other requirements of people, and thus will help for the survival of people. Once their primary needs were achieved, other advantages offered by domestication of animals encouraged that people to settle down, and in this way villages and towns were created. Livestock was reared for the transport purposes, to work in the agriculture fields/farms and for leisure also [4]. Breeding of livestock imply that their feeding needs to be ensured. For this purpose, agriculture procedures were adapted to fulfill the nutrition of animals and humans. Several practices for cropping, pasturing, etc. were developed, making possible the accurate nutrition for livestock and as a consequence their productivity [5].

Towards the 1950 to 1960's, significant modifications in agriculture and farming procedures were applied. Based on the requirements for increasing the inputs, more productive animals were looked for [6, 7]. In this way, genetic selection for milk, meat and/or wool production was stimulated [8]. Genetic selection implies that some characteristics are enhanced and others are lost. Most times, adaptability conditions are misplaced in benefit of production [9]. As a consequence, more attention is needed to ensure a suitable health condition, avoiding the infection by different pathogens. Because of this, preventive care increases the cost for vaccines and the control of parasitic agents. It should be considered that increase in livestock production also increases the food requirements [10]. A significant reduction of the grassland productivity, together with the loss of land, weed ingress or perennial meadow has been associated with increase in grazing pressures [11]. The best option seems to increase the grassland surface [12]. Other consideration to take into account is that the increment in pasture areas for the nutrition of livestock could be achieved by reducing forest areas and/or transformation of unproductive lands by providing water, machinery, etc., favoring thus the exposition of livestock to infective stages of different pathogens. Besides the ecological damage to the environment, animals yielding high production levels are no longer fed on grass alone, and they need to be supplemented with protein, energy, vitamins, minerals and micronutrients from other sources. Another solution to the animal feeding was to implement different management procedures, in most cases by rearing livestock under intensive systems. High inputs are obtained in this way, but the cost for nutrition also increases significantly. The requirements for preventing diseases and disinfection are also higher.

1.1. Typical Farming Procedures

Classically, rearing of livestock consisted of the maintenance of animals grazing on pastures throughout the year [13]. Their nutrition depended only on the grass and animals were seldom stabled. This is the *extensive management*, and the main species kept under this procedure were initially autochthonous breeds, very resistant to adverse weather conditions and insufficient nutrition [14]. Most of the activity is focused to meat production [15].

Nowadays, extensive farming is commonly practiced in regions where grass is the main nutritional resource. Forests and natural pastures are characterized by the presence of *Calluma vulgaris*, *Erica arborea*, *Poa pratensis*, *Dactylis glomerata*, *Plantago major*, *Phleum pratense* and *Chamomilla recutita*. No land-management operations are normally used and there is no rotation of pastures. Free-grazing cattle produced by crossing different breeds are now maintained under extensive management throughout the year; food supplementation, deworming, and some measure for the control of parasites are never applied [16]. In some countries, the main contacts with these animals occur when the Official Veterinary Services impose an obligatory program to survey for diseases.

Other livestock production system based on seasonal pasture and forage production is the *semi-extensive management*, and the main difference in respect to the extensive farming is that the animals usually take advantage of pastures during the day, but in the afternoon they are carried to the stables [17]. Livestock remains indoors when there are adverse weather conditions. Nutritional supplement is normally given to the animals early in the morning and at the end of the day. It currently implies farming a low number of animals on the land available [18]. Partially-selected and autochthonous breeds are maintained under this regime. As mentioned already, the grazing herd feeds on natural pastures, although cultured meadows composed of *Trifolium pratense*, *Trifolium repens*, *Lolium perenne*, *Lolium multiflorum* and *Dactylis glomerata* can also be used.

In response to a high yielding level, some livestock species are reared under an *intensive regime/management*, characterized by keeping them into isolated buildings (stabled) throughout their lives [19]. Feeding is mainly based on concentrates, so fresh pasture or hay is seldom provided [20]. This regime is applied in regions where the surface area for the grazing is much more limited. Pigs, rabbits, hens and chickens are widely maintained under this type of farming [21]. By considering the welfare of farm animals and in attempt to improve their conditions, farmers are changing some of these practices. For example, it is frequent nowadays that animal's walk around in a run outside [22, 23].

1.2. Other Farming Measures

Under non-*grazing farming* the animals are never brought to pastures. Livestock is wholly confined and grass is harvested and fed fresh to them [24]. An increment in the prevalence of diseases and loss of productivity has been reported [25, 26].

From several decades ago, and with the purpose to obtain healthy and quality foods, the *organic farming* has been developed. Animals are maintained under an extensive regime feeding on pastures throughout the year, getting more space and having free access to outside areas [27]. Chemotherapy is not allowed except when the life of the animals is seriously endangered, though these requisites vary with the countries.

Recently, it has been demonstrated that grazing cows produce milk characterized by the presence of beneficial fatty acids, antioxidants and vitamins [28, 29]. Grazing also enhances the contents of other compounds regarded as beneficial in meat, like carotene and tocopherol, which may improve the shelf life of meat [30]. This seems the future for the livestock farming, but it should be remembered that the actual demand for meat/milk can no longer be sustained by traditional livestock production systems. Outside the world's grasslands, most ruminants traditionally are fed on the farms on herbage and crop wastes.

2. Analysis of Parasitic Infections Affecting Livestock under Different Farming Types

During the breeding and rearing, animals might be exposed to infectious agents, according to the type of farming they are receiving. With the goal to gain more knowledge on the risk of parasitic infections affecting livestock under different management regimes, five trials were conducted on cattle, sheep, goats, pigs and horses, as described in the following scheme.

	Livestock species				
Farming type	Cattle	Goat	Sheep	Pigs	Horses
Extensive	*	*		*	*
Semi-extensive	*	*	*		*
Intensive	*		*		*

In all cases, the fecal samples were collected directly from the rectum of the animals. Once in the laboratory, the fecal samples were macroscopically analyzed and then processed by means of the copromicroscopical flotation (detection of coccidian oocysts and eggs of cestodes and nematodes) [31], sedimentation (eggs of trematodes) [32] and migratory (broncho-pulmonary parasites) techniques. The Ziehl-Neelsen staining technique was also employed for the observation of *Cryptosporidium* (coccidia) oocysts [33]. Due to the difficulty in the morphological differentiation of the eggs belonging to some gastrointestinal nematodes, fecal pats were also performed [34, 35].

Statistical Analysis: The prevalence numbers were expressed as the percentage and the 95% confidence interval. The Chi square test was applied to compare the prevalence and differences were considered significant if $P< 0.05$. The possible relationship between the farming system and the prevalence of the parasitism was estimated by calculating the odds ratio (OR) [36]. The OR ranges in value from 0 to infinity. When values near to 1 are achieved, no relationship between the exposure and the outcome is concluded. The OR values minor than 1 indicate a protective effect, whereas values higher than 1 point the negative effect caused by the exposition. Finally, all the data were classified by using the CHAID (Chi-square Automatic Interaction Detector) test [37]. All tests were done using SPSS for Windows (15.0).

2.1. Cattle

A total of 284 animals belonging to 3 herds maintained under different regimes were assayed in the current work (Table 1):

- Extensive: free-grazing cattle obtained by crossing different breeds maintained under extensive management throughout the year.
- Semi-extensive: autochthonous *Rubia gallega* cattle kept under field conditions outside and stabled at night or when adverse weather conditions.
- Intensive: Friesian cows remaining stabled and receiving concentrates and hay.

Eimeria and *Cryptosporidium* specimens were identified among the coccidia excreted in the feces of the sampled cattle. Eggs of the cestode *Moniezia* spp. were observed after the application of the flotation technique. Trematodes belonging to *Fasciola hepatica* and *Paramphistomum* spp. were found in the stools of the ruminants. The gastrointestinal nematodes were identified as *Trichuris*, *Capillaria*, *Nematodirus*, *Chabertia*, *Oesophagostomum*, *Trichostrongylus* and *Ostertagia*.

Table 1. Percentage (95% Confidence Interval) of cattle passing parasites in their feces

Farming type	Coccidia %	Cestode %	Trematode %	Nematode %
Extensive	4 (0, 8)	18 (11, 26)	25 (16, 33)	73 (65, 82)
(*N*= 98)	*OR*= *0.10*	*OR*= *3.96*	*OR*= *0.99*	*OR*= *1.45*
Semi-extensive	0	16 (7, 25)	69 (57, 80)	94 (88, 100)
(*N*= 64)		*OR*= *2.10*	*OR*= *16.41*	*OR*= *9.63*
Intensive	46 (37, 55)	0	2 (0, 4)	51 (42, 60)
(*N*= 122)	*OR*= *33.1*		*OR*= *0.02*	*OR*= *0.23*
χ^2	*179.168*	*23.720*	*101.799*	*37.580*

The percentage of cattle passing coccidian oocysts ranged from 4percent in the extensively managed animals to 46percent in the intensive farming. The highest prevalence of cattle with helminths (cestodes, trematodes and nematodes) was achieved in grazing cattle (extensive and semi-extensive regimes). Statistical differences regarding the farming type in the parasite egg-output were recorded. By the estimation of the OR values, it was proved that extensively managed cattle were at risk of cestodes; semi-extensive regime favored trematodes and nematodes, while the intensive farming was related to coccidian infections.

2.2. Goats

A total of 172 crossbred goats belonging to 2 flocks maintained under extensive or semi-extensive management were assayed (Table 2). One flock was extensively exploited for controlling the unwanted vegetation in a fenced mountainous area focused to wood production (silvo-pasturing system). Supplement was given when necessary. Goats belonging to other flock for dairy farming were also sampled. By the copromicroscopical techniques, the presence of coccidia (*Eimeria*), cestodes (*Moniezia*), trematodes (*F. hepatica*, *Dicrocoelium*) and gastrointestinal nematodes (*Nematodirus*, *Ostertagia*, *Trichostrongylus*, *Chabertia* and *Teladorsagia*) in the fecal samples were detected.

Table 2. Percentage (95% Confidence Interval) of goats passing parasites in their feces

Farming type	Coccidia	Cestode	Trematode	Nematode
Extensive (N= 100)	58 (48, 68)	20 (12, 28)	22 (14, 30) *OR*= *0.25*	56 (44, 67) *OR*= *0.05*
Semi-extensive (N= 72)	61 (50, 72)	22 (13, 32)	53 (41, 64) *OR*= *3.96*	96 (92, 100) *OR*= *19.2*
χ^2	*0.168*	*0.125*	*17.457*	*41.375*
P	*0.682*	*0.724*	*0.001*	*0.001*

As reflected in Table 2, no differences in the percentages of goats shedding coccidian oocysts and cestode eggs were observed. The highest prevalence of trematodes and nematode egg-output were observed in the goats under a semi-extensive regime and the OR values showed the highest risk of infection by trematodes and nematodes among these animals.

2.3. Sheep

Fecal samples were collected from 192 crossbred sheep distributed into two flocks; one was kept on a semi-extensive management and the other on an intensive regime (Table 3).

Table 3. Percentage (95% Confidence Interval) of sheep passing parasites in their feces

Farming type	Coccidia	Cestode	Trematode	Nematode
Semi-extensive (N= 98)	73 (65, 82) *OR*= *0.37*	10 (4, 16)	35 (25, 44) *OR*= *6.6*	75 (67, 84) *OR*= *3.5*
Intensive (N= 94)	88 (82, 95) *OR*= *2.72*	16 (9, 23)	7 (2, 13) *OR*= *0.15*	47 (37, 57) *OR*= *0.3*
χ^2	*6.781*	*1.402*	*21.210*	*16.686*
P	*0.009*	*0.236*	*0.001*	*0.001*

In the fecal samples, there were found *Eimeria* oocysts, eggs of *Moniezia*, *F. hepatica*, *C. daubneyi*, *Trichuris* and *Nematodirus*. The coprocultures showed sheep were also infected by the gastrointestinal nematodes *Oesophagostomum*, *Trichostrongylus*, *Ostertagia*, *Haemonchus*, *Cooperia* and *Chabertia.*

Intensively reared sheep achieved a significantly highest prevalence of coccidiosis (Table 3). The prevalence of ovine trematodes and nematodes were significantly higher under a semi-extensive management. The latter finding was strengthened by the higher odds ratio as well in favor of sheep maintained on semi-intensive management.

2.4. Horses

The presence of parasitic infection was investigated in indigenous Pura Raza Galega horses maintained in a silvo-pastoral system, consisting of combining timber production and livestock grazing. Other herd was composed by Spanish Sport horses pasturing through the day and kept into the sheds in the night. Another group of Spanish Pure Breed horses maintained in sheds and dedicated to the sport and/or leisure was utilized. Eggs of cestodes (*Anoplocephala*) and gastrointestinal nematodes (*Strongylus, Triodontophorus, Gyalocephalus* and *Cyathosthomum*) were only observed, whereas coccidian oocysts or trematode eggs were not found in the feces.

Table 4. Percentage (95% Confidence Interval) of horses passing parasites in their feces

Farming type	Cestode	Nematode
Extensive (silvo-pasturing) (*N*= 69)	0	93 (87, 99) *OR= 14.1*
Semi-extensive (Pasturing by day) (*N*= 72)	18 (9, 27)	62 (51, 74) *OR= 1.01*
Intensive (Box) (*N*= 71)	0	32 (22, 43) *OR= 0.14*
χ^2	*26.929*	*54.262*
P	*0.001*	*0.001*

As shown in Table 4, only the horses under a semi-extensive regime passed cestode-eggs by feces. The nematode egg-output ranged from 32percent in stabled horses (intensive) to 93percent in equines maintained on forests (extensive). An association between extensive farming and nematode infection in the horses was demonstrated (OR> 1).

2.5. Pigs

The analysis of the stools collected from 96 Porco Celta autochthonous pigs maintained in a forest area (extensive management) showed the infection by coccidia (*Eimeria* and *Ballantidium*), and nematodes (*Ascaris suum* and *Oesophagostomum*). No eggs belonging to cestode and/or trematode were observed (Table 5). It is noticeable that swine do not harbor the adult stages of tapeworms and thus egg-output is not possible to detect.

Table 5. Percentage (95% Confidence Interval) of pigs passing parasites in their feces

Farming type	Coccidia	Nematode
Extensive (*N*= 98)	80 (72, 88)	40 (30, 49)

3. Relationship between Farming Type and Livestock Parasitic Infections

Management of livestock can influence the possibilities of endoparasitism. Animals infected by enteric coccidia excrete oocysts to the soil and depending on environmental conditions (including temperature and humidity) these sporulate and become infective (excepting for *Cryptosporidium*). Infection in livestock is higher in the presence of large numbers of oocysts around the animals (soil, water and feed). This explains the high incidence in livestock under intensive farming where ambient conditions favor the development of the oocysts.

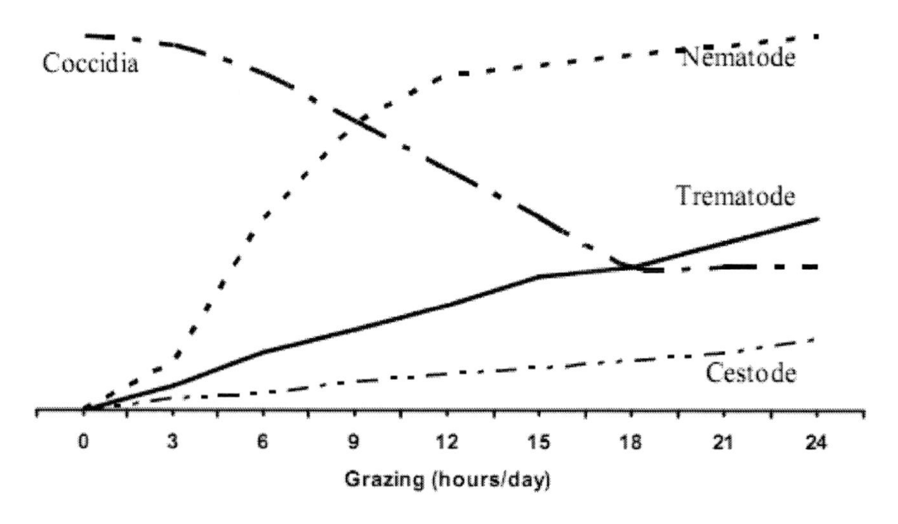

Figure 1. Theoretical relationship between farming type and risk of infection by endoparasites.

Excluding some nematode species (i.e., ascarids do not present free living stages in the soil), most of the helminths (cestode, trematode, nematode) develop a life-cycle with an external phase in the environment, where resistance forms passed by feces (ova, larvae) reach the infective stages (larvae, metacercariae), and are ingested by livestock when grazing on such contaminated pastures. By considering the possible relationship between the farming type (expressed in the time the livestock remain grazing in pastures) and the parasitic infection, animals under an intensive regime are mainly exposed to enteric protozoan, whereas livestock in a semi-extensive or extensive system are more prone to infections by helminths (Figure 1).

Data collected in the current chapter are summarized in Figure 2. Significantly highest prevalence of coccidia (χ^2= 8.831, P= 0.001) were observed in the animals under an intensive regime, supporting a close contact between animals and oocysts passed by the feces in systems where livestock remain stabled for long periods [38]. Although more space is available for semi-extensively reared animals, the presence of feeders and drinkers in the pastures seem responsible for the exposition to both infected animals and/or oocysts [39, 40, 41].

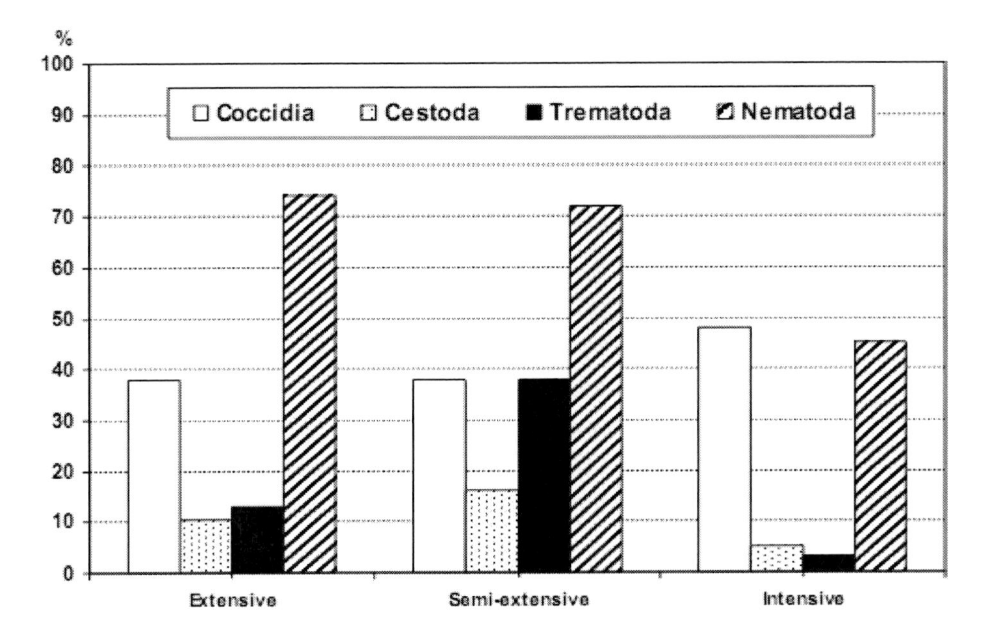

Figure 2. Prevalence endoparasites in livestock under different farming regime.

The prevalence of cestodes and trematodes were significantly higher in semi-extensive managed livestock (χ^2= 18.104, P= 0.001 and χ^2= 132.755, P= 0.001, respectively). Tapeworms are transmitted to grazing animals such as livestock when they feed on grass that harbors oribatid mites containing cysticercus larvae [42]. In forest soils, with thick litter and mycelium-feeding fauna, mites are abundant [43], while in agricultural soils, with bacterial-feeding fauna and quick decay of soil organic matter, the density of mites is low [44, 45]. We found significantly higher percentages of cestodes in intensively managed bovine as well as in horses in a semi-extensive system. Several differences in the efficiency of mites as intermediate hosts, as well as in their preference for meadows or pastures have been pointed [46, 47].

Amphibious lymnaeid snails act as intermediate hosts for both *Fasciola hepatica* and *Paramphistomum* spp., releasing to the environment the cercariae, which finally change into metacercariae, the infective stage. For this reason, livestock feeding on humid pastures exhibits the highest risk of infection by these two trematodes, although stabled animals can also be infected if receiving metacercariae-contaminated hay or fresh pasture [16]. Two are the intermediate hosts involved in the *Dicrocoelium* life-cycle, terrestrial snails and ants, which implies that dicroceliosis is improved in dry and calcareous environments [48, 49]. Infection by trematodes in livestock is mainly associated with grazing systems. The highest prevalence found in semi-extensive maintained livestock suggests that pastures they feed are more than those grazed by extensively reared animals for the survival of intermediate hosts.

The demonstration of moderate to higher prevalence of nematodes in all the management types (Figure 2) confirms that several aspects related to the rearing of the animals must be taken into account. It has been demonstrated that the infection of livestock by endoparasites depends on the management systems. Besides this, one may also consider possibility of a difference in susceptibility in the animal species. For trying to add more information, the simultaneous analysis of the data regarding both factors by means of the CHAID test was carried out.

The occurrence of coccidia was associated with the livestock species (Figure 3), with the highest risk in pigs and sheep. Moreover, it was observed that the farming type influenced the infection by coccidia in bovines and the highest percentages in cattle under intensive farming systems were recorded.

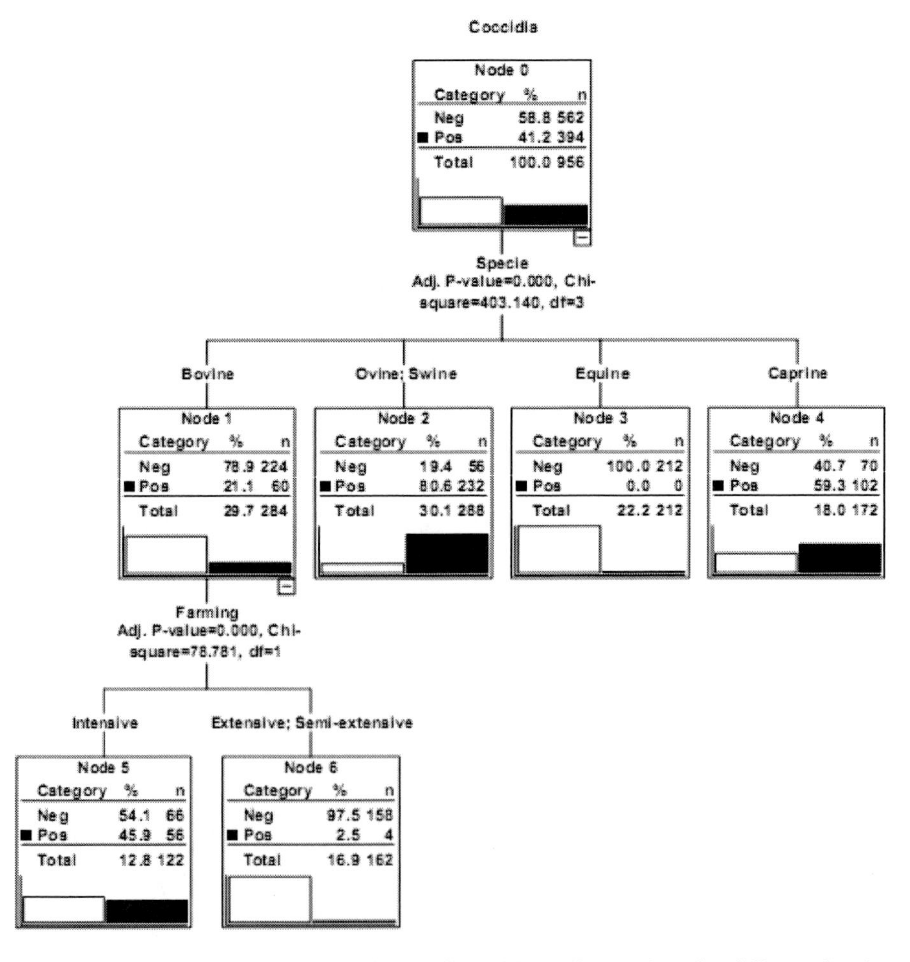

Figure 3. Simultaneous analysis of the prevalence of coccidia in livestock under different farming regimes.

Figure 4 shows that infection by tapeworms depends mainly on the farming system and the highest prevalence among extensive / semi-extensive reared livestock were recorded. The next cluster formed according to the animal species showed that the highest percentages of infection occurred in cattle + goats and the lowest in sheep + horses. Finally, this last group was divided regarding the farming procedures. These results support the risk of exposition to oribatid mites, the intermediate hosts of tapeworms in grazing livestock [50]. Among these animal species, an increased susceptibility in bovine and ovine has also been demonstrated.

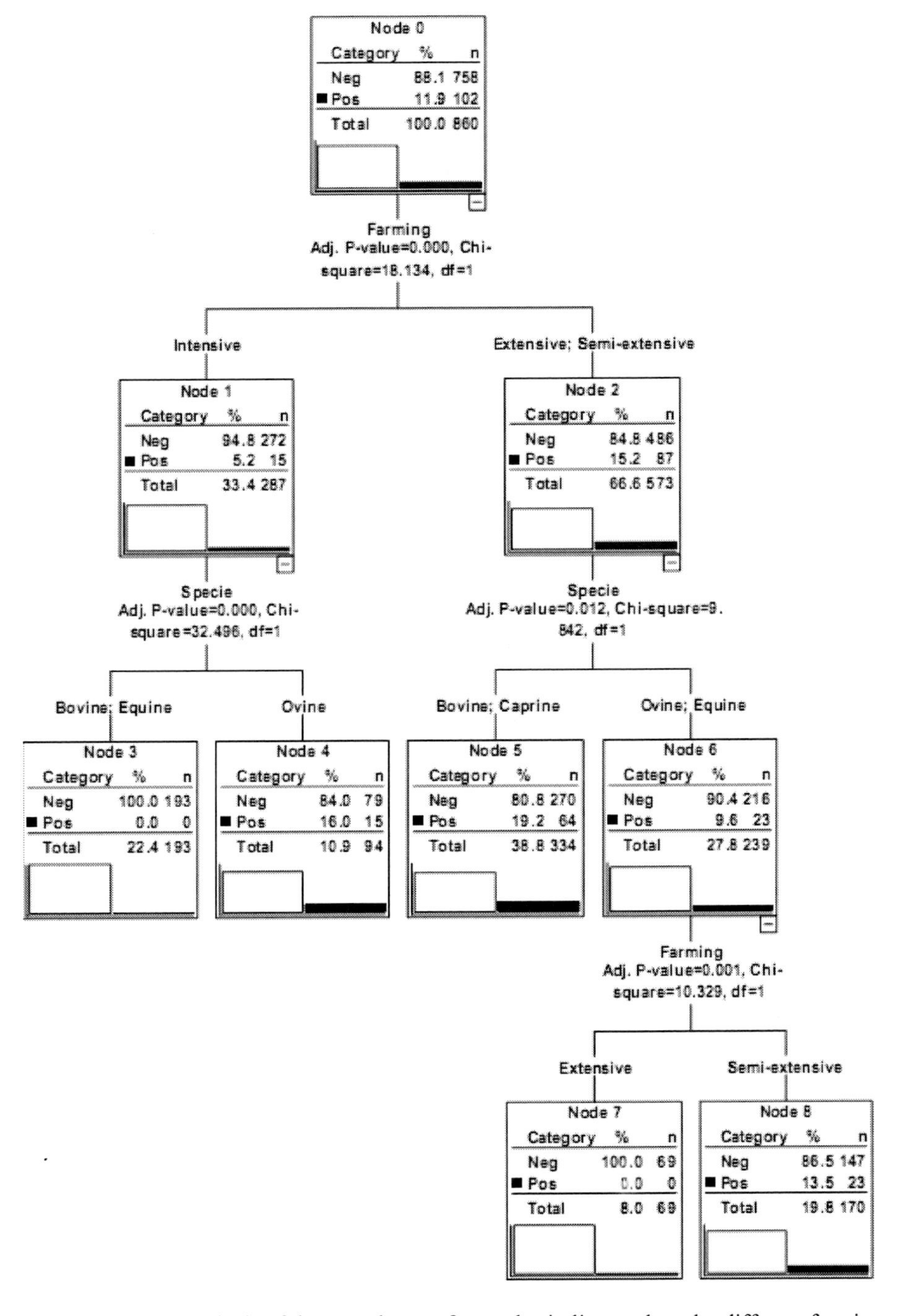

Figure 4. Simultaneous analysis of the prevalence of cestodes in livestock under different farming regime.

In Figure 5 it is shown that livestock management is involved in the risk of infection by trematodes. Semi-extensively maintained animals achieved the highest prevalence by trematodes followed by those under extensive farming. These results indicate that the frequent ingestion of trematode-metacercariae can be enhanced under these regimes (Arias *et al.*, 2010).

The finding of low numbers of positive cases among intensively reared animals reflects that they are fed with herbage, hay or silage contaminated with metacercariae, confirming the results collected by the analysis of cestodes. Livestock under intensive management are focused on the production of milk, and appropriate food and care are provided, including anthelminitics. Likewise, the observation of trematode-positive cases could indicate that these ruminants are not adequately managed [51].

Grazing livestock was divided according to the animal species, and the highest risk of infection was observed in cattle + goats. Infection by the trematodes *Fasciola hepatica* or *Paramphistomum* spp. is associated to areas with abundant vegetation and elevated humidity throughout the year [52]; these are conditions favoring the survival of their intermediate host, i.e., amphibian lymnaeid snails [53]. Infection by *Fasciola* and *Paramphistomum* has also been linked to the ingestion of fresh pasture, which can occur in intensively managed livestock [54, 55]. The presence of *Dicrocoelium* spp. needs the existence of two intermediate hosts, ants (several species) and terrestrial snails [37, 48]. This implies that dicrocoeliosis affects mainly animals maintained under hill and dry areas, as demonstrated in the goats in the current research [49].

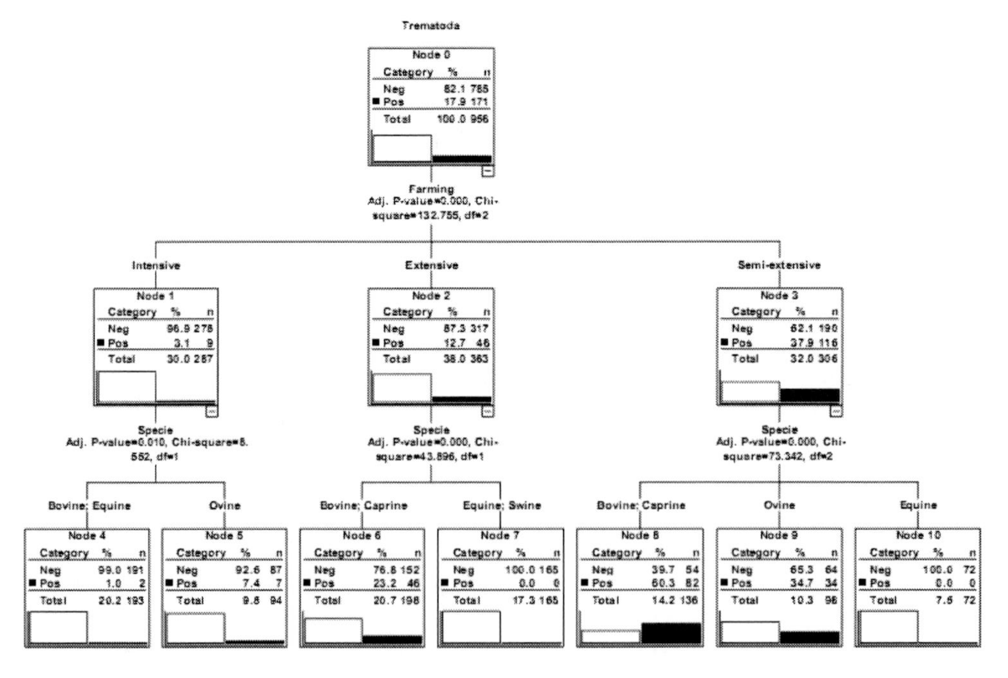

Figure 5. Simultaneous analysis of the prevalence of trematodes in livestock under different farming regime.

Data presented in Figure 6 demonstrates the infection by nematodes depends on the farming system, being the highest risk in extensive and semi-extensive maintained animals.

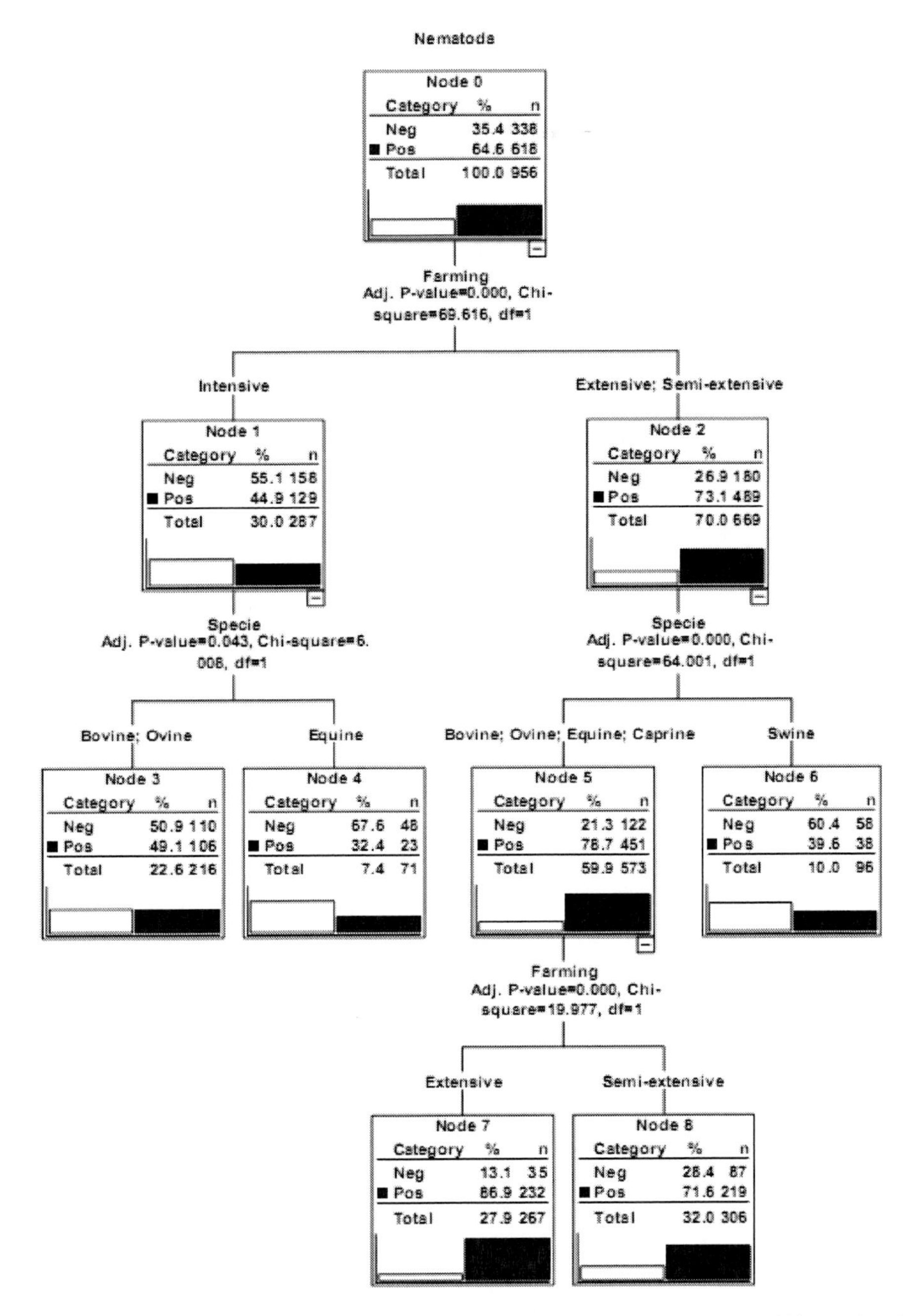

Figure 6. Simultaneous analysis of the prevalence of nematodes in livestock under different farming regime.

Among the pasturing livestock species (extensive or semi-extensive managing), swine exhibited the lowest susceptibility to infection. Finally, the group formed by the rest of the species was divided again in respect to the farming procedure, remarking the importance of management in the livestock nematodes.

4. Possibilities for the Control of Parasitic Infections in Livestock under Different Farming Regimes

In the current chapter the correlation between certain farming types and the development of several endoparasitisms has been demonstrated. Livestock maintained under intensive management exhibit a higher risk for protozoan infections (coccidiosis), while extensively or semi-extensively reared animals are mainly exposed to helminths. Nevertheless, nematodes have been detected in intensive managed animals and a moderate percentage (40%) of grazing livestock is affected by coccidia. The appropriate control of parasites affecting livestock requires accurate methods for their detection, and once identified, development of an integrated program based on the action on the animals and their environment.

4.1. Methods for the Detection of Parasites In Livestock

Accurate detection of parasites depends on many factors, including collection, storage, and transport of the sample, as well as the method of laboratory evaluation [56]. Copromicroscopic techniques are frequently utilized for the observation of parasite stages in the feces of the animals. These are low-cost procedures due to the easiness to get stool samples and to analyze them. Nevertheless, the presence of oocysts, ova or even larvae reflects the animals albeit mature stages, and thus most of the pathological damages have already occurred. Another disadvantage relies in the sensitivity of this method, and infections where the parasite burden is low or when the host is harboring immature parasites that may go undetected [57]. Some serological probes have provided successful results for the detection of early and current infection. Likewise, blood sampling could be difficult from livestock managed under extensive systems. In many cases, these probes are discarded due to the high cost.

4.2. Action on Livestock: Chemotherapy

Most of the control procedures applied on livestock are centered on chemotherapy, and many efficient parasiticides against the endoparasites found in the current chapter are commercially available (Table 6).

A higher efficacy for the toltrazuril against *Eimeria* coccidia has been reported in calves and lambs [58, 59, 60, 61], while the suggested chemotherapy for *Cryptosporidium* infection seems to be halofuginone lactate [62, 63]. Tapeworms have been significantly treated in the final hosts by using praziquantel or benzimidazoles [64, 65, 66]. Benzimidazoles have been proved successful against trematodes and cestodes [53, 67, 68, 69]. Finally, benzimidazoles and macrocyclic lactones are the main anthelminitics administered against nematode parasites [70, 71, 72].

Regardless of the efficacy on the definite hosts, chemotherapy also exhibits several disadvantages, including:

- The cost for the deworming of livestock reduces the afforded benefits.
- Lower dosage or higher frequencies of deworming have been pointed as some of the procedures enhancing the development of resistance of parasites to chemotherapy.
- Chemotherapy does not act against the infective stages of the parasites.

As previously demonstrated, the risk of parasitic infection is important in livestock, especially among those under pasturing systems. There have been formerly shown higher percentages of animals shedding nematode-eggs (≥ 40% prevalence), and the prevalence of livestock passing trematode-eggs in the feces ranged from 3 to 40percent (Figure 2).

Table 6. Control measures currently applied against livestock endoparasites

Farming type	Endoparasite	Chemotherapy
Intensive	Coccidia	Toltrazuril Diclazuril Halofuginone
Extensive Semi-extensive	Cestode	Benzimidazoles Praziquantel
Semi-extensive	Trematode	Benzimidazoles
Extensive Semi-extensive Intensive	Nematode	Benzimidazoles Macrocyclic Lactones Pyrantel salts

In view of this, some actions against the parasite infective stages living in the environment for preventing infection in livestock are required.

4.3. Action to the Environment

Most of the measures should be directed to prevent environmental contamination where livestock is reared. Though these procedures can be easily applied against coccidian parasites, the presence of free-living stages from cestode, trematode or nematode makes difficult to avoid infection to livestock maintained under semi-extensive or extensive regimes.

Coccidia: it is predominantly a problem among the youngest animals. The frequent presence of stools in stabled livestock enhances coccidial outbreaks. The most appropriate environments for these protozoa are warm and humid ambient conditions.

Outbreaks on pasture depend on seasonal changes. Other factors to consider are those related to the stress (shipping, weaning).

By taking into account the risk factors, exposition to coccidia could be limited if the hygiene in all the facilities is increased. It is necessary to keep watching for feeders/drinkers that they are not in direct contact to the ground; to avoid that animals cannot walk in or even defecate into them. Finally, an adequate sanitation is required.

Trematode: to avoid pasturing in humid areas, the latter provide adequate environment for the amphibious snails (their intermediate hosts) to develop and survive.

Fencing or drainage of the meadows to reduce the habitat of the intermediate host is a very useful measure for limiting the risk of infection.

Cestode and Nematode: some pasturing strategies are recommended, in combination with other management measures to reduce the application of anthelminitics. One of these strategies consists of combining rotation and haying, due to that once animals ended the rotation, they are moved to regrowth following a hay crop. Another measure can be alternation in the animal species. For example, one pasture is firstly pastured by horses, and the next year by small ruminants. Since several years ago, the possibilities of the biological control of several helminths have been underlined. Actually, successful results against gastrointestinal nematode with free-living stages in the pasture by using some nematophagous fungus as *Duddingtonia flagrans* are being reported [73].

5. Conclusion

Rearing and farming practices can increase infection by different parasites in livestock. While coccidiosis depends on animal species, the appearance of helminth-infected livestock has been associated with the farming management. Besides this, the animals maintained under intensive systems demonstrated more susceptibility for coccidia infections, which can be controlled by the administration of appropriate chemotherapy accompanied by several measures on the farm facilities to reduce the viability of oocysts.

Extensive or semi-extensive systems favor the infection by helminths (cestodes, trematodes, nematodes). The presence of free-living stages reduces the efficacy of any chemotherapy considered for their control. Some actions on the environment are required for limiting the risk of helminthoses and thus to increase the successful control of the parasites. Among the advantages provided by grazing animals, it should be underlined the possibility of control of pastures and forests free of shrubs and bushes, which becomes an ecological contribution to reduce the risk of fire. Grazing enhances tree growth because the competition for moisture, nutrients, and sunlight is reduced. The presence of pasturing animals also provides organic fertilizer. Finally, it should be considered that the biggest limitation in raising organic livestock is the need to control internal parasites. More attention is needed to offer measures different to the pharmaceutical based.

Acknowledgments

This work was supported, in part, by the Projects XUGA 10MDS21261PR and 10MDS261023PR, and complies with the current laws for Animal Health Research in Spain. We are very thankful to Mrs. B. Valcárcel for preparing and editing the manuscript.

REFERENCES

[1] Calub AD, Anwarhan H, Roder W. Livestock production systems for Imperata grasslands. *Agroforest. Syst.,* 1997; 36: 121-128.

[2] Kemp DR, Michalk DL. Towards sustainable grassland and livestock management. *J. Agr. Sci.,* 2007; 145:543-564.

[3] Schitanz S. *Ethical consideration of the human-animal relationship under conditions of asymmetry and ambivalence.* Eursafe Conference Science, ethics and society, Leuven, Belgium, 2004.

[4] GrahamRK. Future of Man. *Foundation for the Advancement of Man*, California, USA, 1981.

[5] Tirel JC. Extensive farming: opportunity or challenge for agriculture? *INRA Prod. Anim.*, 1991; 4: 5-12.

[6] Montano-Bermúdez M, Nielsen MK, Deutscher GH. Energy requirements for maintenance of crossbred beef cattle with different genetic potential for milk. *J. Anim. Sci.,* 1990; 68: 2279-2288.

[7] Gosey J. Composites: A beef cattle breeding alternative. *Proceedings of the Beef Improvement Federation Annual Meeting.* June 1-4, W. Des Moines, Iowa, 1994.

[8] Steinfeld H, de Haan C, Blackburn H. Livestock-environment interactions: issues and options. *Report of a study coordinated by the Food and Agriculture Organisation of the United Nations,* United States Agency for International Development and the World Bank,1997. http://www.fao.org/ag/aga/LSPA/LXEHTML/policy/index.htm.

[9] Miah A. Genetic selection for enhanced health characteristics. *J. Int. Biotech. Law,* 2007; 4:239-264.

[10] Taberlet P, Valentini A, Rezaei HR, Naderi S, Pompanon F, Negrini R, Ajmone-Marsan P. Are cattle, sheep, and goats endangered species? *Mol. Ecol.*, 2008; 17: 275-284.

[11] Stafford M, Buxton R, McKeon GM, Ash AJ, Seasonal climate forecasting and the management of rangelands: do production benefits translate into enterprise profits?. In: G.L. Hammer, N. Nicholls and C. Mitchell(Eds), *Applications of Seasonal Climate Forecasting in Agricultural and Natural Ecosystems,* Kluwell Academic Pulishers, London,UK, 2000.

[12] Zhou D, Lin J, Qin M. Extensive grazing and designed feeding with supplemented legume forages on natural grassland. *Ying Yong Sheng Tai Xue Bao*, 2004; 15: 1187-1193.

[13] Li-Menglin, Bo-Hua Y, Suttie JM. Winter feed for transhumant livestock in China: the Altai experience. *World Anim Rev.*, 1996; 87: 38-44.

[14] Gaspar P, Escribano M, Mesías FJ, Rodríguez de Ledesma A, Pulido F. Sheep farms in the Spanish rangelands (dehesas): Typologies according to livestock management and economic indicators. *Small Ruminant Res.*, 2008; 74: 52-63.

[15] Pulido F, Escribano M, Mesías FJ, Rodríguez A. Use of energetic resources in sheep farms in dehesas of south-west Extremadura. *Cahiers Options Méditerranéennes. Centre International de Hautes Etudes Agronomiques Méditerranéennes*, 1999, 39: 269-272.

[16] Arias M, Piñeiro P, Hillyer GV, Suárez JL, Francisco I, Cortiñas FJ, Díez-Baños P, Morrondo P, Sánchez-Andrade R, Paz-Silva A. An approach of the laboratory to the field: assessment of the influence of cattle management on the seroprevalence of fascioliasis by using polyclonal- and recombinant-based ELISAs. *J. Parasitol.*, 2010; 96: 626-631.

[17] Delgado-Pertiñez M, Alcalde MJ, Guzmán-Guerrero JL, Castel JM, Mena Y, Caravaca F. Effect of hygiene-sanitary management on goat milk quality in semi-extensive systems in Spain. *Small Ruminant Res.*, 2003; 47: 51–61.

[18] Murwira KH, Swift MJ, Frost PGH. Manure as a key resource in sustainable agriculture. In: J.M. Powell, S. Fernández-Rivera, T.O. Williams and C. Renard (Eds.), *Livestock and Sustainable Nutrient Cycling in Mixed Farming Systems of Sub-Saharan Africa.* Addis Ababa, ILCA, 1995; pp. 131–148.

[19] Atkinson D, Watson CA. The environmental impact of intensive systems of animal production in the lowlands. *Anim. Sci.*, 1996; 63: 353-361.

[20] Stookey JM. Is Intensive Dairy Production Compatible With Animal Welfare? Advances in Dairy Technology Vol. 6. *Proceedings of the 1994 Western Dairy Canadian Dairy Seminar*,1994.

[21] Turnpenny JR, Parsons DJ, Armstrong AC, Clark JA, Cooper K, Matthews AM. Integrated models of livestock systems for climate change studies. 2. Intensive systems. *Global Change Biol.*, 2001; 7: 163-170.

[22] Sundrum A. Organic livestock farming - a critical review. *Livest. Prod. Sci.*, 2001; 67: 207-215.

[23] Gottardo F, Brscic M, Contiero B, Cozzi G, Andrighetto I. Towards the creation of a welfare assessment system in intensive beef cattle farms. *Ital. J. Anim. Sci.*, 2009; 8: 325-342.

[24] Haskell MJ, Rennie LJ, Bowell VA, Bell MJ, Lawrence AB. Housing system, milk production and zero grazing effects on lameness and leg injury in dairy cows. *J. Dairy Sci.,* 2006; 89: 4359-4266.

[25] Somers JGCJ, Frankena K, Noordhuizen-Stassen EN, Metz JHM. Prevalence of claw disorders in Dutch dairy cows exposed to several floor systems. *J. Dairy Sci.*, 2003; 86: 2082-2093.

[26] Bradley AJ, Leach KA, Breen JE, Green LE, Green MJ. Survey of the incidence and aetiology of mastitis on dairy farms in England and Wales. *Vet. Rec.,* 2007; 160: 253-257.

[27] Barton Gade P. Welfare of animal production in intensive and organic systems with special reference to Danish organic pig production. *Meat Sci.*, 2002; 62: 353-358.

[28] Noci F, Monahan FJ, French P, Moloney AP. The fatty acid composition of muscle fat and subcutaneous adipose tissue of pasture-fed beef heifers: Influence of the duration of grazing. *J. Anim. Sci.*, 2005; 83: 1167-1178.

[29] Alfaia CMM, Castro MLF, Martins SIV, Portugal APV, Alves SPA, Fontes CMGA, Bessa RJB, Prates JAM. Effect of slaughter season on fatty acid composition, conjugated linoleic acid isomers and nutritional value of intramuscular fat in Barrosã-PDO veal. *Meat Sci.,* 2007; 75: 44-52.

[30] Lourenço CF, Gago F, Barbosa RM, de Freitas V, Laranjinha J. LDL isolated from plasma-loaded red wine procyanidins resist lipid oxidation and tocopherol depletion. *J. Agr. Food Chem.*, 2008; 56: 3798-3804.

[31] Francisco I, Arias M, Cortiñas FJ, Francisco R, Mochales E, Sánchez JA, Uriarte J, Suárez JL, Morrondo P, Sánchez-Andrade R, Díez-Baños P, Paz-Silva A. Silvopastoralism and autochthonous equine livestock: analysis of the infection by endoparasites. *Vet. Parasitol.*, 2009; 164: 357-362.
[32] Paz-Silva A, Sánchez-Andrade R, Suárez JL, Pedreira J, Arias M, López C, Panadero R, Díaz P, Díez-Baños P, Morrondo P. Prevalence of natural ovine fasciolosis shown by demonstrating the presence of serum circulating antigens. *Parasitol. Res.*, 2003; 4: 328-331.
[33] Ortolani EL. Standardization of the modificied Ziehl-Neelsen technique to stain oocysts of *Cryptosporidium* sp. *Braz. J. Vet. Parasitol.*, 2000; 9: 29-31.
[34] Osterman E, Uggla A, Waller P, Höglund J. Larval development assay for detection of anthelmintic resistance in cyathostomins of Swedish horses. *Vet. Parasitol.*, 2005; 128: 261-269.
[35] Kuzmina TA, Kuzmin YI, Kharchenko VA. Field study on the survival, migration and overwintering of infective larvae of horse strongyles on pasture in central Ukraine. *Vet. Parasitol.,* 2006; 141: 264-272.
[36] Thrusfield M. *Veterinary Epidemiology*. Wiley-Blackwell (3rd Edition), Iowa, USA, 2005.
[37] Arias M, Lomba C, Dacal V, Vázquez L, Pedreira J, Francisco I, Piñeiro P, Cazapal-Monteiro C, Suárez JL, Díez-Baños P, Morrondo P, Sánchez-Andrade R, Paz-Silva A. Prevalence of mixed trematode infections in an abattoir receiving cattle from northern Portugal and north-west Spain. *Vet. Rec.*, 2011; 168: 408-412.
[38] Bandara NWRVN, Rajakaruna RS, Rajapakse RPVJ. Identification and prevalence of *Eimeria* spp. causing coccidiosis in goats in selected sites from Kandy and Nuwara Eliya districts. *Proceedings of the Peradeniya University Research Sessions*, Sri Lanka, Vol. 12, Part I, , 2007.
[39] Díaz Fernández P. Epidemiology of the main parasitoses in Rubia Gallega autochthonous cattle. *PhD Thesis*, University of Santiago de Compostela, Spain, 2006.
[40] Jiménez AE, Montenegro VM, Hernández J, Dolz G, Maranda L, Galindo J, Epe C, Schnieder T. Dynamics of infections with gastrointestinal parasites and *Dictyocaulus viviparus* in dairy and beef cattle from Costa Rica. *Vet. Parasitol.*, 2007; 148: 262-271.
[41] Theodoropoulos G, Peristeropoulou P, Kouam MK, Kantzoura V, Theodoropoulou H. Survey of gastrointestinal parasitic infections of beef cattle in regions under Mediterranean weather in Greece. *Parasitol. Int.*, 2010; 59: 556-559.
[42] Shimano S. Oribatid mites (Acari: Oribatida) as an intermediate host of *Anoplocephalid* cestodes in Japan. *Appl. Entomol. Zool.*, 2004; 39: 1-6.
[43] Seniczak S. Juvenile stages of moss mites (Acari, Oribatei) as an essential component of agglomerations of these mites transforming the organic matter of soil. *Rozprawy UMK Toruñ*, 1978.
[44] Bardgett RD, Cook R. Functional aspects of soil animal diversity in the agricultural grasslands. *Appl. Soil Ecol.*, 1998; 10: 263-276.
[45] Behan-Pelletier V. Oribatid mite biodiversity in agroecosystems: role for bioindication. *Agr. Ecosys Environ.*, 1999; 74: 411-423.
[46] Shimano S, Kaminura K. A Cysticercoid in *Mixacarus exilis* (Acari: Oribatida) nymph as an intermediate host of *Anoplocephalid* cestodes. *J. Acarol. Soc. Jap.*, 2005; 14: 31-34.

[47] Chachaj B, Seniczak S. The influence of sheep, cattle and horse grazing on soil mites (Acari) of lowland meadows. *Folia Biol.-Prague.*, 2005; 53: 127-132.

[48] Manga-González MY, González-Lanza C. Field and experimental studies on *Dicrocoelium dendriticum* and dicrocoeliasis in northern Spain. *J. Helminthol.*, 2007; 79: 291-302.

[49] Díaz P, Paz-Silva A, Sánchez-Andrade R, Suárez JL, Pedreira J, Arias M, Díez-Baños P, Morrondo P. Assessment of climatic and orographic conditions on the infection by *Calicophoron daubneyi* and *Dicrocoelium dendriticum* in grazing beef cattle (NW Spain). *Vet. Parasitol.*, 2007; 149: 285-289.

[50] Schuster R, Coetzee L, Putterill JF. Oribatid mites (Acari, Oribatida) as intermediate hosts of tapeworms of the family *Anoplocephalidae* (Cestoda) and the transmission of *Moniezia expansa* cysticercoids in South Africa. *Onderstepoort J. Vet.*, 2000; 67: 49-55.

[51] Stromberg B, Vatthauer R, Schlotthauer. J, Myers G, Haggard D, King V, Hanke H. Production responses following strategic parasite control in a beef cow/calf herd. *Vet. Parasitol.,* 1997; 68: 315-322.

[52] Díaz P, Pedreira J, Sánchez-Andrade R, Suárez JL, Arias MS, Francisco I, Fernández G, Díez-Baños, P, Morrondo P, Paz-Silva A. Risk periods of infection by *Calicophoron daubneyi* (Digenea:Paramphistomidae) in cattle from oceanic climate areas. *Parasitol. Res.,* 2007; 101: 339-342.

[53] Arias MS, Suárez JL, Hillyer GV, Francisco I, Calvo E, Sánchez-Andrade R, Díaz P, Francisco R, Díez-Baños P, Morrondo P, Paz-Silva A. A recombinant-based ELISA evaluating the efficacy of netobimin and albendazole in ruminants with naturally acquired fascioliasis. *Vet. J.,* 2009; 182: 73-78.

[54] Szmidt-Adjidé V, Abrous M, Adjidé CC, Dreyfuss G, Lecompte A, Cabaret J, Rondelaud D. Prevalence of *Paramphistomum daubneyi* infection in cattle in central France. *Vet. Parasitol.*, 2000; 87: 133-138.

[55] Díaz P, Lomba C, Pedreira J, Arias M, Sánchez-Andrade R, Suárez JL, Díez-Baños P, Morrondo P, Paz-Silva A. Analysis of the IgG antibody response against Paramphistomidae trematoda in naturally infected cattle. Application to serological surveys. *Vet. Parasitol.,* 2006; 140: 281-288.

[56] Ballweber LR. Diagnostic methods for parasitic infections in livestock. *Vet. Clin. N Am.-Food. An.,* 2006; 22: 695-705.

[57] Gorman T, Aballay J, Fredes F, Silva M, Aguillón JC, Alcaíno HA. Immunodiagnosis of fasciolosis in horses and pigs using western blots. *Int. J. Parasitol.*, 1997; 11: 1429-1432.

[58] Epe C, von Samson-Himmelstjerna G, Wirtherle N, von der Heyden V, Welz C, Beening J, Radeloff I, Hellmann K, Schnieder T, Krieger K. Efficacy of toltrazuril as a metaphylactic and therapeutic treatment of coccidiosis in first-year grazing calves. *Parasitol. Res.*, 2005; 97: S127-S133.

[59] Le Sueur C, Mage C, Mundt HC. Efficacy of toltrazuril (Baycox 5% suspension) in natural infections with pathogenic *Eimeria* spp. in housed lambs. *Parasitol. Res.*, 2009; 104: 1157-1162.

[60] Mundt HC, Dittmar K, Daugschies A, Grzonka E, Bangoura B. Study of the comparative efficacy of toltrazuril and diclazuril against ovine coccidiosis in housed lambs. *Parasitol. Res.*, 2009; 105: S141-S150.

[61] Veronesi F, Diaferia M, Viola O, Fioretti DP. Long-term effect of toltrazuril on growth performances of dairy heifers and beef calves exposed to natural *Eimeria zuernii* and *Eimeria bovis* infections. *Vet. J.* 2010 Dec 6. doi:10.1016/j.tvjl.2010.10.009.

[62] Klein P. Preventive and therapeutic efficacy of halofuginone-lactate against *Cryptosporidium parvum* in spontaneously infected calves: a centralised, randomised, double-blind, placebo-controlled study. *Vet. J.,* 2008; 177: 429-431.

[63] De Waele V, Speybroeck N, Berkvens D, Mulcahy G, Murphy TM. Control of cryptosporidiosis in neonatal calves: use of halofuginone lactate in two different calf rearing systems. *Prev. Vet .Med.*, 2010; 96: 143-151.

[64] Bauer C. Comparative efficacy of praziquantel, albendazole, febantel and oxfendazole against *Moniezia expansa. Vet. Rec.*, 1990; 127: 353-354.

[65] Ward JK, Ferguson DL, Parkhurst AM, Berthelsen J, Nelson MJ. Internal parasite levels and response to anthelmintic treatment by beef cows and calves. *J. Anim. Sci.*, 1991; 69: 917-922.

[66] Southworth J, Harvey C, Larson S. Use of praziquantel for the control of *Moniezia expansa* in lambs. *N. Zeal. Vet.* J., 1996; 44: 112-115.

[67] Sánchez-Andrade R, Paz-Silva A, Suárez JL, Panadero R, Pedreira J, Díez-Baños P, Morrondo P. Effect of fasciolicides on the antigenaemia in sheep naturally infected with *Fasciola hepatica. Parasitol. Res.*, 2001; 87: 609-614.

[68] Keiser J, Engels D, Büscher G, Utzinger J. Triclabendazole for the treatment of fascioliasis and paragonimiasis. *Expert Opin. Invest. Drug.*, 2005; 14: 1513-1526.

[69] Martínez-Valladares M, Cordero-Pérez C, Castañón-Ordóñez L, Famularo MR, Fernández-Pato N, Rojo-Vázquez FA. Efficacy of a moxidectin/triclabendazole oral formulation against mixed infections of *Fasciola hepatica* and gastrointestinal nematodes in sheep. *Vet. Parasitol.*, 2010; 174: 166-169.

[70] Hidalgo-Argüello MR, Díez-Baños N, Rojo-Vázquez FA. Efficacy of moxidectin 1% injectable and 0.2% oral drench against natural infection by *Dictyocaulus filaria* in sheep. *Vet. Parasitol.*, 2002; 107: 95-101.

[71] Díez-Baños P, Pedreira J, Sánchez-Andrade R, Francisco I, Suárez JL, Díaz P, Panadero R, Arias M, Painceira A, Paz-Silva A, Morrondo P. Field evaluation for anthelmintic-resistant ovine gastrointestinal nematodes by in vitro and in vivo assays. *J. Parasitol.,* 2008; 94: 925-928.

[72] Sargison ND. Pharmaceutical control of endoparasitic helminth infections in sheep. *Vet. Clin. N. Am-Food An.*, 2011; 27: 139-156.

[73] Paz-Silva A, Francisco I, Valero-Coss RO, Cortiñas FJ. Sánchez JA, Francisco R, Arias M, Suárez JL, López-Arellano ME, Sánchez-Andrade R, Mendoza de Gives P. Ability of the fungus *Duddingtonia flagrans* to adapt to the cyathostomin egg-output by spreading chlamydospores. *Vet. Parasitol.,* 2011; 179: 277-282.

In: Livestock: Rearing, Farming Practices and Diseases
Editor: M. Tariq Javed
ISBN 978-1-62100-181-2

Chapter 6

Animal Trypanosomosis: An Important Constraint for Livestock in Tropical and Sub-Tropical Regions

***Marc Desquesnes*[1] *and Carlos Gutierrez*[2*]**
[1]Centre de Coopération Internationale en Recherche Agronomique pour le Développement (CIRAD), Montpellier, France
[2]University of Las Palmas de Gran Canaria, Canary Islands, Spain

Abstract

Trypanosomosis is an important constraint to the livestock production in many parts of Africa, Asia and Latin America. Tsetse-transmitted trypanosomosis (nagana) is a disease complex caused particularly by *Trypanosoma vivax*, *T. congolense* and *T. brucei brucei*. Non tsetse-transmitted trypanosomosis, on the other hand, is principally caused by *T. evansi* (surra), a widely distributed pathogenic trypanosome affecting livestock, but *T. equiperdum* and *T. cruzi* are also relevant pathogenic trypanosomes. From an economic viewpoint, it has been estimated that trypanosomosis reduces the cattle population between 30percent and 50percent and the production of milk and meat by at least 50percent in those infected areas of Africa. The presence of *Trypanosoma vivax* in and out of Africa transmitted by mechanical vectors rather than tsetse flies or the recent descriptions of *T. evansi* and *T. equiperdum* in European countries could pose a new threat for animal production in those territories. The purpose of this chapter is to review the current knowledge of the pathogenic trypanosomes that affect livestock, including the economic impact and control programs.

* E-mail address: cgutierrez@dpat.ulpgc.es.

1. INTRODUCTION

Trypanosomosis are protozoan diseases, affecting both human and animals, mainly found in tropical Africa, Latin America and Asia. Species of trypanosomes infecting mammals fall into two distinct groups and, accordingly, have been divided into two sections [1]:

- Stercoraria (subgenera *Schizotrypanum, Megatrypanum* and *Herpetosoma*), in which trypanosomes are typically produced in the hindgut and are then passed on by contaminative transmission from the posterior end of the digestive tract.
- Salivaria (subgenera *Duttonella, Nannomonas* and *Trypanozoon*), in which transmission occurs by the anterior station and is inoculative.

Characteristically, salivarian species, by virtue of variant surface glycoprotein (VSG) genes, are the only trypanosomes to exhibit antigenic variation [2].

Within section Stercoraria, trypanosomes of medical and veterinary importance are *T. theileri*, *T. lewisi* and *T. cruzi*; while within section Salivaria are included *T. vivax*, *T. congolense*, including at least 3 types (type Forest, Savannah, Kilifi, Tsavo) or species (*T. godfreyi*) [3, 4], *T. simiae* (including Tsavo-type) [3], *T. brucei brucei*, *T. b. gambiense*, *T. b.rhodesiense*, *T. equiperdum* and *T. evansi* and possibly other species such as *T. suis* [4].

From a geographical viewpoint, Africa is a particularly affected continent because the tsetse-transmitted trypanosomosis is circumscribed to sub-Saharan Africa, in an area of approximately 9 million km^2. *Trypanosoma vivax, T. congolense, T. simiae, T. brucei* are transmitted by this vector. In Africa, a daily mortality of about 100 people and 10,000 cattle is estimated to be occurring as a consequence to trypanosomosis [5]. Other trypanosomes present in the continent are *T. evansi*, which is transmitted by hematophagous flies, and *T. equiperdum*, a sexually transmitted protozoon.

In Latin America, Chagas' disease (*Trypanosoma cruzi*) is also a major human disease that particularly affects poor populations living in dilapidated dwellings; of the 90 million people exposed to the risk, 20 million are estimated to be infected, with a mortality rate of about 10percent [6]. In Latin America, four species of trypanosomes are of medical and economic importance, or can interfere with the diagnosis of livestock hemoparasitoses: *Trypanosoma equiperdum*, *Trypanosoma vivax, Trypanosoma evansi*, and *Trypanosoma cruzi*. Only *T. cruzi*, in the Stercoraria group, is indigenous to America, while humans imported the other three salivarian species together with their main domestic hosts [6].

In Asia, *Trypanosoma evansi* is endemic affecting Southeast countries, where it poses important economic losses to livestock especially to smallholders [7]. *Trypanosoma equiperdum* has also been described in Asia [8].

Trypanosomes are widely distributed in many areas of these continents; however, some trypanosomes that have been introduced into new continents have adapted to the new environments and infecting new areas. Some species of trypanosomes are transmitted by hematophagous insects with worldwide distribution, and outbreaks can easily occur if sanitary measures are not exercised. Thus, *Trypanosoma evansi* is a typical example; transmitted by hematophagous flies belonging to Genus *Tabanidae, Stomoxys* and *Liperosia;* it has recently caused acute outbreaks in many countries previously considered as *T. evansi*-free such as metropolitan France [9] and metropolitan Spain [10], or countries where the disease was rarely reported such as Israel [10] and Iran [12]. Another not so typical is *Trypanosoma vivax*; exclusively transmitted by tsetse flies in Africa, was introduced in South America possibly in the late 19th centuries, adapting to new vectors and affecting nowadays 10 of the 13 South-American countries [13]. This adaptive capacity can pose a threat to the non-endemic areas. Thus, free areas close to infected areas are at risk to be infected, which is linked to the movement of the animals from infected areas.

The present chapter, for that matters, aims at reviewing the current knowledge of the pathogenic trypanosomes that affect livestock including clinical features, diagnostic methods, treatments, preventions, control programs and economic impact on livestock.

2. Section Stercoraria

2.1. *Trypanosoma theileri*

Background: *Trypanosoma theileri* can be considered as a type-species of the subgenus *Megatrypanum*. In 1902, this parasite was described for the first time in cattle by Laveran (in South Africa) and by Bruce (in East Africa) almost simultaneously. *Trypanosoma theileri* has a wide distribution and presents a high incidence in all continents excluding Antarctica [14].

Morphology: Trypanosoma theileri belongs to subgenus *Megatrypanum*, which contains the largest mammalian trypanosomes. This parasite is one of the largest trypanosome with mean length between 60 and 70μm, occasionally up to 100 μm, and presents a small kin-etoplast which is located close to the nucleus.

Host range and vectors: Based on their morphology and that *Megatrypanum* spp. are mammalian host-specific, all isolates from cattle are considered as *Trypanosoma theileri* [15]. Wells [16] also suggested that only isolates from cattle could be considered synonyms of *T. theileri*. However, further work based on the gene sequence encoding for the cathepsin L-like (CATL) cystein protease showed that isolates from cattle, buffalo and deer could also be characterized [17]. Thus, it has been proposed to classify as *T. theileri*–like the *Megatrypanum* isolates from *Artiodactyla* species other than cattle (including isolates from goats and sheep), until their genetic characterization. *Megatrypanum* spp. is cyclically transmitted by hematophagous insects and the infection to new hosts is by contamination route. Tabanids are the most important vector implied in the transmission of *T. theileri* [18].

Pathogenicity: The pathogenicity of *T. theileri* within the mammalian host is not well understood. *Trypanosoma theileri* may persist in cattle for many years without any evidence of clinical disease. In most of the infected animals, parasite cannot be detected in blood smears. It is thought to be of very low pathogenicity [19].

Clinical features: *T. theileri* is commonly associated with mild or no clinical signs at all. However, clinical disease has been reported in a calf [20], in a cow with a history of abortion and diarrhea and in a dairy cow with severe regenerative anemia [21]. The *T. theileri* is also commonly isolated along with other microorganisms [22].

Diagnosis: Most infections caused by *T. theileri* cannot be diagnosed by blood examination, even using some concentration tests (micro-hematocrit method); however, prevalence observed by using this method can be as high as 12 percent [23]. The possibility to use laboratory rodents to isolate *T. theileri* may be of no use because of the host restriction to this parasite [15]. Thus, the diagnosis requires isolation using blood culture, which is a time-consuming, difficult and expensive method; however the rate of success of this method is very high which is up to 70percent [24] and in some cases it may be 100percent [1]. Serologic tools have not yet described for *T. theileri*. However, this species can be detected by means of PCR-amplified spliced leader transcript [25] and single PCR based on internal transcribed spacer 1 of rDNA [26].

Treatment: Specific treatment against *T. theileri* infection has not been developed.

Another close *Megatrypanum* can also be observed in antelope and bovines. *T. ingens*, can easily be recognized in blood smears, due to its deeply stained body and clear nucleus presenting as a transversal clear band [1, 23]; alike *T. theileri,* it is not suspected to have any significant medical impact.

2.2. *Trypanosoma cruzi*

Background: Most of the literature concerning *T. cruzi* infection (Chagas disease) is focused to human infections; however, this protozoon presents a wide host range. Many species of mammals, including bats, have been reported to be infected with *T. cruzi*; thus, all mammals are considered to be susceptible. Distinct phylogenetic lineages of *T. cruzi* appear to have preferential associations with different groups of hosts [2]. Chagas disease affects exclusively Latin American continent where most of the countries have reported this disease. *Trypanosoma cruzi* is not a livestock trypanosome as such, but it is sometimes found in domestic ruminants and horses [27]. Although the human disease is confined to South and Central America (25° N-38° S), the parasite is spreading towards the North deep inside the USA, up to California and Virginia where the main reservoirs seems to be dogs, opossums and raccoons [28]. The spreading of *T. cruzi* is speculated to continue and sometime if not already, might lead to the further spread of Chagas disease in the USA.

Morphology: In mammalian blood, *T. cruzi* is described as small (mean length including flagellum 16.3-21.8 μm), typically C-shaped trypanosome with a large kinetoplast (diameter 1.2 μm, approximately) present near to the posterior end, has a free flagellum, and an undulating membrane with 2-3 shallow convolutions [2].

Transmission: Its transmission is essentially cyclical through bugs that belong to the *Reduviidae* family: triatomines, or reduviid bugs, of the genera *Rhodnius, Panstrongylus* and *Triatoma*. The metacyclic trypomastigote infective form (metatrypanosome) is present in the excreta of the bugs that contaminate bite wounds or the mucous membranes, particularly the eye. Bugs may be both a reservoir and vector for *T. cruzi*. Transmission of the infective forms found in the feces of the bugs can also take the oral route. Ingestion of infected bugs is also thought to be the cause of contamination of dogs and cats and possibly livestock [6]. The life cycle described in the gut of triatomines has also been observed in *Didelphis marsupialis* [29], and probably the parasite implements its cycle in *Didelphis virginiana* [30].The human infections by oral route after drinking the fruit juice (sugar cane or palm) contaminated by bug's feces are increasingly recorded in Venezuela and Brazil [31, 32]; this may also be the cause of Chagas disease in the USA. *Trypanosoma cruzi* could colonize other countries and continents such as Asia for example, where *Triatoma rubrofasciata* is present. Human carriers travelling to Asia could allow the establishment of new endemic areas. Diagnosis of such infection in Asian countries may not be evident due to epidemiological and geographical considerations; serological surveys in humans could allow insuring free status of such countries.

Pathogenicity: The current knowledge about the pathogenicity of *T. cruzi* in livestock is scarce, but it appears to be fairly low according to the few experimental findings available. Alcaino *et al.* [33] reported that no symptoms were visible in an experimentally infected goat kid although they did detect ventricular hypertrophy using ECG. Unfortunately little research has been conducted either in the field, where no specific diagnostic tools are available. *T. cruzi* is markedly pathogenic in dogs and produces cardiac signs with a potentially fatal outcome [6]. Since receptivity and susceptibility of livestock appeared to be low in the experimental studies, with low and transient parasitemia, livestock infections probably play a minor epidemiological role in the Chagas disease. However, in poor rural populations, cohabitation with animals might foster peri-domestic zoonotic spread of the parasite. Veterinary practitioners who deal with pets and livestock in endemic areas should be careful. A positive diagnosis in an animal may be an indication of risk for humans. Pathogenicity of *T. cruzi* in livestock is assumed to be low but it requires further investigations to determine the real role that animals can play in the epidemiology of the disease.

*Diagnosis:*In acute phase, when parasitemia is normally high conventional parasitological tests including stained smear examination and micro-hematocrit centrifugation techniques are usually applied. In chronic phases, indirect immunofluorescence assay (IFA), indirect hemagglutination assay (IHA) and enzyme-linked immunosorbent assay (ELISA) have been used. It has been observed that the antibodies against *T. cruzi* in humans can cross react with antigens from *T. evansi*; interferences between different *Trypanosoma* species are then highly expected in serological diagnosis [34].

Treatment: A nitroimidazole compound, benznidazole (Rochagan, Radanil, Roche) and a nitrofuran compound, nifurtimox (Lampit, Bayer) are currently used to cure human infections. However, the differences in drug susceptibility of *T. cruzi* strains have been described and even the existence of strains naturally resistant to both the drugs. These drugs are also administered to infected dogs with unsatisfactory results [35].

3. Section Salivaria

3.1. Tsetse Transmitted Animal Trypanosomes:*Trypanosoma vivax, Trypanosoma congolense, Trypanosoma brucei brucei*

Background: *Trypanosoma vivax, T. congolense* (including several types) and *T. brucei brucei*, constitute a tsetse-transmitted group that causes a disease complex known as "nagana". It is extended over 10 million km^2 and affects 37 countries, mostly in Africa. Nagana is a very important disease in cattle, in which it produces enormous economic losses. The disease affects also camels and poses a natural barrier avoiding the introduction of these animals into the southern Sahel area. Equines are also considered highly sensitive and it is generally believed that the historical Arabian invasion was stopped by tsetse flies because of their impact on the Arabian horses. Human infection caused by animal species of *Trypanosomes*, notably *T. vivax* and *T. congolense*, have rarely been observed [1, 36]. However, tsetse-transmitted trypanosomes can also infect humans, in particular *T. brucei gambiense* or *T. brucei rhodesiense*, the causal agents of sleeping sickness. There is a large range of domestic and wild animals which can act as reservoirs of those human trypanosomes [37].

Morphology: *Trypanosoma vivax* differs from other salivarian trypanosomes by its length (means from 21 µm to 25.4 µm). However, length is not a reliable taxonomic character since many longer forms of this trypanosome have been described in rodents [2]. Its characteristics are a large terminal kinetoplast (1-1.2µm) located at a rounded posterior extremity; a free flagellum and a medium developed undulating membrane.

Trypanosoma congolense: It is a small parasite (mean 12-17 µm), with sub-terminal and marginal medium sized kinetoplast (0.7-0.8µm), with an inconspicuous undulating membrane, while the free flagellum is most often absent [1]. However, there have been described at least three different forms based on the mean length, width and presence of free flagellum [38]. Significantly, these forms have not been associated with biological, biochemical or molecular characteristics [1, 2]. Only the rate of these forms are different, but variable, inside the types [38]; now these types are better defined and characterized using molecular tools [39].

Trypanosoma brucei brucei: Bloodstream forms of *T. brucei* ranges between 11 to 42 µm in length. They are typically polymorphic, being represented by three forms: long slender form (23-30 µm); short stumpy form (17-22 µm) presenting usually lacking of free flagellum and intermediate form (20-25 µm) [2]. The most representative blood form of the subgenus *Trypanozoon* is the slender form which characteristics are: the length (25µm), a small kinetoplast (0.6 µm) sub-terminal and sub-marginal located far from the narrow posterior extremity (sometime truncated), a large undulating membrane (3-5 waves) and a free flagellum [1].

Vectors: Nagana is mostly transmitted by *Glossina* spp. (tsetse flies), but mechanical transmission by biting flies is also possible. *Trypanosoma vivax* is a particular case which is also transmitted by biting flies such as tabanids and stomoxes and it explains its presence in Central and South America, as well as in those African areas free or cleared of tsetse [37]. Mechanical transmission of *T. congolense* and *T. vivax* were demonstrated [40, 27], but only *T. vivax* seem to be of epidemiological importance.

Clinical features: Clinical signs of nagana would include intermittent fever, anemia, edema, decreased fertility and abortions, loss of milk and meat production and work capacities and emaciation. Anemia is a typical finding in infected animals and is commonly followed by weight loss, decreased productivity and mortality [37]. Clinical signs are very acute in camels and horses which may not survive under the pressure of tsetse transmitted trypanosomosis. In cattle, the clinical signs are strong in early infection and may be fatal, or decrease and lead to chronic or subclinical infection. The animals may recover for a period of time, but the clinical signs may relapse due to a peculiar stress such as seasonal high parasitic pressure, drought, lack of food, interference of inter-current diseases (anaplasmosis, babesiosis, etc.), vaccination or transportation of the animals, etc. Clinical signs are also variable with the type of cattle; thus the so-called trypano-tolerant *Bos taurus* exhibit milder signs than the trypano-sensible zebu cattle [41]. This is also true in sheep and goats to a lesser extent; the southern Djallonke sheep and the West African Dwarf (WAD) goat being trypano-tolerant compared to the large size sheep and goat originating from the northern Sahelian area [42].

Diagnosis: There is no clinical sign or post-mortem lesion that can be considered as pathognomonic of nagana. For that, diagnosis must be based on the direct detection of trypanosomes by microscopic visualization or by means of polymerase chain reaction (PCR) for which positive results mean active infection, or can be diagnosed indirectly by serological techniques [37]. To increase the sensitivity of the tests, the blood can be enriched by capillary centrifugation, and the buffy coat can be observed microscopically [43], or collected for further use in PCR [44]. PCR can detect various levels of specificity: sub-genus (*Trypanozoon*), species (*vivax, simiae*), sub-species (*T. brucei brucei, T. b. gambiense* and *T. b. rhodesiense*) or types (Savannah, Forest, Kilifi, Tsavo). Thus, at least 5 different types and/or species have been described in the subgenus *Nannomonas* [45, 3, 39]. *T. congolense* Savannah-type is the most pathogenic for livestock and the most common in cattle; *T. congolense* Forest-type is less pathogenic and is frequently found in dogs, while Kilifi-type is rare and of low pathogenicity to cattle [44, 46]. Tsavo-type has been described but little is known about its pathogenicity and epidemiology [3, 47]. Finally, *T. simiae* is a Nannomonas parasite found in pigs and monkeys, which showed a high pathogenicity in the former. Serological tests can detect immunoglobulins directed against the trypanosomes; the most employed is the ELISA technique which is useful for epidemiological studies since it is adapted to large scale sampling and can be standardized at the regional level [48]. Sensitivity and specificity of the ELISA for *T. vivax*, *T. congolense* and *T. brucei* are generally above 90 to 95percent, although the 3 parasites can cross react amongst themselves [49]. Positivity by serological test does not mean active infection; however, based on the history of treatment and the time for clearance of antibodies, active infection or cure can be inferred from the serological results.

Treatment: Treatment against tsetse-transmitted trypanosomes is based on several drugs. Currently available trypanocides are the followings [Based on 50]:

- Diminazene aceturate (Berenil® and others), used as therapeutic at dose rate 3.5 to 7 mg/kg, by intramuscular route, is recommended at 3.5 mg/kg against *T. congolense* in cattle, *T. vivax* in large and small ruminants and at 7 mg/kg against *T. brucei* in livestock and dogs. Diminazene aceturate cannot be used in camels due to its high toxicity [51]. Nowadays, it seems that the theoretical dose of 3.5 mg/kg is able to get rid of the clinical signs, at least temporarily, but is most often unable to cure the infection [6].
- Homidium chloride (Novidium®), Homidium bromide (Ethidium®), used as preventive and therapeutic, at the dose rate of 1 mg/kg by IM route is effective against *T. congolense* in cattle and *T. vivax* in pigs, small ruminants and horses; however it is not advised to use it because of the well know carcinogenic activity of ethidium bromide [52, 6].
- Isometamidium chloride can be used at dose rate of 0.25 to 0.5 mg/kg (Samorin®) or 0.5 to 1 mg/kg (Trypamidium®), by intramuscular route as preventive and therapeutic for *T. vivax* and *T. congolense* in cattle and small ruminants and *T. brucei* in equines.
- Quinapyramine dimethylsulphate (Trypacide sulphate®) at dose rate of 3 to 5 mg/kg by subcutaneous route is recommended as therapeutic treatment against *T. congolense* in camels.

- Quinapyramine dimethylsulphate/chloride (Trypacide pro-Salt®), at dose rate of 3 to 5 mg/kg by subcutaneous route is suggested as prophylactic use for *T. vivax* in equines and *T. brucei* in pigs

3.2. Non Tsetse Transmitted Animal Trypanosomes

3.2.1. Trypanosoma evansi

Background: *Trypanosoma evansi*, which causes a disease known as surra, is widely distributed and affects domestic livestock and wildlife in Africa, Asia and Latin America [53]. The presence of the parasite causes an important economic impact in the endemic areas.

Morphology: In the mammalian host, *Trypanosoma evansi* is typically monomorphic and occurs almost exclusively as long-slender forms (trypomastigotes) indistinguishable from the intermediate and slender forms of *Trypanosoma brucei*. The overall length ranges between 15 and 36 µm [2].

Vectors: The transmission of *T. evansi* is mechanical by biting flies particularly belonging to genus *Tabanus*, *Stomoxys* and *Liperosia*. Vampire bats are biological vectors in South America, and, since they can be infected and transmit to other bats and other hosts, they are considered as host, reservoir and vector of the parasite [6]. Immediate mechanical transmission of *T. evansi* by hematophagous vectors has been demonstrated in experimental conditions [54], but delayed transmission may also be suspected since *T. evansi* can survive in the crop of stable flies [55] and these insects may have several meals at hours or even at day interval [56].

Host range: *T. evansi* presents the widest host range of the pathogenic trypanosomes. Surra can affect a large number of domestic animals and the main host species varies depending on the geographical location (camels in Africa, horses in Latin America and bovines in Asia). Camels, horses, buffalo, cattle, and dogs are highly susceptible species, but other species such as sheep, goat, pig, deer, elephant, rhinoceros, cat and rodents, including other wildlife, can also be infected [57].

Clinical features: The disease is normally manifested by pyrexia, directly associated with parasitemia, progressive anemia, weight loss and lassitude. Such recurrent episodes lead to parasitemia and intermittent fever during the course of the disease. Edema (in particular affecting the lower parts of the body), rough coat in camels, urticarial plaques or petechial hemorrhages of the serous membranes are commonly detected. In advanced cases, parasites invade the central nervous system (CNS), which can lead to nervous signs (progressive paralysis of the hind quarters and exceptionally paraplegia), especially in horses, but also in other host species before complete recumbence and death. Abortions and immunodeficiency associated to *T. evansi* infection have also been reported, notably in bovines and pigs [57].

Diagnosis: The clinical signs of *T. evansi* infection are indicative but are not pathognomonic; thus, laboratory methods are required to confirm the diagnosis. Identification of the agent is needed, which can be reached by direct, but also by indirect parasitological methods since *T. evansi* can grow in laboratory rodents. Mice inoculation is still in use at least for diagnosis, or for strain isolation or production [58]. Within serological tests, enzyme-linked immunosorbent assay (ELISA), card agglutination tests (CATT/*T. evansi*), indirect immunofluorescent antibody test (IFAT) and immune trypanolysis tests are recommended [57].

Treatment: Current available drugs against *T. evansi* are the followings [Based on 50]:

- Diminazene aceturate (Berenil® and others), at dose rate of 7 mg/kg by intramuscular route is recommended as therapeutic in equines.
- Isometamidium chloride (Trypamidium®), at dose rate of 0.5 to 1 mg/kg by intramuscular route is indicated as prophylactic in camels.
- Quinapyramine dimethylsulphate/chloride (Trypacide pro-Salt®), at dose rate of 3 to 5 mg/kg by subcutaneous route is indicated as prophylactic in dogs.
- Suramine (Naganol®), at dose rate of 7 to 10 g per animal by intravenous route is recommended as a therapeutic and prophylactic agent in camels and equines.
- Melarsomine (Cymelarsan®), at dose rate of 0.25 mg/kg by intramuscular or subcutaneous route is used as therapeutic in camels, but also in horses. Although melarsomine has been scarcely used in other animal species, some studies have demonstrated its efficacy but only with higher doses (0.5-0.75mg/kg) in goats, pigs, cattle and buffaloes [59, 60, 61].

3.2.2. Trypanosoma equiperdum

Background: Trypanosoma equiperdum produces a disease called "Dourine", an acute or more frequently chronic contagious disease of equines that is transmitted directly by coital contact during mating, and possibly from the mare to the foal by eye or nose mucous membrane [62]. Infected equids are the natural reservoir of the parasite, which is not transmitted by invertebrate vectors. This trypanosome is considered primarily a tissue parasite and infrequently invades the blood [37].

Clinical features: The natural host of *Trypanosoma equiperdum* is equines, no other animal species is affected. In horses, the disease is chronic and persists for 1-2 years. Clinically, dourine is categorized into three phases, although the evolution of the disease can vary depending on certain conditions. The first phase occurs in 1 to 2 weeks after the infection and is characterized by tumefaction, edema and effects on the genitalia. The second phase is clearly pathognomonic for dourine. Typical cutaneous plaques or skin thicknesses can be observed, which can range from very small to hand sized; the name of the disease originates from the shape of this skin lesion similar to the coin named "duro" (ancient Spanish money). The third stage is recognized by anemia, impairments of the nervous system, in particular paraplegia and paralysis of the hind legs and finally death [63].

Diagnosis: Typical clinical signs of the disease can be useful for the diagnosis of dourine, but clinical diagnosis needs to be confirmed by laboratory methods. Given the extreme difficulty to discover the protozoa in the body fluids of affected horses, diagnosis of dourine is based on serological tests. The antibody and antigen-ELISAs have been developed for *T. equiperdum* but complement-fixation test (CFT) is the test internationally recommended by the World Animal Health Organization (WAHO/OIE). However, this test does not distinguish among *T. evansi, T. equiperdum* and *T. b. brucei* [63].

Treatment: There is not any available drug for dourine. The disease is considered to be incurable and, for that, seropositive horses should be killed.

4. Animal Trypanosomosis: Economic Impact in the Infected Areas

Little information is available about economic impact of trypanosomosis and most of it is limited to local or regional areas. We have subdivided the available information by continents, in an attempt to evaluate the impact of trypanosomosis on animal production according to trypanosomes present in each one and their estimated consequences.

4.1. Africa

Tsetse-transmitted trypanosomosis is one of the most important constraints to agricultural production in the tropical and subtropical areas of Africa. It has been reported that in an area of 9 million km^2 about 50 million people and 46 million cattle are exposed to the African tsetse transmitted trypanosomosis [64].

Animals kept in moderately infected areas show low calving rates, high rates of calf mortality, low milk yields, and require periodical preventive and curative trypanocide treatments compared to animals kept in free areas. Susceptible herds can dramatically be affected by sudden exposure to high risk to contract trypanosomosis [65].

Trypanosomosis also affects human populations living in the same affected areas. In these areas, the disease decreases the milk and meat productions about 50percent. Thus, the overall benefits for herdsmen are seriously reduced and, on the other hand, the direct negative effects on the agriculture are evident because trypanosomosis reduces cultivated area and the benefits as well as the efficiency of resource allocation. It has been estimated that the total value of agricultural production can be increased about 10percent with an increase about 50percent of the livestock population [65].

4.2. Latin America

Latin America, which includes Mexico, Central and South America as well as several Caribbean islands, is home to approximately a quarter of the world bovine population, i.e., 280 million heads, most of which are exposed to *T. cruzi, T. vivax* and *T. evansi* infections, while the horses, that are mainly used for herding cattle, are exposed to *T. equiperdum, T. evansi* and *T. cruzi* [6] and even to *T. vivax* as recently observed in Brazil [66]. Studies carried out in a beef farm in French Guiana showed that a *T. vivax* outbreak lead to a mean loss of 10 kg/animal, for a total cost of US$40-50 /animal [6]. Some studies carried out in Bolivian lowlands and Brazilian Pantanal have shown that 11 million cattle, with an approximate value of US$3 billion, are at risk to contract *T. vivax* infection; thus economic impact can be more than US$160 million [67]. For cattle ranchers in the Pantanal region, *Trypanosoma evansi* can produce economic losses of about US$2.4 million, and about 6,462 per year [68].

4.3. Asia

Official information about *Trypanosoma evansi* in the Philippines revealed estimated losses of more than US$1.1 million in nine years those are caused only by mortality. However, many cases frequently go unreported and it has been estimated that the true losses due to the parasite may be higher and are about US$7.9 million for nine years [69]. In Indonesia, it has been reported that the annual losses from mortality and morbidity associated with *T. evansi* were US$28 million [7]. Recently, models were developed in the Philippine to evaluate the losses due to surra in buffaloes and other livestock [70]; for a typical village, the total net-benefit derivate from an effective *T. evansi* control program was US$158,000/year, and the value added to cattle, horses, buffaloes, sheep/goats and pigs as a result of this control was estimated at US$84, $151, $88 $7, $114 per animal/year, respectively [71].

5. Control Programs

Control programs against trypanosomosis should include both, the use of chemical agents as therapeutic and/or preventive and, at the same time, a control plan to eradicate or minimize the vector.

5.1. Trypanocide Drugs

Currently there are four trypanocides available for livestock in the market. Three of them (diminazene, isometamidium and homidium) are used against tsetse transmitted trypanosomosis, while the fourth one (melarsomine) is specifically used for *T. evansi.*

- Diminazene aceturate (Berenil®, Veriben®, Ganaseg®, etc.) is a sanative treatment of trypanosomosis in cattle, sheep and goats, especially when it is caused by *T. vivax* or *T. congolense*, but it has no prophylactic effect. For this reason, it is recommended in Africa in areas where parasite loads are low, or for the purposes of controlling epizootic or sporadic outbreaks. For a long time, no resistance to diminazene aceturate was observed, but starting in 1960, resistant strains of trypanosomes were identified in many African and South American countries [6]. The recommended dose for treatment of *T. vivax* or *T. congolense* infections is 3.5 mg/kg; however, almost all strains tested are cleared only for a transient period before the infection relapse [6]; for that reason, nowadays, it is advisable to treat with 7 mg/kg diminazene aceturate, which in some instance may even not be sufficient.
- Isometamidium chloride (Trypamidium®, Samorin®) is recommended for the treatment and prevention of cattle and sheep trypanosomosis, in particular those due to *T. vivax.* In cattle, at dose rate of 1 mg/kg of bodyweight, isometamidium chloride affords protection for 17 to 28 weeks [72], five months on an average. That is why it is recommended in Africa in enzootic areas and/or periods of very high parasite load. Isometamidium chloride is, therefore, the therapy of choice in most cases, especially in the case of *T. congolense* and *T. vivax* infections that are resistant to diminazene

aceturate [6]. However, some strains are resistant to isometamidium chloride, e.g., a Nigerian *T. congolense* strain that was resistant to homidium was also found to be resistant to isometamidium chloride. Initially, this was a rare occurrence but is apparently becoming more frequent [73].

- Homidium chloride (Novidium®) and homidium bromide (Ethidium®) are closely related derivatives of isometamidium chloride. Their sanative effect is good and they are prophylactic for one to three months. They have been recommended for treating cattle, sheep and goats against trypanosomosis due to *T. vivax* and *T. congolense*, but their use is not advised due to their potential carcinogenic effect [6].
- Melarsomine (Cymelarsan®) is one of the few trypanocides those have been developed in the last 30 years. It is active only against *T. evansi* and *T. brucei*. Its metabolism is rapid and, as such, its action is purely curative. Specifically, it is effective against parasites that are resistant to suramin and quinapyramine [6]. The dose of 0.25 mg/kg can be applied to horses and camels; however, in other species (cattle, buffalo, goat, sheep, pig), higher doses (0.5-0.75mg/kg) are required.

5.2. Control Programs against Vectors

Vectors that transmit trypanosomes can be specific at local, regional or even continent level or, on the contrast, widely distributed elsewhere. An example of the first case is the vampire bat (*Desmodus rotundus*) in South America which is a vector for *T. evansi*, or, tsetse flies in tropical Africa for the trypanosomes those are responsible of Nagana, while hematophagus flies like *Tabanus* spp. or *Stomoxys* spp. are found in many parts of the world.

In Africa, before 1940s the destruction of bush and the extermination of game were the main measures of tsetse control. In the 1940s, the development of organochlorine insecticides allowed their use as an effective and relatively inexpensive approach to tsetse control in the 1950s. Organochlorides such as DDT, dieldrin or γ-BHC (lindane; gamma isomer of benzene hexachloride) were applied to tsetse areas either from the air or from the ground. These toxic molecules were really persistent and killed many generations of flies [74]. However, for many decades the wildlife in the continent was seriously affected due to the indiscriminate use of residual doses of broad-spectrum insecticides.

In Latin America, some attempts have been made to control the transmission of trypanosomes by vampire bats using net screens, but the results were insufficient for large scale application. Attempts to control horseflies with insecticide spray on cattle were somehow successful, but their use could only be restricted to isolated farms, which limited the extension of the method [6].

At farm level, the use of fly-trap was adequate to reduce the vector population. Against tsetse flies, different models have been developed (Nzi, Vavoua, tetra trap) presenting several structures, colors (in particular blue and black) and visual and olfactory substances that can contain insecticides or sterilizing agents. For other vectors such as horseflies, for example, some similar traps have been experimented: Manitoba, Vavoua, Malaise or Nzi traps [6]. The use of insecticides is normally required and can spread in the environment, applied to special stands or directly on the animals. The most suitable methods for Tabanid control can be by the use of sprays and pour-on insecticides. Pour-on to cattle containing deltamethrin has successfully been used against tsetse flies [75]. However, little information is available and further evaluations about efficacy of the insecticides and cost-effectiveness are required.

As for the stable flies, integrated control can be achieved by the combined use of fly-traps (type Vavoua), sticky traps (lines with glue), spray insecticides and the larvicides (cyromazine/neporex), but the most efficient is the elimination or the management of breeding sources by removal of landing areas (medium size vegetation in the surrounding of the stables) and manure, silage, wastes and decaying plants which are the substrates used for egg laying and larval development [76].

6. Conclusions

Trypanosomosis affects severely the livestock production in many parts of the world, in particular tropical and sub-tropical regions. Trypanosomosis causes enormous economic impact in those regions affecting not only the animal production but also the agriculture and even the human beings by several mechanisms, in particular because livestock can be reservoirs of some human trypanosomes and, secondly, because infected herds can become nonproductive. Control programs must be established in those infected areas in order to minimize and/or eradicate the trypanosomes present, when possible. Environmental, geo-climatic and many other epidemiological features must be taken into consideration to achieve better results.

References

[1] Hoare CA. The trypanosomes of mammals. *A zoological monograph*. Oxford: Blackwell Scientific Publications, 1972.

[2] Stevens JR, Brisse S. Systematic of Trypanosomes of Medical and Veterinary Importance. In: Maudlin, I; Holmes, PH; Miles, MA, editors. *The Trypanosomiases*. Trowbidge: Cabi publishing, 2004; pp. 1-23.

[3] Majiwa PA, Maina M, Waitumbi JN, Mihok S, Zweygarth E. Trypanosoma (Nannomonas) congolense: molecular characterization of a new genotype from Tsavo, Kenya. *Parasitology,*1993; 106: 151-162.

[4] Gibson WC, Stevens JR, Mwendia CM, Ngotho JN, Ndung'u JM. Unravelling the phylogenetic relationships of African trypanosomes of suids. *Parasitology,*2001; 122: 625-631.

[5] Hursey BS. The programme against African trypanosomiasis: Aims, objectives and achievements. *Trends Parasitol.,*2001; 17: 2-3.

[6] Desquesnes M. *Livestock Trypanosomoses and their Vectors in Latin America*. Paris, OIE (World Organisation for Animal Health), http://www.oie.org/, 2004.

[7] Reid SA. Trypanosoma evansi control and containment in Australasia. *Trends Parasitol.,*2002; 18: 219-224.

[8] Lun ZR, Brun R, Bibson W. Kinetoplast DNA and molecular karyotypes of Trypanosoma evansi and Trypanosoma equipedum from China. *Mol. Biochem. Parasitol.,*1992; 50: 189-196.

[9] Desquesnes M, Bossard G, Patrel D, Herder S, Patout O, Lepetitcolin E, Thevenon S, Berthier D, Pavlovic D, Brugidou R, Jacquiet P, Schelcher F, Faye B, Touratier L, Cuny G. First outbreak of Trypanosoma evansi in camels in metropolitan France. *Vet. Rec.,*2008; 162: 750-752.

[10] Tamarit A, Gutierrez C, Arroyo R, Jimenez V, Zagala G, Bosch I, Sirvent J, Alberola J, Alonso I, Caballero C. Trypanosoma evansi infection in mainland Spain. *Vet. Parasitol.,* 2010; 167: 74-76.

[11] Berlin D, Loeb E, Baneth G. Disseminated central nervous system disease caused by Trypanosoma evansi in a horse. *Vet. Parasitol.,*2009; 161: 316-319.

[12] Derakhshanfar A, Mozaffari AA, Zadeh AM. An outbreak of trypanosomiasis (surra) in camels in the Southern fars province of Iran: Clinical, hematological and pathological findings. *Res. J. Parasitol.,*2010; 5: 23-26.

[13] Jones TW. Davila AM. Trypanosoma vivax Out of Africa. *Trends Parasitol.,*2001; 17: 99-101.

[14] Hoare CA. Morphological and taxonomic studies on mammalian trypanosomes. X. Revision of the systematics. *J. Protozool.,*1964; 11: 200–207.

[15] Rodrigues AC, Campaner M, Takata CS, Dell' Porto A, Milder RV, Takeda GF, Teixeira MM. Brazilian isolates of Trypanosoma (Megatrypanum) theileri: diagnosis and differentiation of isolates from cattle and water buffalo based on biological characteristics and randomly amplified DNA sequences. *Vet. Parasitol.,*2003; 116: 185–207.

[16] Wells EA. *Biology of the Kinetoplastida*. London: Academic Press, 1976.

[17] Rodrigues AC, Garcia HA, Ortiz PA, Cortez AP, Martinkovic F, Paiva F, Batista JS, Minervino AH, Campaner M, Pral EM, Alfieri SC, Teixeira MM. Cysteine proteases of Trypanosoma (Megatrypanum) theileri: cathepsin L-like gene sequences as targets for phylogenetic analysis, genotyping diagnosis. *Parasitol. Int.,*2010; 59: 318-325.

[18] Rodrigues AC, Paiva F, Campaner M, Stevens JR, Noyes HA, Teixeira MM. Phylogeny of Trypanosoma (Megatrypanum) theileri and related trypanosomes reveals lineages of isolates associated with artiodactyl hosts diverging on SSU and ITS ribosomal sequences. *Parasitology,* 2006; 132: 215-224,.

[19] Greco A, Loria GR, Dara S, Luckins T, Sparagano O. First isolation of Trypanosoma theileri in Sicilian cattle. *Vet. Res. Comm.,*2000; 24: 471–475.

[20] Doherty ML, Windle H, Voorheis HP, Larkin H, Casey M, Murray M. Clinical disease associated with Trypanosoma theileri infection in a calf in Ireland. *Vet. Rec.,*1993; 132: 653–656.

[21] Ward WH, Hill MWM, Mazlin ID, Foster CK. Anaemia associated with a high parasitaemia of Trypanosoma theileri in a dairy cow. *Aust. Vet. J.,*1984; 61: 324.

[22] Villa A, Gutierrez C, Gracia E, Moreno B, Chacón G, Valera P, Büscher P, Touratier L. Presence of Trypanosoma theileri in Spanish cattle. *Ann. N. Y. Acad. Sci.,*2008; 1149: 352-354.

[23] Van Vlaenderen G. In search of cattle trypanosomiasis in Suriname. *Master of Science Thesis Prince Leopold Institute of Tropical Medicine*, Antwerp, Belgium, 88p., 1996.

[24] Toure SM. Diagnostic des trypanosomiases animales. *Rev. Elev. Med. Vet. Pays Trop.,*1977; 30: 1-10.

[25] Gibson W, Bingle L, Blendeman W, Brown J, Wood J, Stevens J. Structure and sequence variation of the trypanosome spliced leader transcript. *Mol. Biochem. Parasitol.,*2000; 107: 269–277.

[26] Desquesnes M, McLaughlin G, Zoungrana A, Davila AMR. Detection and identification of Trypanosoma of African livestock through a single PCR based on internal transcribed spacer 1 of rDNA. *Int. J. Parasitol.*,2001; 31: 610–614.

[27] Desquesnes M, Dia ML. Mechanical transmission of Trypanosoma vivax in cattle by the African tabanid Atylotus fuscipes. *Vet. Parasitol.*,2004; 119: 9-19.

[28] Desquesnes M, de Lana M. Veterinary aspects and experimental studies. In: Telleria J, Tibayrenc, M, editors. *American Trypanosomiasis Chagas disease: One Hundred Years of Research.* Burlington: Elsevier; 2010; 277-317.

[29] Urdaneta-Morales S, Nironi I. Trypanosoma cruzi in the anal glands of urban opossums.I. Isolation and experimental infections. *Mem. Inst. Oswaldo Cruz.*,1996; 91: 399-403.

[30] Yabsley MJ, Noblet GP. Seroprevalence of Trypanosoma cruzi in raccoons from South Carolina and Georgia. *J. Wildl. Dis.,*2002; 38, 75-83.

[31] Cardoso AV, Lescano AS, Amato Neto V, Gakiya E, Santos SV. Survival of Trypanosoma cruzi in sugar cane used to prepare juice. *Rev. Inst. Med. Trop.*, 2006; 48: 287-289.

[32] Steindel M, Kramer Pacheco L, Scholl D, Soares M, de Moraes MH, Eger I, Kosmann C, Sincero TC, Stoco PH, Murta SM, de Carvalho-Pinto CJ, Grisard EC. Characterization of Trypanosoma cruzi isolated from humans, vectors, and animal reservoirs following an outbreak of acute human Chagas disease in Santa Catarina State, Brazil. *Microbiol. Infect. Dis.*,2008; 60: 25-32.

[33] Alcaino TV, Lorca M, Nunez F, Issota A, Gorman T. Chagas' disease in goats from the Metropolitan region (Chile): Seroepidemiological survey and experimental infection. *Parasitología al Día,*1995; 19: 30-36.

[34] Desquesnes M, Bosseno MF, Breniere SF. Detection of Chagas infections using Trypanosoma evansi crude antigen demonstrates high cross-reactions with Trypanosoma cruzi. Infect. *Genet. Evol.*,2007; 7: 457-462.

[35] Guedes PM, Urbina JA, de Lana M, Afonso LC, Veloso VM, Tafuri WL, Machado-Coelho GL, Chiari E, Bahia MT. Activity of the new triazole derivative albaconazole against Trypanosoma (Schizotrypanum) cruzi in dog hosts. *Antimicrob. Agents Chemother.,*2004; 48: 4286-4292.

[36] Truc P, Jamonneau V, N'Guessan P, N'Dri L, Diallo PB, Cuny G. Trypanosoma brucei ssp. and T. congolense: mixed human infection in Cote d'Ivoire. *Trans. Royal Soc. Trop. Med. Hyg.,*1998; 92: 537-538.

[37] OIE. Trypanosomosis (Tsetse-transmitted), World Organisation for Animal Health, Paris, http://www.oie.org/ 2008.

[38] Toure S, Kebe B, Seye M, Sa N. Biométrie, morphologie et virulence de Trypanosoma (Nannomonas) congolense à travers 640 passages sur souris en 10 ans. 1976. *Rev. Elev. Med. Vet. Pays Trop.,*1976; 29: 17-22.

[39] Garside LH, Gibson WC. Molecular characterization of trypanosome species and subgroups within subgenus Nannomonas. *Parasitology,*1995; 111: 301-312.

[40] Desquesnes M, Dia ML. Mechanical transmission of Trypanosoma congolense in cattle by the African tabanid Atylotus agrestis. *Exp. Parasitol.*,2003; 105: 226-231.

[41] Naessens J. Bovine trypanotolerance: A natural ability to prevent severe anaemia and haemophagocytic syndrome?. *Int. J. Parasitol.*,2006; 36: 521-528.

[42] Geerts S, Osaer S, Goossens B, Faye D. Trypanotolerance in small ruminants of sub-Saharan Africa. *Trends Parasitol.*,2009; 25: 132-138.

[43] Woo PTK. The heamatocrit centrifuge technique for diagnosis of African trypanosomiasis. *Acta Trop.*,1970; 27: 384-386.

[44] Lefrancois T, Solano P, de la Rocque S, Bengaly Z, Reifenberg JM, Kabore I, Cuisance D. New epidemiological features on animal trypanosomiasis by molecular analysis in the pastoral zone of Sideradougou, Burkina Faso. *Mol. Ecol.*,1998; 7: 897-904.

[45] Masiga DK, Smyth AJ, Hayes P, Bromidge TJ, Gibson WC. Sensitive detection of trypanosomes in tsetse flies by DNA amplification. *Int. J. Parasitol.*,1992; 22: 909-918.

[46] Bengaly Z, Sidibe I, Ganaba R, Desquesnes M, Boly H, Sawadogo L. Comparative pathogenicity of three genetically distinct types of Trypanosoma congolense in cattle: Clinical observations and haematological changes. *Vet. Parasitol.*,2002; 108: 1-19.

[47] Njiokou F, Simo G, Nkinin SW, Laveissiere C, Herder S. Infection rate of Trypanosoma brucei s.l., T. vivax, T. congolense "forest type", and T. simiae in small wild vertebrates in south Cameroon. *Acta Trop.*,2004; 92: 139-146.

[48] Desquesnes M. Standardisation internationale et régionale des épreuves immuno-enzymatiques: méthode, intérêts et limites. *Rev. Sci. Tech./OIE,*1997; 16: 809-823.

[49] Desquesnes M, Bengaly Z, Millogo L, Meme Y, Sakande H. The analysis of the cross-reactions occurring in antibody-ELISA for the detection of trypanosomes can improve identification of the parasite species involved. *Ann. Trop. Med. Parasitol.*,2001; 95: 141-155.

[50] Holmes PH, Eisler MC, Geerts S. Current chemotherapy for trypanosomiasis. In: Maudlin I, Holmes PH, Miles MA, editors. *The Trypanosomiases.* Trowbidge: Cabi publishing, 2004; 431-444.

[51] Homeida AM, El Aminb EA, Adamb SEI, Mahmoudb MM. Toxicity of diminazene aceturate (Berenil) to camels. *J. Comp. Pathol.*,1981; 91: 355-360.

[52] McCann J, Choi E, Yamasaki E, Ames BN. Detection of carcinogens as mutagens in the Salmonella I microsome test: assay of 300 chemicals. *Proc. Natl. Acad. Sci.,*1975; 72: 5135-5139,.

[53] Luckins AG, Dwinger RH. Non-tsetse-transmitted Animal Trypanosomiasis. In: Maudlin I, Holmes PH, Miles MA, editors. *The Trypanosomiases.* Trowbidge: Cabi publishing, 2004; 269-281.

[54] Sumba AL, Mihok S, Oyieke FA. Mechanical transmission of Trypanosoma evansi and T. congolense by Stomoxys niger and S. taeniatus in a laboratory mouse model. *Med. Vet. Entomol.,*1998; 12: 417-422.

[55] Coronado A, Butler JF, Becnel J, Hogsette J. Artificial feeding in the stable fly Stomoxys calcitrans and their relationship with the blood meal destination. *Proceedings of the 1st international symposium and 2nd national symposium on Hemoparasites and their vectors*, Caracas (Venezuela), 2004; 50-51.

[56] Kunz SE, Monty J. Biology and ecology of Stomoxys nigra Macquart and Stomoxys calcitrans (L.) (Diptera, Muscidae) in Mauritius. *Bull. Entomol. Res.*,1976; 66: 745-755.

[57] OIE. *Trypanosoma evansi infection (Surra).* OIE (World Organisation for Animal Health), Paris, http://www.oie.org/ 2010.

[58] Monzon CM, Mancebo OA, Roux JP. Comparison between six parasitological methods for diagnosis of Trypanosoma evansi in the subtropical area of Argentina. *Vet. Parasitol.,*1990; 36: 141-146.

[59] Lun ZR, Min ZP, Huang D, Liang JX, Yang XF, Huang YT. Cymelarsan in the treatment of buffaloes naturally infected with Trypanosoma evansi in South China. *Acta Trop.*,1991; 49: 233-236.

[60] Dia M, Desquesnes M. Infections expérimentales de bovins par Trypanosoma evansi: pathogénicité et efficacité du traitement au Cymelarsan®. Revue AFSSA de Santé et de *Productions Animales,*2007; 5: 37-41.

[61] Gutierrez C, Corbera JA, Bayou K, van Gool F. Use of cymelarsan in goats chronically infected with Trypanosoma evansi. *Ann. N. Y. Acad. Sci.*,2008; 1149: 331-333.

[62] Luckins AG. Trypanosoma evansi and T. equiperdum an overview. In: Proceeding Premier Séminaire International sur les Trypanosomose Animales non Transmises par les Glossines, Annecy (France), 18, 1992.

[63] Claes F, Büscher P, Touratier L, Goddeeris BM. Trypanosoma equiperdum: Master of disguise or historical mistake?. *Trends Parasitol.*,2005; 21: 316-321.

[64] Shaw APM. Economics of African trypanosomiasis. In: Maudlin I, Holmes PH, Miles MA, editors. *The Trypanosomiases*. Trowbidge, Cabi publishing, 2004; 369-402,.

[65] Swallow BM. Impacts of trypanosomiasis on African agriculture. PAAT Technical and Scientific Series 2. Roma: Food and Agriculture Organization (FAO), 2000.

[66] Da Silva AS, Garcia Perez HA, Costa MM, Franca RT, De Gasperi D, Zanette RA, Amado JA, Lopes ST, Teixeira MM, Monteiro SG. Horses naturally infected by Trypanosoma vivax in Southern Brazil. *Parasitol. Res.*,2011; 108: 23-30.

[67] Dávila AM, Silva RA. Animal trypanosomiasis in South America. Current status, partnership, and information technology. *Ann. N. Y. Acad. Sci.*,2000; 916: 199-212.

[68] Seidl A, Moraes AS, Silva RAMS. A financial analysis of treatment strategies for Trypanosoma evansi in the Brazilian Pantanal. *Prev. Vet. Med.*,1998; 33: 219–234.

[69] Manuel MF. Sporadic outbreaks of Surra in the Philippines and its economic impact, *J. Protozool. Res.*,1998; 8: 131–138.

[70] Dargantes AP, Mercado RT, Dobson RJ, Reid SA. Estimating the impact of Trypanosoma evansi infection (surra) on buffalo population dynamics in Southern Philippines using data from cross-sectional surveys. *Int. J. Parasitol.*, 2009; 39: 1109-1114.

[71] Dobson RJ, Dargantes AP, Mercado RT, Reid SA. Models for Trypanosoma evansi (surra), its control and economic impact on small-hold livestock owners in the Philippines. *Int. J. Parasitol.*,2009; 39: 1115-1123.

[72] Toro M, Leon E, Lopez R, Pallota F, Garcia JA, Ruiz A. Effect of isometamidium on infections by Trypanosoma vivax and Trypanosoma evansi in experimentally infected animals. *Vet. Parasitol.*,1983; 13: 35-43.

[73] Silber AM, Bua J, Porcel BM, Segura EL, Ruiz AM. Trypanosoma cruzi: specific detection of parasites by PCR in infected humans and vectors using a set of primers (BP2) targeted to a nuclear DNA sequence. *Exp. Parasitol.*,1997; 85: 225-232.

[74] Grant IF. Insecticides for tsetse and trypanosomiasis control: Is the environmental risk acceptable?. *Trends Parasitol.*,2001; 17: 10-14.

[75] Bauer B, Amsler-Delafosse S, Clausen P, Kabore I, Petrich-Bauer J. Successful application of deltamethrin pour on to cattle in a campaign against tsetse flies, Glossina

spp in the pastoral zone of Samorogouan Burkina Faso. *Trop. Med. Parasitol.*,1995; 46: 183-189.

[76] Gilles J, David JF, Duvallet G, De La Rocque S, Tillard E. Efficiency of traps for Stomoxys calcitrans and Stomoxys niger niger on Reunion Island. *Med. Vet. Entomol.*,2007; 21: 65-69.

In: Livestock: Rearing, Farming Practices and Diseases
Editor: M. Tariq Javed
ISBN 978-1-62100-181-2

Chapter 7

SURVEILLANCE AND MANAGEMENT OF TRYPANOSOMIASIS IN CATTLE HERDS IN KAURU AREA, KADUNA STATE, NIGERIA

***Felicia. N. C. Enwezor,*[a,*] *Balaraba Bello,*[a] *Abdullahi Kalgo,*[a] *and Lamido Tanko Zaria*[a, b]**

[a]Nigerian Institute for Trypanosomiasis and Onchocerciasis Research (NITOR),P.M.B. 2077, Kaduna, Kaduna State, Nigeria

[b]Present address: Faculty of Veterinary Medicine, Department of Veterinary Microbiology, University of Maiduguri, Nigeria

ABSTRACT

Surveillance for tsetse and cattle trypanosomiasis distribution aimed at instituting control measures was undertaken in parts of Kauru Local Government Area in Kaduna State, Nigeria between late February to early March, 2007 and October, 2008 following claims of severe trypanosomiasis which had disrupted farming activities, devastated communal cattle industry and flight of 32 cattle herders with over 70 herds. A total of 964 cattle selected at random in nine sampling sites were each bled from the jugular vein and blood collected. Dark ground buffy coat and Giemsa stained thin film were methods used in diagnosis of trypanosomes. Anemia was quantified by packed cell volume assessment using the hematocrit method. Traps were set for tsetse catch. The knowledge, attitude, perception and treatment seeking behaviors (KAPTSB) and economic losses were assessed using questionnaire administered on 36 cattle herders, focus group discussions held with selected cattle herders and in-depth interview with the Head, Agriculture Department. Overall, 210 cattle were found positive with an infection rate of 22percent. The infection rates in the dry and wet seasons were 23.6 and 20.1percent, respectively, indicating increased trypanosome prevalence in the dry season. We detected 203 *Trypanosoma vivax* and 7 *T. congolense* infections and these differed significantly ($P< 0.05$) across the sampling sites. Out of 964 cattle, 395 (41%) were found anemic with significant variation across the sampling sites ($P < 0.05$). The traps captured few *Glossina*

* Correspondence: Dr. Felicia N.C. Enwezor. Telephone: +2348077252297. E-mail address: feliciaenwezor@yahoo.com.

palpalis palpalis and *G. tachinoides* with lots of Tabanidae. The tsetse flies did not harbor trypanosomes. From the qualitative data, it became apparent that the cattle herders, though very much aware of trypanosomiasis problem, were faced with the challenge of inappropriate health care seeking behavior due to ignorance about where to seek for the service as a result of unavailability of veterinary clinics in the study districts. The 61percent of the respondents procured veterinary drugs from the open markets and resorted to quacks, while the rest said they relied on traditional remedies. The Local Government on its part lacked reliable information for planning, management and decision making against the disease. This research provided data on the magnitude of trypanosomiasis and linkage between community and Government. The Government treated cattle in all the districts and brought down the incidence of trypanosomiasis. It recommended the strategies for sustainable control to enhance livestock productivity, which is critical to poverty reduction and creation of wealth.

1. Introduction

Pathogenic African trypanosomes present a very significant group of parasitic infections and each year the parasites cause huge morbidity and mortality [1]. The disease in bovines, equines and small ruminants is caused by three species of *Trypanosomes*, viz, *Trypanosoma brucei brucei*, *T. congolense* and *T. vivax*. The *T. simiae* affect pigs. These parasites are transmitted by tsetse flies of the genus *Glossina* spp. The word ‘tsetse’ is generally believed to have originated from the Tswana language in Botswana, meaning ‘fly that kills livestock’ [2]. *T. vivax* and *T. simiae* could also be transmitted mechanically by horseflies (Diptera, Tabanidae) and other bloodsucking insects. *T. evansi* mainly infects camels in which the horse flies spread the disease. It causes the disease named Surra that is endemic in parts of Africa, Asia and South America. Besides camels, other domesticated livestock are affected [3].

Their pathologic effects in livestock have been reviewed and grouped as either hematic (congolense, vivax, simiae) or humoral (*brucei*) [4]. The diseases caused by the hematic group are largely due to anemia. The anemia also occurs with *brucei* infection; the pathology has more often been associated with tissue degeneration and inflammation [4]. *T. brucei* infections in horses and small ruminants (sheep and goats), could cause central nervous systems involvement, including staggering gait and paralysis [3]. Trypanosomiasis results in severe wasting disease of cattle (Nagana).

Tsetse termed the ‘poverty insect’ [5] is found in the African continent covering an approximated area of 10,000,000 km^2. The tsetse and trypanosomiasis problems constitute serious impediments to both human beings and their livestock. The impact on livestock not only reduces the availability of beef and dairy produce but also affects work efficiency of draft animals and mixed farming [6]. The implication is vulnerability to food shortages, starvation and famine. Studies indicated that if draught animals were available, a family dependent on manual labor alone could increase its income from agricultural work by 45percent per unit of land and 143percent per unit of labor [2, 6]. Estimates by Steelman and FAO showed that overall agricultural losses due to trypanosomiasis for the whole Africa were more than US$ 4 billion per annum [7, 8]. This figure was later corroborated by Budd in which he projected that agricultural benefits accruing to tsetse elimination could reach US$4.5 billion per annum [9]. Animal trypanosomiasis, therefore, impedes livestock and mixed farming production systems including socio-economic development in much of sub-Saharan Africa [1, 10].

It is evident from the above that the threat to health and developmental constraints imposed by trypanosomiasis justifies the need for continued epidemiological study of the disease. So also, the current efforts on a regional basis aimed at achieving a complete annihilation of tsetse and trypanosomiasis from Africa by the Pan African Tsetse and Trypanosomiasis Eradication Campaign (PATTEC) of the African Union needs to be applauded. Before now, there has not been any study conducted in Kauru area that assessed the status of trypanosomiasis. This chapter reports the results of a longitudinal study, which showed how research could be used to impact positively towards improving cattle industry and alleviating poverty in a village community in Northern Nigeria.

2. Rationale for the Study

Over 66percent of the populace, the nation's main food producers, live in rural areas where they engage in agricultural production for their livelihood. Animal trypanosomiasis makes livestock farming extremely difficult and has remained a major constraint on production [2, 11]. The disease is aptly described as the most singular disease that has had tremendous influence on farmers deciding where to settle for mixed agriculture [1]. This scenario was the case for Kauru area, Kaduna State, Nigeria, where 32 cattle owners deserted their farmlands and fled with their families and over 70 herds for settlement elsewhere. The worst hit districts were Kadage, Kware and Kagadama where tsetse and trypanosomiasis allegedly disrupted farming activities and had devastated effects on communal cattle industry. No fewer than 450 cattle died of suspected trypanosomiasis from these districts in the past five years (Chief of Kadage, per. Communication). These developments no doubt have serious implications on both local and national economies given the agrarian nature of the state with great potentials for livestock industry. Information generated provided data for planning, developing strategies and implementing control interventions required to curb the problem of cattle trypanosomiasis on agricultural production and avert further disruption of socio-economic activities in the area. This research work fits into the present efforts of the Nigerian Government to eradicate extreme poverty, create jobs and ensure food security for the teaming populace as well as better economy for the affected rural communities of Nigeria.

3. African Trypanosomiasis

African trypanosomiasis is included on the list of 13 Neglected Tropical Diseases (NTDs), is caused by tsetse-transmitted African trypanosomes. The disease is peculiar to Africa affecting both humans and animals and causes fatal sleeping sickness in people and 'nagana' in livestock (a fatal wasting disease).

African sleeping sickness is scientifically called as human African trypanosomiasis (HAT). The disease occurs in 36 African countries where 60 million people are at risk [12]. Each year, between 300,000 and 500,000 individuals are affected with 100,000 deaths [13, 14]. The endemicity levels in the afflicted countries have been provided [15]. Nigeria is now classified as low endemic area with 146 HAT cases reported between 1999 and 2000 [16] with epidemiological shift to the South of Nigeria, especially in parts of Delta State [17, 18].

Retrospective accounts showed that HAT had caused three severe pandemics; one about the end of the nineteenth century, the second in the 1920s and Nigeria was among the countries severely affected [19, 20]. The third occurred in the 1970s in which Angola, Democratic Republic of Congo (DRC) and Sudan were affected [21]. Other countries in recent years were Central African Republic, Congo Republic, Tanzania and Uganda [12]. Contributing factors to the upsurge were somehow linked to a collapse of infrastructure arising from political instability or health policy change [12]. These had dismantled the infrastructure of disease surveillance and vector control in many regions of sub-Saharan Africa and resulted in great exposure to infected tsetse. Reversal of this trend in these countries may be facilitated by a return of civil order and rehabilitation of disease surveillance and treatment programs [21].

Nigeria through NITOR and the Federal Ministry of Health received funding from WHO and the Foundation for Innovative New Diagnostics (FIND) in 2010, to improve sleeping sickness awareness, diagnosis and treatment in four selected states aimed at enhancing the local disease control infrastructure. These activities considered important in forestalling a possible future outbreak of sleeping sickness began first in Delta State where 20 communities were surveyed in July 2010.

3.1. Nagana

Nagana or African animal trypanosomiasis (AAT) is endemic throughout the humid and semi-humid zones of sub-Saharan Africa co-incident with the distribution of tsetse, which infest an area covering 10 million km^2 [22, 23, 24]. In Nigeria, AAT remains a major threat to livestock in all the agro-ecological zones with cases recorded from the zones [25]. The situation was not different from other African countries including Uganda [25, 26]; Kenya [27] and Burkina Faso [28] where severe trypanosomiasis had been reported. Overall infection rates in the range of 3 to 18percent had been reported in Nigeria by several workers [29, 30, 31, 32, 33, 34, 35, 36, 37]. However, other studies of outbreak situations had reported a much higher prevalence of 40 to 50percent in parts of North Central [38] and South East [39] Nigeria. The problem of AAT had been on the increase due to bites of tsetse and other hematophagous flies and lack of active surveillance of the disease, which was increasingly jeopardized by lack of funds [40].

4. Trypanosomal Parasites

Trypanosomes in the family Trypanosomatidae belong to the group of flagellated protozoa called kinetoplastids. Three distinct kinetoplastids cause human disease, viz, *T. brucei gambiense* and *T. b. rhodesiense* (the causative agents of sleeping sickness); *Trypanosoma cruzi* (Chagas' disease or American trypanosomiasis) and *Leishmania* spp. (leishmaniasis). Also three distinct kinetoplastids are responsible for trypanosomiasis in large and small ruminants, viz, *T. brucei brucei, T. congolense* and *Trypanosoma vivax* producing single or co-infections [41]. *Trypanosoma simiae* and *T. suis* cause trypanosomiasis in pigs and *T. equiperdum* in equines. All the pathogenic African trypanosomes, the causative agents of trypanosomiasis belong to the genus Trypanosoma and are heteroxenous, i.e., transmitted by vectors except *T. equiperdum* the cause of dourine that is sexually transmitted. Much research has been undertaken on this genus because of its extreme importance to the health of man and livestock. The morphology, life cycle, epidemiology, pathology and symptomatology have been described in many standard text books [41, 42] while the phenomenon of antigenic variation by trypanosomes and their suppression of hosts' immune responses have been the subject of several reviews [43, 44]. Control by vaccination has not been possible owing to the problem of antigenic variation and prospects seem poor [44]. The Table 1 below highlights the major morphological features of the three important trypanosome species that cause disease in cattle.

Table 1. Morphological characteristics of trypanosomes

Species	Site of development in the tsetse fly	Free flagellum	Kinetoplast	Undulating membrane	Size (μm)	Motility
T. vivax	Proboscis	Large, present	terminal	Not prominent	20-26	Large, extremely active, traverses the whole field very quickly
T. brucei	Mid gut/salivary gland	Present in all but stumpy forms	Small, sub terminal or central	Prominent	13-35	Large, rapid movement in confined areas.
T. congolense	Mid gut/ proboscis	Absent	Medium, sub terminal	Not prominent	9-18	Small, sluggish, adheres to red blood cells by anterior end
Adapted from	Hoare	1972.				

5. Study Objectives

5.1. Main Objective

To determine the extent and impact of tsetse and trypanosomiasis on socio-economic activities with a view to instituting and recommending sustainable intervention measures.

5.2. Specific Objectives

a. To determine tsetse species and their apparent density in the study area.
b. To determine trypanosome prevalence in cattle in the study area.
c. To document farmers knowledge, attitude and perception and treatment seeking behavior (KAPTSB) and socio-economic impacts using questionnaire .
d. To institute intervention measures for control of the disease.

e. To evaluate intervention effectiveness.
f. To formulate strategies and opportunities for sustainable control of trypanosomiasis.

6. METHODOLOGY

6.1. Study Design

The study was a longitudinal one designed to determine the status of tsetse and cattle trypanosomiasis through quantitative and qualitative data collection as a basis for developing a strategy for control in order to increase livestock productivity and poverty alleviation. Approaches to the study included engaging the Veterinary section of the Agriculture Department at the Local Government (LG) and the study districts and communities through advocacy visits. Community participation in sensitization and mobilization helped to secure acceptance and consent for the study. Field surveys to obtain quantitative data on tsetse and trypanosome prevalence and qualitative data on KAPTBS by interviews with the Head, Veterinary Section Unit, Agriculture Department and village leaders of the study communities and cattle herders were undertaken. Tsetse and trypanosomiasis surveillance were conducted during the dry (late February, 25th to early March, 4th) and rainy (October 2008) seasons.

6.2. Study Area

Kaduna State is one of the 36 States of Nigeria and is located in the North West part of Nigeria. The State shares boundaries with Katsina and Kano States to the north, Plateau State to the east, Nassarawa to the southeast, Niger State to the west and the Federal Capital Territory Abuja (FCT) to the south. Kaduna State is made up of 23 Local Government Areas (LGAs) with a population of 6.1 million (2006, Census). The State is essentially an agrarian society, which has the potentials for the livestock industry and fisheries. It also has the potentials to grow arable (crops) and tree. However, there is increasing poverty where an estimated 70percent of the populations, which are the rural dwellers, live on less than one US dollar a day [45].

The study LGA is Kauru with headquarters in the town of Kauru. It has an area of 2,810 km^2 and a human population of 170,008 as at the 2006 census. The animal population size as at 2001 was 106,470 cattle; 1800 sheep and 1665 goats [46]. There are numerous rivers and streams, which provide sources of water for livestock. The cattle were owned by pastoralists who had settled in the study area for several years and cattle management was traditional. The Kauru LGA was once in Saminaka LGA, a known human African trypanosomiais (HAT) / sleeping sickness endemic focus before the splitting of the latter into two to create Kauru and Lere LGAs. Both LGAs are also endemic for Onchocerciasis [47]. The rationale for the study has been provided and formed the basis for selection of the LGA and the districts of study, namely Kadage and Kware. Only Kagadama was selected at random from a list of other districts that expressed interest to participate in the study during the wet season survey.

6.3. Study Population

The study populations were made up of the key officials of Agriculture Department and were the Head of Department of Agriculture and the Head, Veterinary Section. At the district level, the district heads and at the community level, the village heads, the herd owners selected and their herds. The herds studied were resident. Prior to the study; we got the ethical clearance from Kaduna State Ministry of Agriculture. We also obtained informed consent for specimen collection and interviews from the Chairman, Kauru LGA; district heads, village leaders, Fulani leaders (the ardos) and herders.

6.4. Entomological Surveys

Entomological surveys were carried out in selected sites along the streams in the study area in order to assess the tsetse species, their apparent density and other biting flies in relation to the occurrence of trypanosomiasis. Entomological data was collected twice during the study period; in the dry and wet seasons. Tsetse flies were caught with traps and hand net when approaching people or cattle or while resting on vegetation. A total of 40 traps baited with acetone (20 each in dry and wet seasons) were deployed before sunrise in the morning alongside streams of uwargurjiya, wawan rafi, likarbu, rafin magaji, rafin kabari, rafin kogi-kaduna-fulani-jin-biri, etc. and kept in positions for 48 hours before removing and mounting to other locations. The flies caught in each trap were counted, identified, analyzed and dissected. Tsetse fly was identified based on the characteristic morphology [48] and other biting flies, especially on wing venation structure and proboscis at the genus level [49]. Sex was determined by observing the posterior end of the ventral aspect of the abdomen by hand lenses. Consequently, male flies were identified by enlarged hypophygium in the posterior ventral part of the abdomen. The apparent density of tsetse fly was given as number of tsetse catch per trap per day [50]; and determined as follows:

$$\text{Apparent density} = \frac{\text{No. of flies caught}}{\text{No. of traps set x No. of days of trapping}}$$

6.5. Animal Sampling and Diagnosis

The study LGA and districts were purposively selected based on the reported presence of tsetse and trypanosomiasis problems, while random sampling was used in selecting the herdsmen and their herds. During the dry season survey, four communities of which two were purposively selected (Likarbu and Uwargurjiya) and the other two (Kabari and Kagura) were selected at random. In the wet season, five communities were studied, namely, Likarbu, Uwargurjiya and the other three Kagadama, Matakarko 1 and Matakarko 2 were selected at random. The communities of Kabari and Kagura could not be visited due to inaccessibility occasioned by flood and swamps. Sample size was determined based on the expected prevalence of 9percent, allowable error of 3percent and confidence level of 95percent using the already described method [51]. Following which a minimum total sample size of 696 cattle were required with the equation below [51].

$$n = \frac{D^* Z^2_{1-\&/2} P(1-P)N}{d^2(N-1) + Z^2_{1-\&/2} P(1-P)};$$

where D = design effect =

$$\frac{\textit{variance of cluster sampling}}{\text{var. of simple random sampling}}$$

$$D = 2.0, Z^2_{1-\&/2} = (1.96)^2,\ P = 9\% = 0.09;\ N = 106,470,\ d = 3\ (\%) = 0.03;$$
$$= \frac{2 * (1.96)^2 * 0.9 * 0.91 * 106470}{(0.03)^2 * 106469 + (1.96)^2 * 0.9 * 0.91} = 695.76;\ n = 696.$$

Cluster sampling was applied in which herds were considered as clusters and selection was randomly drawn from a list of cattle owners who consented to participate. Based on the calculated sample size, n, a total of 964 samples were taken comprising 461 and 503 in the dry and wet seasons, respectively. During sampling, information on age, sex, breed, herd composition, grazing patterns and treatment seeking behaviors was obtained using questionnaire for each cattle bled. Within the herd, the cattle were drawn at random without replacement. Between 10 and 15percent of individual herd was sampled depending on the herd size and predisposition of the herd owner. From each animal, five milliliters of blood was taken through the jugular vein into specimen bottles containing anticoagulant (EDTA) dispensed as one-milligram powder per milliliter of blood and sent in cold boxes with ice packs to the field laboratory for analysis. The samples were analyzed using the dark ground buffy coat technique [52] and thin blood films made and examined for trypanosome species differentiation using morphology [53]. Packed cell volume (PCV) was also determined. The age range of the sampled animals was 0 to 10 years and essentially white Fulani breeds, i.e., trypano-susceptible.

6.6. Treatment of Cattle

Cattle with trypanosomes received treatment with diminazene aceturate (dimna vet) at the recommended dose of 3.5 mg/ kg, during the research work with the consent of the owners.

6.7. Focus Group Discussion

This aspect involved a maximum of ten selected herd owners from each study district. It assessed the perception of the herd owners on the occurrence of tsetse and trypanosomiasis, socio-economic status, livestock constraints, herd composition and source and use of trypanocidal drugs, among others. Suggestions were made on strategies to adopt to improve livestock development.

6.8. Data Analysis

Data were analyzed using STATA version 9 in Microsoft Access® database (Microsoft Corporation, USA) [54, 55] and level of significance of 5percent (2-sided) was used to interpret the results.

7. RESULTS

7.1. Entomological Data

The distribution of the tsetse flies in the locations where traps were set is shown in Table 2. The traps captured *Glossina palpalis palpalis* and *G. tachinoides*with lots of biting flies. Altogether, 72 tsetse flies were caught comprising 20 *G. palpalis palpalis* and 52 *G. tachinoides*. Other biting flies caught were 70 stomoxys, 194 chrysops and 8 hematopota, giving a total of 272. Flies were also seen following cattle and most of the catches were made where cattle tracks and foot path cross the streams. The tsetse flies caught were dissected but none was found infected with trypanosomes. The absence of trypanosomes in the dissected flies might not exclude transmission as *T. congolense* infections observed showed evidence of biological transmission, more so the surveyed area lies within the tsetse belt. The mean apparent density was 0.63 and 1.18fly/trap/day for the dry and wet seasons, respectively. Limited days of trapping due to logistics and the effect of human activities, for instance, bush burning could account for the low tsetse density observed. Both *G. palpalis* and *G. tachinoides* have been shown to be good transmitters of human and animal trypanosomiasis [50]. These flies together with other biting flies, which were in abundance, could portend a great danger in the area. These biting flies being blood feeders have been considered to be of enormous economic importance due to their roles in the mechanical transmission of animal trypanosomiasis and their nuisance value [56, 57]. The biting flies, when in large numbers, among man and his livestock could cause intense provocation and may contribute to loss of weight, high mortality, migration and disruption of farming activities.

Table 2. Tsetse Catch (Kauru districts, February-March and October 2008)

Trapping site	No. of traps		Days of trapping		No. tsetse caught		Apparent density (F/T/D)	
	Feb/March	Oct	Feb/March	Oct	Feb/March	Oct	Feb/March	Oct
Uwargurjiya/wawan rafi	4	4	2	2	7	10	0.88	1.25
R. Likarbu	4	6	2	2	5	15	0.63	1.25
R. Magaji	3	2	2	2	4	5	0.67	1.25
R.Kogi/Kaduna/Fula ni jin Biri	3	4	2	2	4	7	0.67	0.88
R. Kaberi	3	2	2	2	3	4	0.50	1.00
R. Dudu	3	2	2	2	2	6	0.33	1.50
Total	20	20	2	2	25	47	0.63	1.18

7.2. Trypanosome Prevalence

Of the 964 cattle examined, 210 were found positive for trypanosomes with an overall prevalence of 22percent. On herd basis, among the 36 cattle herds sampled, trypanosome infections were found in all the herds. The severity of the infection was such that 50percent infection rates were recorded in three of the 36 herds. Seasonal prevalence showed that of 461 cattle examined in the dry season, 109 were found positive with infection rate of 23.6percent, while in the wet season, 101 cattle out of 503 were found positive, giving an infection rate of 20.1percent. Analysis, based on sampling site showed high infection rates in the range of 15 and 40percent, indicative of severity of trypanosomiasis in the area (Table 3). The higher prevalence in the dry season may be related to the continuous exposure of cattle to tsetse at river banks where pastures and access to drinking water provided contacts given the traditional system of husbandry whereby cattle trekked distances in search of pastures and in so doing enter the riverine tsetse areas. This could contribute to stress, which in itself is a predisposing factor to increased susceptibility and higher infection rate. Season as found out contributed to a higher trypanosome infection rate as more infections were observed in the dry compared to the wet season in the area. Some workers have reported similar findings and showed that the dry season effect was influenced by the nature of the production system, which relies on the traditional mixed crop-livestock pattern [58]. Thus farming system is important consideration in planning disease intervention. Trypanosome species encountered were *Trypanosoma vivax* (203) and *Trypanosoma congolense* (7) for which the infection rates differed significantly ($P< 0.05$) in all the sampling sites. Similar findings with *T. vivax* accounting for majority of all infections have been reported [35, 37, 58]. *T. vivax* spread cyclically and mechanically. Therefore, the predominance of *T. vivax* infection could be due to a greater ease of mechanical transmission of this species of parasite by other hematophagous Diptera, such as encountered in this chapter. The team noticed that cattle were infected with ticks. Ticks being ectoparasites are important because of the diseases they transmit.

Table 3. Seasonal prevalence of trypanosomes in the sampling sites (Feb/March, Oct. 2008)

Seasonal distribution	No. of samples	Positive cases[%]	Trypanosome species *T. vivax*,	*T. congolense*
Dry:				
Likarbu	110	46 [41.82]	46	1
Uwargurjiya	123	23 [23]	23	
Kagura	90	17 [15]	15	2
Kaberi	138	23 [16.67]	21	
Sub total	461	109 [23.64]	105	3
Wet:				
Likarbu	195	32 [16.41]	32	
Uwargurjiya	157	39 [24.84]	32	2
Kagadama	41	11 [26.83]	11	
Matarkako 1	70	13 [18.57]	13	2
Matarkako 2	40	10 [25]	10	
Subtotal	503	101 [20.08]	94	4
Total	964	210 [21.78]	203	7

7.3. Packed Cell Volume

Of the 964 cattle examined, 395 (41%) were found anemic, while569 (59%) had normal PCV values (> 24%). Mean total PCV values were 22.19 ± 4.82percent for trypanosome positive cattle and 25.14 ± 4.55percent for trypanosome negative. The mean PCV values for dry season were 23.35 ± 4.49percent and 25.62 ± 4.65percent, respectively for trypanosome positive and negative cattle, while in the wet season, the mean PCV recorded were 20.94 ± 4.87percent for trypanosome positive cattle and 25.62 ± 4.65percent in trypanosome negative cattle. Analyzing the variability of PCV of the cattle across the sampling sites showed significant difference at 95% Confidence limit ($F = 9.031$; $df = 8$, $P = 0.000$; $P < 0.05$). Anemia characterized by depressed PCV (%) values is a common feature of hemo-protozoan infections including trypanosomiasis as a result of destruction of red blood cells and inhibition of hemopoiesis [59, 60]. The trypanosome species recorded in this chapter are grouped as hematic, as diseases they cause are largely due to anemia [4]. Mechanisms of trypanosomal anemia have been advanced; prominent among which is immunologic mechanism arising from cleavage of erythrocyte surface sialic acid by neuraminidase produced by trypanosomes. This renders the red blood cells to erythro-phagocytosis [61]. This can explain the findings whereby 41percent of the cattle had anemia. It is also known that *T. vivax* produces virulent infection in cattle and coupled with anemia, which is frequently a significant cause of morbidity and mortality in trypanosomiasis [60], that could explain the huge deaths and devastation it caused to the cattle industry in the area.

7.4 Knowledge, Attitude, Perception and Treatment Seeking Behaviour of Respondents

A total of 36 cattle herders were interviewed using questionnaire. All (100%) of the respondents expressed knowledge of trypanosomiasis which they called 'ciwon barci' in Hausa or 'samore' in Fulfudae and said that the disease was a major cause of re-location of cattle herders out of the area. On how they got information about the disease, the majority (72.22%) of the respondents said they experienced it in their animals 15 years ago and that the disease had persisted since then, causing high mortality. As to the cause, the majority of the respondents (77.78%) said it could be tsetse, which resides by the river side, a few others (13.9 and 8.3%) said overcrowding and fertilizer poisoning, respectively. Most of the respondents (83.33%) perceived trypanosomiasis as a disease problem that has to come to stay in the area without any hope of elimination. From the qualitative data, it was apparent that the cattle herders were faced with the challenge of inappropriate health care seeking behavior due to ignorance about where to seek for the service as a result of unavailability of veterinary clinics in the study districts. A good number of the respondents (61.1%) procured veterinary drugs from the open markets and resorted to quacks. The rest said they used local remedy, which they administered thrice for three days but were reluctant to disclose the composition/nature of the local treatment. These practices might have contributed to the high risk of trypanosomiasis and the persistence of the disease in the area. Other disease problems mentioned were fasciolasis and helminthiasis in young cattle.

7.5 Socio-Economic effects

Most of the respondents (80.44%) recounted high mortality in cattle and desertion of community by some cattle herders and their families, fleeing with them with over 70 herds. The focus group discussion also corroborated the respondents' assertion by providing data on number of families that migrated from the area as follows: 20, 7 and 5 families from Likarbu, Matarkako and Uwargurjiya, respectively. This they pointed out brought untold social misery and inexplicable psychological effects. Data on the number of trypanosome-attributed deaths in cattle in the past five years were also provided: Kabari 110, Kagura 45, Likarbu 183, Matarkako 48 and Uwargurjiya 75. In Likarbu, a herd owner who said he lost 50 cattle to samore, i.e. trypanosomiasis, named one of his sampled cattle samore. When probed, he said that since youth, the cattle had been afflicted with samore and had had only one calving with much difficulty. Clinical symptoms mentioned during the FGD included: weakness, severe weight loss, geophagia (soil eating) even when apparently healthy, watery stool with blood at the onset, watery discharge from the eyes, eat bones and metals, enlarged lungs and spleen when slaughtered at autopsy, etc. Cattle with these symptoms they said could not add value or be of marketable quality. The anemia was found in 41percent of the cattle sampled and could account for not only the huge deaths reported but also in weakness and extensive tissue degeneration/weight loss with negative economic impacts. The socio-economic effects of trypanosomiasis include:

i). Direct losses due to death of animals including costs of trypanosomiasis control.
ii). Decrease in livestock farming through desertion of the area.
iii). Limits livestock productivity and agricultural production.
iv). Wasting disease which affects quantity and quality of meat, milk and dairy products.
v). Hinders both rural and national economy.

7.6 Interview with Head of Agriculture Department

The interview made it clear that the Local Government was not aware of the trypanosomiasis problem in the area and lacked reliable information for planning, management and decision making against the disease. The impact of the government was felt in crop agriculture by making available fertilizer and other farming implements to farmers as disclosed by the HOD Agriculture Department. Prior to this study, the veterinary section in the Agriculture Department appeared to exist in name only with no serious consideration on animal health care.

7.7. Intervention Measures for Control of Trypanosomiasis

The prevalence of trypanosomiasis greater than 20percent was observed in both dry and wet seasons and that showed the magnitude of trypanosomiasis. Data including implications were made available to the Agriculture Department. Armed with this information, the Government immediately planned and mobilized resources to the sum of $22,580 used in procuring veterinary drugs- trypanocides, acaricides, etc. and got all cattle in the entire districts treated. The veterinary staff divided themselves into groups to effectively conduct the task of treatment. Therefore, following this surveillance the treatment intervention was implemented.

7.8. Evaluating Effectiveness of the Intervention Measures

The evaluation was conducted one month after the mass treatment of cattle. All treated cattle were reported by their owners as having recovered and feeding well. Their perception of wellbeing was expressed largely in terms of the tangible improvements seen in their cattle looking very healthy, decreased mortality and increase in herd size through calving. The intervention was effective because the Government supported treatment of all cattle in the districts and trypanosomiasis was brought down.

8. Strategies and Opportunity for Sustainable Control of Trypanosomiasis

The Kauru area is very suitable for cattle farming in terms of availability of fodder round the year. There is hardly any forage shortage and numerous water bodies are scattered in the area. The Zebu breed being the dominant breed requires improved management strategies, parasite and vector control to drastically curb trypanosomiasis and other disease problems in the area towards improved cattle health and production. Good sanitation of stables is essential to avoid chances of breeding of flies. Control measures should target destroying breeding places by regular removal of moist hay and feces. Insecticide spraying, pour-on or dipping methods can be applied to animals to kill the flies and spraying of stables with insecticides at frequent intervals can effectively control these flies and ticks and reduce the density in the area. For parasite control, the Government should assume a greater role in monitoring and control of major animal diseases, such as trypanosomiasis; and provision of veterinary clinics. So far, there are no veterinary clinics around and the Veterinary Section of the Local Government is not readily accessible, which is at a more than 10 km distance from the study area. Lack of contact could adversely affect outbreak reporting and disease surveillance in general. This in turn can complicate the planning and implementation of effective disease control policies.

The opportunity afforded by the research was the acquisition of data on prevalence and the burden of trypanosomiasis and linkage between the community and the Government. The synergy thus provided became manifested by the willingness of the Government to support targeted veterinary care for improved animal health in the area and community's acceptance to seek appropriate livestock health interventions. The relationship established between the veterinary section of the Government and cattle owners is good and need to be encouraged. The Government through the Ministry of Agriculture and Rural Development in conjunction with PATTEC Nigeria-Office should take a leading role to effectively control animal trypanosomiasis in Kauru area and to completely eradicate the tsetse and other heamato-phagous Diptera insect populations to ensure sustainable control of livestock diseases in the area. Improved cattle health will translate to improved family incomes and poverty alleviation through better economy of the affected communities.

9. Conclusions and Recommendations

9.1. Conclusions

The research work has clearly suggested that:

i) Trypanosomiasis constitutes a serious health problem to cattle rearing in Kauru area.
ii) Trypanosomiasis is a major cause of migration and causes devastation to the cattle industry in the area.
iii) Awareness about trypanosomiasis is high and cattle owners know the symptoms of the disease.
iv) Tsetse flies (*Glossina palpalis palpalis* and *G. tachinoides)*are low in addition, other haematophagous flies –stomoxys, chrysops, haematopota are in abundance.
v) Though there was no trypanosome infection seen in the flies dissected, cattle /man fly contact is high.
vi) *Trypanosoma vivax* infection predominate in the area.
vii) Transmission is both cyclical and mechanical with some infections occurring due to *T. congolense*.
viii)Cattle owners practiced inappropriate health care seeking behavior due to ignorance about where to seek for the service as a result of unavailability of veterinary clinics in the area.
ix) Linkage and information provided by this study were vital in getting the Local Government to respond to the needs of the cattle herders through procurement of drugs and mass treatment of cattle in the districts.
x) Infection by trypanosomes is brought down in the districts based on reports of the herd owners.

9.2. Recommendations

The study recommends:

i) Targeted treatment of cattle against trypanosomiasis in the dry season. Restricted treatment is valuable to avoid outbreak of the disease in the area.
ii) Increased funding to the Local Government to be able to assume a greater role in planning and monitoring, surveillance and control of major animal diseases.
iii) Provision of veterinary clinics around the study districts to promote good health care seeking behavior by cattle owners.
iv) Ground spraying of insecticides at contact points and cattle routes to protect both humans and cattle during routine daily activities.

v) Cattle owners should practice good sanitation of stables, insecticide spraying, pour-on or dipping methods can be applied to animals to protect them from the bites of these haematophagous insects.
vi) PATTEC-Nigeria office and the Ministry of Agriculture and Rural Development should assume a leading role in the eradication of tsetse and trypanosomiasis problems for sustainable agriculture and rural development.

ACKNOWLEDGMENTS

This work was financed by the Nigerian Institute for Trypanosomiasis and Onchocerciasis Research (NITOR). We acknowledge support and assistance of Dr. Yusufu Audi, Director, Kaduna State, Veterinary Services, the Chairman, Dr. Paul R. Wani, HOD Agriculture, Alhaji Suleiman Aminu and staff of Kauru Local Government, in particular, Linus Akut, Dr. Billy La'ah and Usman Hassan. We also duly acknowledge the consent, interest and cooperation of the village leaders, the ardos and the cattle owners/herders. We appreciate the excellent technical assistance provided by Mr. Francis Doro. Finally, we thank the Director General and Chief Executive of NITOR, Prof. Mohammed Mamman for revision and permission to publish.

REFERENCES

[1] FAO. Food, *Agriculture and food security: The global dimension*, WFS02/Tech/ Advance unedited version, FAO 2002, Rome pp. 19-28.
[2] FAO. Impacts of trypanosomiasis on African agriculture. *PAATTechnical and Scientific Series* 2, 2000.
[3] WHO. *Strategic Review of traps and targets for tsetse and African trypanosomiasis control.* TDR/IDE/ 2005.1
[4] Losos GJ, Ikede BO. Review of pathology of diseases in domestic and laboratory animals caused by Trypanosoma congolense, T. vivax, T. brucei, T. rhodesiense and T. gambiense. *Vet. Pathol.* (suppl), 1972; 9: 1-71.
[5] FAO. *Food and Agriculture Organization Meeting Report*, Rome, 2009.
[6] Govereh J. Impacts of animal disease on migration, livestock adoption and farm capital accumulation: *Zambesi valley, Zimbabwe. Ph. D th*esis, Michigan State University, 1999.
[7] Steelman CD. Effects of external and internal arthropod parasites on domestic livestock production. *Ann. Rev. Entomol.*, 1976; 21: 155-178.
[8] FAO. A systematic approach to tsetse and trypanosomiasis control. *FAO Animal and Production Health paper,* 121, 1994.
[9] Budd L. Vol. 2: Economic analysis, In: DFID-funded research and development since 1980. Chattam, UK, Department for International Development: *Livestock Production Programme, Animal Health Programme/ Natural Resources Systems Programme*, 1999.

[10] Michel JF, Dray S, De La Rocque S, Desquesnes M, Solano P, De Wispelaere G, Cuisance D. Modelling bovine trypanosomosis spatial distribution by GIS in agro-Pastoral zone o0f Burkina Faso. *Prev. Vet. Med.,* 2002; 56: 5-18.
[11] FAO. Animal Production and Health paper, Rome. *Animal health year book*, 1996, FAO Rome.
[12] *WHO Expert Committee on strategic emphasis for African trypanosomiasis research.* In: TDR 2002.
[13] *World health report: Health systems improving performance*, WHO 2000, Geneva.
[14] *TDR Scientific Working Group*, June 2001. World Health Organization, Geneva.
[15] Meda HA, Pepin, J. Epidemiology and control of human African trypanosomiasis, in: *African trypanosomiasis (sleeping sickness) Scientific Working Group*, 2001, TDR/SWG/01, Geneva, Switzerland.
[16] Simmaro PP, Jannin J, Cattand P. Eliminating human African trypanosomiasis: Where do we stand and what comes next. *Plos Med.* 2008; 5: e55.
[17] Edeghere H, Olise PO, Olatunde DS. Human African Trypanosomiasis (Sleeping sickness). A new endemic foci in Bendel State, Nigeria. *Trop. Med. Parasitol.*, 1989; 40: 16-40.
[18] Elhassan EO, Ukah JCA, Sanda SA, Ikenga MA. *Integration of recent diagnostic techniques into Primary Health Carein Ethiope East Local Government Area.* Paper presented at the annual conference of the Nigerian Society of Parasitologyheld at the Obafemi Awolowo University, Ile-Ife, Nigeria.
[19] Duggan AJ. A survey of sleeping sickness control in Northern Nigeria from the earliest time to the present day. *Trans. Roy. Soc. Hyg.,* 1962; 55: 439-480.
[20] McLetchie JL. The control of sleeping sicknessin Nigeria. *Tran. Roy. Soc. Hyg.*, 1948; 41: 445-451.
[21] WHO Expert Committee on Control and Surveillance of African trypanosomiasis Geneva, World Health Organization, 1998, *WHO Tech. Rep. Ser.*, No. 881.
[22] Hommel PH. New opportunities for the diagnosis and control of animal diseases in the tropics. *Trans. Roy. Soc. Med. Hyg.*, 1991; 85: 163.
[23] FAO Meeting report on harmonization of the activities of PAAT and PATTEC. In: *Tsetse and trypanosomiasis Information Quarterly*, 2002, 5: 12204-12287.
[24] Catley A, Irungu P, Simiyu K, Dadye J, Mwakio W, Kiragu J, Nyamwaro SO. Participatory investigations of bovine trypanosomiasisin Tana River district, Kenya. *Med. Vet. Entomol.*, 2002; 16 (1): 55-66.
[25] Picozzi K, Tilley A, Fevre EM, Coleman PG, Magona JW, Odiit M, Eisler MC, Welburn SC. The diagnosis of trypanosome infections: Applications of novel technology for reducing disease risk. *Afr. J. Biotechnol.,* 2002; 1: 39-45.
[26] Waiswa C, Katungunka-Rwakishaya E. Bovine trypanosomosis in south- western Uganda: Packed cell volumes and Prevalences of infection in the cattle. *Ann. Trop.Med. Parasitol.,* 2004; 98: 21-27.
[27] Njiru ZK, Constantine CC, Guya S, Crowther J, Kiragu JM, Thompson RCA, Davilla AMR The use of ITS1 rDNA PCR in detecting pathogenic African trypanosomes. *Parasitol. Res.*, 2005; 95:186-92.
[28] Lefrancois T, Solano P, De la Rocque S, Bengaly Z, Reifenberg JM, Kabore I, Cuisance D. New epidemiological featurers on animal trypanosomiasis by molecular

analysis in the pastoral zone of Sideradougou, Burkina Faso. *Mol. Ecol.*, 1998; 7: 897-904.

[29] Agu WE, Kalejaiye JO, Olatunde AO. Prevalence of bovine trypanosomosis in some parts of Kaduna and Plateau States, Nigeria. *Bull. Anim. Prod. Afr.*, 1989, 37: 161-166.

[30] Kalu AU. An outbreak of trypanosomiasis on the Jos Plateau, Nigeria. *Trop. Anim. Hlth. Prod. Afr.*, 1991; 23: 215-216.

[31] Daniel AD, Joshua RA, Kalejaiye JO, Dada, AJ. Prevalence of trypanosomosis in sheep and goats in a region of northern Nigeria. *Rev. Elev. Med. Vet. T rop*, 1992; 47: 295-297.

[32] Ahmed, MI, Osiyemi TIO, Ardo MB. Prevalence of bovine trypanosomosis in Damboa Local Government Area, Borno State. *Nig. J. Anim. Prod.*, 1994; 21(1): 186-187.

[33] Ogunsanmi AO, Ikede BO, Akpavie SO. Effects of management, season, vegetation zone and breed on the prevalence of bovine trypanosomiasis in Southwestern Nigeria. *I. J. Vet Med.,* 2000; 55: 54-61.

[34] Abenga JN, Enwezor FNC, Lawani FAG, Ezebuiro O, Sule J, David KM.Prevalence of trypanosomosis in trade cattle at slaughter in Kaduna, Nigeria. *Nig. J. Parasitol.*, 2002; 23: 107-110.

[35] Abenga JN, Enwezor FNC, Lawani FAG, Osue HO, Ikemereh ECD. Trypanosome prevalence in cattle in Lere area in Kaduna State, North central Nigeria. *Rev. Elev. Med. Vet. Trop.*, 2004; 57: 45- 48.

[36] Enwezor FNC, Umoh JU, Esievo KN, Anere JI. Prevalence of trypanosomes in sheep and goats in the Kachia Grazing Reserve of Kaduna State, Northwest Nigeria. *Bull. Anim. Hlth. Prod. Afr.*, 2006; 54: 306-308.

[37] Enwezor FNC, Umoh JU, Esievo KAN, Halid I, Zaria LT, Anere JI.Survey of bovine trypanosomosis in the Kachia Grazing Reserve, Kaduna State, Nigeria. *Vet. Parasitol.,* 2009; 159: 121- 125.

[38] Maikaje D. *Some aspects of the epidemiology and drug sensitivity of bovine trypanosomosis in Kaura Local Government Area of Kaduna State*, Nigeria. Ph. D thesis, Ahmadu Bello University, Zaria, Nigeria, pp 147.

[39] Anene BM, Ezekwe AG. Trypanosomiasis in intensively reared Muturu calves in Nigeria. *Trop. Anim. Hlth. Prod.*, 1995; 27: 229-230.

[40] Onyia JA. African animal trypanosomosis: An overview of the current status in Nigeria. *Trop. Vet.,* 1997; 15: 111-116.

[41] Mulligan HW. *The African trypanosomes*. George Allen and Unwin Ltd London.

[42] *Text book of Parasitology*, Bhatia BB, Pathak KML, Banergie DP.

[43] Steinert M, Pays E. Genetic control of antigenic variation in trypanosomes. *Br. Med. Bull.*, 1985; 41: 149-155.

[44] McCulloch R. Antigenic variation in African trypasnosomes, monitoring progress. *Trends Parasitol.*, 2004; 20: 117-121.

[45] *National Statistics Report*, 2004.

[46] Kaduna State Ministry of Agriculture and Rural Development document.

[47] NOCP. *National Onchocerciasis document.*

[48] Ford J. *The role of the trypanosomiasis in the African ecology*. Oxford, Clarendon Press, 1971.

[49] Haeselbarth E, Segerman J, Zumpt F. The arthropod parasites of vertebrates in Africa south of the Sahara (Ethiopian region), 3 (Insecta excl. Phthiraptera). Pub. S. *Afr. Inst. Med Res.,* 1966; 13: 1-283.

[50] Davies H. *Tsetse flies in Nigeria.* 3rd ed. Oxford University Press, Ibadan. P 340.

[51] WHO.Immunization coverage cluster-survey. Reference Manual (WHO Immunization, *Vaccines and Biologicals* 04.23., 2006.

[52] Murray M., Murray PK, McIntyre WIM: An improved parasitological technique for the diagnosis of African trypanosomiasis. *Trans. Roy. Soc. Trop. Med. Hyg*., 1977; 71: 325-326.

[53] Baker FJ, Silverton RE, Pallister CJ. *Introduction to Medical Laboratory Technology*, 7ed. Arnold, Bounty Press Ltd. 2000.

[54] Statacorp. *Stata statistical software: intercooled stataversion 7.* Stata corporation,College station, Texas.

[55] Gachohi JM, Bett B, Murilla GA. Factors Influencing prevalence of trypanosomosis in Orma Boran (trypanotolerant) and Teso Zebu (trypanosusceptible) crosses in Teso district, Western Uganda. *Bull. Anim. Hlth. Prod. Afr*., 2009; 57: 327-338.

[56] Rahman AHA, Abdon AMO, El-Khidir MF, Hall MJR. Studies on tabanid flies in Sudan. *Sud. J. Vet Res.,* 1990; 10: 1-12.

[57] Dijiteye A, Moloo SK, Foua BI, Coulibaly E, Diarra M, Quattara I, Traore D, Coulibaly Z, Diarra A. Control trial on G. p. gambiensis in the Sudanese zone of Mali using delthamethrin-impregnated traps and screens. *Rev. Elev Med Vet Pay Trop*., 1998; 51(1): 37-45.

[58] Kaikabo AA, Salako MA. Effects of vitamin A supplementation on anemia and tissuepathology in rats infected with Trypanosoma brucei (Federe strain). *Trop. Vet.,* 2006; 24: 46-51

[59] Toma I, Shinggu DY, Ezekiel W, Barminas JT. Effect of intrasperitoneal administration of vitamin C (ascorbic acid) on anaemia in experimental Trypanosoma congolense infected rabbits. *Afr. J. P.Appl. Chem*., 2008; 2: 37-40.

[60] [Jennings FW, Murray PK, Murray M, Urquhart GM. Anaemia in trypanosomiasis: Studies in rats and mice infected with Trypanosoma brucei. *Res. Vet. Sci*., 1974; 16: 70-76.

[61] Esievo KAN, Saror DI, Ilemobade AA, Hallaway MH. Variation in erythrocyte surface and free serum sialic acid concentrations during experimental Trypanosoma vivax infection in cattle. *Res. Vet. Sci*., 1985; 32: 1-5.

In: Livestock: Rearing, Farming Practices and Diseases
Editor: M. Tariq Javed
ISBN 978-1-62100-181-2

Chapter 8

ANTHELMINTIC RESISTANCE A GIANT OBSTACLE FOR LIVESTOCK WORM CONTROL IN CURRENT ERA: A CHALLENGE

Shahid Prawez,[1]* Azad Ahmad Ahanger[2] and Shafiqur Rahman[3]

[1]Division of Pharmacology & Toxicology and
[3]Division of Veterinary Pathology, Faculty of Veterinary Sciences and Animal Husbandry, Sher-e-Kashmir University of Agricultural Sciences and Technology of Jammu,R. S. Pura, JAMMU (INDIA)-181102
[2]Division of Pharmacology and Toxicology, Faculty of Veterinary Sciences and Animal Husbandry, Sher-e-Kashmir University of Agricultural Sciences and Technology of Kashmir, Kashmir

ABSTRACT

Antibiotics are in wide use for the control of microbial infections and as growth promoters in food animals. The indiscriminate use of these antibiotics has given rise to the menace of resistance in bacteria against these antibiotics. In similar fashion, wide spread use of anthelmintics resulted in resistance among parasites which has become a challenge for the clinician. The anthelmintic resistance in parasites has wide implications in terms of animal health and their productivity. Resistance is said to be present when there is a large frequency of microbes within a population able to tolerate doses of a compound than in a normal population. The resistance against a particular antimicrobial agent is also heritable. Among the parasites, nematode infestations are of major economic concern, particularlyin domesticated animals throughout the world. Anthelmintics have been developed with objective to control the parasites and finally reduce the production loss. Unplanned and injudicious use of anthelmintics has resulted in the development of resistance. Anthelmintic resistance has been noted against all classes of anthelmntics. Parasites (nematodes) have developed resistance against most prescribed groups of anthelmintics like benzimidazole through changes in primary target β-tubulin,whereas,

*Corresponding Author: shahidprawez@gmail.com.

resistance against levamisole, morantel, and pyrantel occur through loss or changes in nicotinic-acetylcholine receptors (nAChR).The P-glycoprotein, a drug efflux protein, was over expressed in avermectin resistant nematodes. Main goal of clinician is to minimize the resistance against the anthelmintics in planned ways which include correct dose administration, cyclical changeover of anthelmintics, regular monitoring on development of resistance by *in vivo* as well *in vitro* tests, controlling of livestock density on particular pasture area and use of combination of anthelmintics.

1. History

Anthelmintic resistance by parasites has nowbecome a global issue, and therefore, requiresa sound approach to combat this menace. Modern era of anthelmintics was started with the discovery of phenothiazine in 1940 which is considered as first modern anthelmintic with good efficacy against gastro-intestinal nematodes of sheep [1]. In 1954, another anthelmintic piperazine was introduced and phenothiazine with piperazine was considered as the first generation broad spectrum combination of anthelminitics. In 1960s, the 2nd generation broad spectrum anthelmintic benzimidazole was introduced, showing less toxic organotropic effects. Frequent and injudicious use of anthelmintics in livestock has given rise toresistance [2].

Amongst the parasites, *Haemonchuscontortus* was the first nematode that showed resistance problem to different anthelmintics. It was also observed that once a parasite has developed resistance against one group of anthelmintic, cross resistance against other members of the group would invariably ensue.

2. Introduction

Parasite infestations in domestic animals are responsible for monetary loss as it affects the health and production of animals worldwide. To minimize this economic loss, use of anthelmintics against the parasites was adopted. Intensive injudicious use of anthelmintics results in development of resistantparasites (nematodes), either genetically or physiologically. Anthelminticresistance is defined as “Greater frequency of parasite within a population able to tolerate doses of a compound than in a normal population of the same species and is heritable” [3]. Later on this definition wasexpanded as "A change in the gene frequency of a population, produced by drug selection, which renders the minimal effective dosage previously used to kill a defined portion (e.g., 95%) of the population no longer effective"[4].

Anthelmintic resistance was mainly reported in ruminants and horses infested with gastro-intestinal nematodes and most serious threat were especially in sheep and goats [5, 6]. The resistance problem becomes complex when one parasite may show resistance against more than one anthelmintics [7]. Anthelmintic resistance has been reported against most of the frequently prescribed drugs and, therefore, this is a challenge for clinician to tackle emerging global problem. The main aim, now, is to deal with the parasitic resistance problem against the existing anthelmintic drugs and also to develop new anthelmintic drugs [8, 9].

3. How to Test Anthelmintic Resistance Developed in Parasites

As a consequence of development of resistance and spread to other susceptible parasites, different tests are required to detect existence of resistance problem. These methods include conventional *in vivo,in vitro* and more reliable molecular Polymerase Chain Reaction (PCR) tests [10, 11]. To detect anthelmintic resistance, *in vivo*fecal eggs count reduction test (FECRT) and *in vitro* larval development assay (LDA) are good tests.

3.1. Fecal Egg Count Reduction Test (FECRT)

This test is very simple, practicallyapplicable and most usable one, even at field site. To see the resistance against an anthelmintic, fecal egg count has to perform, before and 10 days after the drugs administration in an experimental animal. Fecal egg count in untreated group has to compare with drug treated group, giving the idea of resistance development [12]. After ten days recommended period of administration of an anthelmintic, 80-95 % reduction in eggs count should be consider as drug is clinically effective.

3.2. Larval Development Assay (LDA)

This test is very similar to culture and sensitivity tests for antibiotics against bacteria [13]. In this, eggs have to be isolated from fecal samples for which we have to see the resistance against a particular anthelmintic. The eggs have to hatch and produce most resistant stage larvae (L3).To check whether the L3 larvae is showing the susceptibility to anthelmintic or not, an anthelmintic has to be poured over agar plate seeded with eggs and to watch the further development to L3 stage.

3.3. Polymerase Chain Reaction (PCR)

The said technique has very high precision and accuracy but the test is exclusively limited for research work. For this technique genetic probe as markers are used, to observe the expression of P-glycoprotein for a particular parasite, and to compare the allelic frequency in resistant isolates versus susceptible isolates against the anthelmintic ivermectin and moxidectin [14].

4. Anthelmintics for the Parasitesof Livestock

Most of the frequently used anthelmintics belong to one of the groups such as benzimidazoles, imidazothiozoles and macrocyclic lactones. The benzimidazoles group from which most of the drugs are in use as anthelmintics, viz. albendazole, fenbend-azole, mebendazole, oxibendazole, cambendazole, etc. and thiabendazole is considered as a prototype drug of the group. The benzimidazoles possess potent anthelmintic activity against gastrointestinal nematodes. Whereas the levamisole belongs to imidazothiazoles, is most frequently in use to check the parasite infestation in livestock as well to improve their immunity. Macrocyclic lactones and their analogs are also potent anthelmintics such as abamectin, milbemycin, milbemectin, doramectin, etc. The ivermectin is a broad spectrum synthetic lactones and act against both endoparasites (nematodes) and ectoparasites (arthropods).

5. Mechanisms of Resistance Developed by Parasites against Anthelmintics

The parasites have developed resistance by reducing the susceptibility against a particular anthelmintic through modifications in drug target, increase in number of drug target, increase of drug efflux, and increase in rate of drug metabolism. Anthelmintic resistance associated with specific target site change is considered as specific mechanism of resistance. Prior to understand the molecular changes occurring in resistant parasites, the molecular mechanism of action of an anthelmintic should be known. Once resistance has been developed against anthelmintic, the mechanistic pathway will help to illustrate the changes that occur in the target of resistant parasites. In the eighties, resistance against benzimidazoles was reported throughout the world and at the same time levamisole resistance also appeared [15, 16].

Benzimidazole produces anthelmintic action through multiple targets such as inhibition of mitochondrial enzyme fumurate reductase, reduces glucose transport and also uncouples oxidative phosphorylation, but the primary anthelmintic activity is binding to β-tubulin leads to inhibition of microtubule polymerization[17, 18]. The development of resistance against benzimidazoles is associated with change in β-tubulin (primary target). Resistant parasites reveal changes at amino acid position such as phenylalanine200 tyrosine, phenylalanine167 tyrosine and glutamic acid198 alanine [19, 20, 21, and 22]. The change in respective position of amino acid is considered to be due to change in genetic sequence known as single nucleotide polymorphism (SNP) which results in amino acid changes in genetic sequences. The single nucleotide polymorphism (SNP) in *Haemoncus contortus* is the sequential change from TTC to TAC, leads to benzimidazoles resistance [23]. The changes persuade the expression of amino-acid tyrosine in place of phenylalanine and as a result reduction in binding capability of drugs to their respective target β-tubulin [20]. It has been seen that resistance against benzimidazoles is largely related to allelic changes in β-tubulin having low binding capacity for the drugs. The resistant parasites also modulate ATP-binding cassette (ABC) transporters such as P-glycoprotein efflux mechanism that contributesto development of resistance against benzimidazoles [24, 25].

The macrocyclic lactone anthelmintic like melbemycins (moxidectin), ivermectin, and abamectin inhibit the parasites through glutamate gated chloride channels expressed in pharyngeal muscle cells of nematodes resulting in paralysis [26]. Single nucleotide polymorphism (SNP) is also responsible for resistance against anthelmintics, macrocyclc lactone (e.g., ivermectin) and endogenous glutamate after change in gene encoding CTT_{256} to TTT_{256} which resultsin reduction of susceptibility to anthelmintics. Resistant *Haemoncus contortus* exhibit resistance against ivermectin and moxidectin after alteration in gene encoding ATP-dependent P-glycoprotein transporters, which lead to over expression of drugs efflux proteins [14, 27].

Anthelmintics those act through nicotinic acetylcholine receptors (nAChR) are levamisole, morantel, pyrantel and oxantel causing spastic paralysis of the worms [28]. The nicotinic acetylcholine receptors in nematodes have three subtypes which areexpressed mainly in the pharynx and head muscles [29]. The resistance against levamisole develops after the loss of L-subtype nicotinic acetylcholine receptors, which reduces the levamisole sensitivity in *Oesophagostomum dentatum* [30].

6. How to Combat Resistance Problems

The major drawback of using antibiotics and anthelmintics is development of resistance. To control microbial is perhaps not that difficult as to control the anthelmintic resistance. Also vaccines for bacterial infections are easy to prepare but these are not so simple to produce against worms, because nematodes have genetic diversity, large population size and high genetic mutation rates. The central idea is to control the resistance problems against existing anthelmintics through proper management such as:

1. Only therapeutic dose of anthelmintics should be used by weighing the animals. The recommendation is to avoid the under dosing of drugs, which otherwise increases the survivability of resistant parasites thus creating a congenial environment for the proliferation of resistant strains [31].
2. It is better to start therapy with narrow spectrum anthelmintics, and broader spectrum should be attempted only when narrow spectrum fails.
3. The existing anthelmintics should be used on rotation basis to slow down the resistance development. The annual rotation of anthelmintics should be exercised, because frequent change of drugs may retard the development of resistant parasites.
4. Only to increase the milk production in animals, unnecessary use of anthelmintics should be avoided.
5. In first year after calving, enough exposure should be recommended for the development of immunity.
6. Quarantine drenching management should be included, before introducing newly purchased animals in existing herd; proper deworming of animals should be done to avoid entry of resistant parasites [32, 33].
7. Pasteur management is one of the important preventive measures to check parasite infestation and, therefore, helps to block development of resistant problem. Grazing of animals is also very important management practice and care should be takento avoid grazing on same pasture throughout the year [34]. A pasture is said to be safe pasture on which animals had not grazed for about 6 months during the cool weather or minimum for about 3 months during hot and dry season. Pasture area should be divided into smaller parts, and the grazing duration of animals should also be short with rotational practices.
8. Biological control is also an important way to control the parasite (nematodes) infestation using fungus *Duddingtonia flagrans* that halt the growth of larval stage of

parasites. The fungus isfed to ruminants that excrete it out through feces. In the feces, fungus gets colonized and uses the larva as nutrient.

9. Even feed ingredient present in plants such as tannins also helps to cut short the fecal egg counts and decrease the L3 larvae.
10. Nutritional status of animals infested with parasites is also an important factor, as nutrition not only alleviate reduction in productivity but also decrease the mortality percentage. Phosphorus supplementation to a certain extent helps to restrict the worms. Copper helps to improve immunity and leads to reduction in the worm load. The most important component of the nutrition is protein, which helps to improve immunity and reduction in the worm load in an animal.
11. Weather condition also plays an important role to tackle this problem. After a considerable rainfall, the formerly existing inactive larvae get hatched. So the strategy is to use the anthelmintics in animals 1 to 2 weeks after the rainfall.

References

[1] Forsyth BA. The anthelmintic effects of a phenothiazine-benzimidazole mixture.*Aust. Vet. J.,1962;* 38: 398.

[2] Drudge JH, Leland SE, Wyant ZN. Strain variation in the response of sheep nematodes to the action of phenothiazine. 1. Studies of mixed infections in experimental animals. *Am. J. Vet.Res.*, 1957; 133-141.

[3] Prichard R. Anthelmintic resistance.*Vet.Parasitol.*, 1994; 54:259-268.

[4] Shoop WL, Mrozik H, Fisher M.H. Structure and activity of avermectins and milbemycins in animal health. *Vet.Parasitol.*, 1995; 59:139-156.

[5] Kaplan RM. Drug resistance in nematodes of veterinary importance: a status report.*Trends Parasitol.*,2004; 20: 477-481.

[6] Wolstenholme AJ, Fairweather I, Prichard RK, Von Samsonhimmelstjerna G, Sangster NC.Drug resistance in veterinary helminths.*Trends Parasitol.*, 2004;20: 469-476.

[7] Van Wyk JA, Malan FS. Resistance of field strains of Haemonchus contortusto ivermectin, closantel, rafoxanide and the benzimidazoles in South Africa. *Vet. Rec.,* 1988*;* 123: 226-228.

[8] Waller PJ. Anthelmintic resistance and the future for roundworm control. *Vet.Parasitol.,*1987; 25: 177-199.

[9] Waller PJ. Anthelmintic resistance.*Vet. Parasitol.*, 1997; 72: 391-412.

[10] Beech RN, Prichard RK, Scott ME.Genetic variability of the beta-tubulin genes in benzimidazole-susceptible and -resistant strains of Haemonchus contortus.*Genetics,*1994; 138: 103-110.

[11] Geerts S, Gryseels B. Anthelmintic resistance in human helminths: *a review. Trop.Med. Int. Health.,* 2001;6: 915-921.

[12] Gill JH, Kerr CA, Shoop WL,Lacey E. Evidence of multiple mechanisms of avermectin resistance in Haemonchus contortus - comparison of selection protocols. *Int. J. Parasitol.,* 1998; 28: 783-789.

[13] Varady M, Bjorn H, Nansen P. In vitro characterization of anthelmintic susceptibility of field isolates of the pig nodular worm Oesophagostomum spp. susceptible or resistant to various anthelmintics. *Int. J.Parasitol.,*1996; 26: 733-740.

[14] Blackhall WJ, Liu HY, Xu M, Prichard RK, Beech RN. Selection at a P-glycoprotein gene in ivermectin- and moxidectin-selected strains of Haemonchus contortus. *Mol.Bioch.Parasitol.*, 1998; 95: 193-201.

[15] Brunston RV. Principles of helminth control. *Vet. Parasitol.*, 1980; 6:185-215.

[16] Morley FHW, Donald AW. Farm management and systems of helminth control. *Vet. Parasitol.,* 1980; 6:105-134.

[17] Prichard RK.Anthelmintic resistance.*Vet. Parasitol.*, 1994; 54: 259-68.

[18] Sangster N, Dobson RJ. Anthelmintic Resistance. In: *The Biology of Nematodes* (Lee, D.L. 2ndedn.), Taylor and Francis, London and New York, 2002; 531-567.

[19] Kwa MSG, Veenstra JG, Roos MA. Benzimidazole resistance in Haemonchus contortus is correlated with a conserved mutation at amino-acid-200 in β-tubulin isotype 1. *Mol. Bioch. Parasitol.*, 1994;63:299-303.

[20] Drogemuller M, Schnieder T, Von Samson-Himmelstjerna G. Beta-tubulin cDNA sequence variations observed between cyathostomins from benzimidazole-susceptible and -resistant populations. *Int. J. Parasitol.*, 2004; 90:868-870.

[21] Ghisi M, Kaminsky R, Maser P. Phenotyping and genotyping of Haemonchus contortus isolates reveal a new putative candidate mutation for benzimidazoleresistance in nematodes. *Vet. Parasitol.,* 2007; 144: 313-320.

[22] Mottier ML, Prichard RK.Genetic analysis of a relationship between macrocyclic lactone and benzimidazole anthelmintic selection on Haemonchuscontortus.*Pharmacogenet. Genom.*, 2008; 18: 129-140.

[23] Silvestre A, Cabaret J. Mutation in position 167 of isotype 1 b-tubulin gene of trichostrongylid nematodes: role in benzimidazole resistance? *Mol. Biochem. Parasitol.*, 2002; 120: 297-300.

[24] Kerboeuf D, Blackhall W, Kaminsky R, Von Samson-Himmelstjerna G. P-glycoprotein in helminths: function and perspectives for anthelmintic treatment and reversal of resistance. *Int. J. Antimicrob.Agents.*, 2003; 22: 332-346.

[25] Blackhall WJ, Prichard RK, Beech RN.P-glycoprotein selection in strains of Haemonchus contortus resistant to benzimidazoles.*Vet.Parasitol.*, 2008; 152: 101-107.

[26] Sangster NC, Gill J. Pharmacology of anthelmintic resistance.*Parasitol.Today,* 1999; 15: 141-146.

[27] Prichard RK, Roulet A. ABC transporters and β-tubulin in macrocyclic lactone resistance: prospects for marker development. *Parasitology,* 2007; 134: 1123-1132.

[28] Robertson SJ, Pennington AJ, Evans AM, Martin RJ. The action of pyrantel as an agonist and an open-channel blocker at acetylcholine receptors in isolated Ascarissuum muscle vesicles.*Eur. J.Pharmacol.*, 1994;271: 273-282.

[29] Qian H, Martin RJ, Robertson AP. Pharmacology of N-, L-, and B- subtypes of nematode nAChR resolved at the single-channel level in Ascarissuum. *FASEB J.*, 2006; 20: 2606-2608.

[30] Robertson AP, Bjorn HE, Martin RJ. Resistance to levamisole resolved at the single-channel level. *The FASEB J.,* 1999; 13:749-760.

[31] Besier RB, Hopkins DL, Anthelmintic dose selection by farmers.*Aust. Vet. J.,* 1988; 65:193-194.

[32] Varady M, Praslicka J, Corba J, Vesely L. Multiple anthelmintic resistance of nematodes in imported goats. *Vet. Rec.*,1993; 132: 387-388.
[33] Himanos C, Papadopoulos E, Anthelmintic resistance in imported sheep.*Vet. Rec.*, 1994; 134: 456.
[34] Barger IA, Miller JE,Klei TR.The role of epidemiological knowledge and grazing management for helminth control in small ruminants.*Int. J. Parasitol.*, 1999; 29: 41-47.

In: Livestock: Rearing, Farming Practices and Diseases ISBN 978-1-62100-181-2
Editor: M. Tariq Javed

Chapter 9

SALMONELLA AND SALMONELLOSIS IN ANIMALS AND HUMANS: EPIDEMIOLOGY, PATHOGENICITY, CLINICAL PRESENTATION AND TREATMENT

Sonia Téllez, María Concepción Porrero and Lucas Domínguez
VISAVET Health Surveillance Centre, Universidad Complutense de Madrid, Spain

ABSTRACT

Salmonellosis is one of the most common and widely distributed foodborne diseases. It is a major public health problemin many countriesand requires a significant amount of money to deal with. Millions of human cases are reported worldwide every year and the disease results in thousands of deaths. Salmonellosis is caused by the bacteria *Salmonella.* According to contemporary classification, the genus *Salmonella* contains two species; *Salmonella bongori* and *Salmonella enterica*, but there are more than 2,500 serotypes of *S. enterica*. *Salmonella* serovars can be divided into host restricted, host adapted, and ubiquitous serotypes with important implications for epidemiology and public health. Most of these cause acute gastroenteritis characterized by a short incubation period and a predominance of intestinal over systemic symptoms. Only a small number of serotypes typically cause severe systemic disease in man or animals, characterized by septicemia, fever and/or abortion, and such serotypes are often associated with one or few host species.

Since the beginning of the 1990s, antimicrobial-resistant *Salmonella*strains have emerged and are threatening to become a serious animal and human health problem. This resistance results from the use of antimicrobials, both in humans and animals. The global spread of multi-drug resistantstrains to critically important antimicrobials; including the first-choice agents used for the treatment of humans are of great concern.

1. Introduction

Salmonella spp. is a facultative anaerobe, gram negative rod-shaped flagellated bacterium belonging to the family *Enterobacteriaceae* (Figure 1). The genome of *Salmonella spp.* has a G+C content of 50-52 mol%, similar to that of *Escherichia, Shigella,* and *Citrobacter* [1, 2, 3].

The principal habitat of *Salmonella* spp. is the intestinal tract of warm-blooded and cold-blooded vertebrates,the fecal-oralbeing the main route of transmission. *Salmonella* are disseminated in the natural environment through human or animal excretion. *Salmonella* do not seem to multiply significantly in the natural environment, but they can survive several weeks in water and several years in soil if conditions of temperature, humidity, and pH are favorable [1, 2].

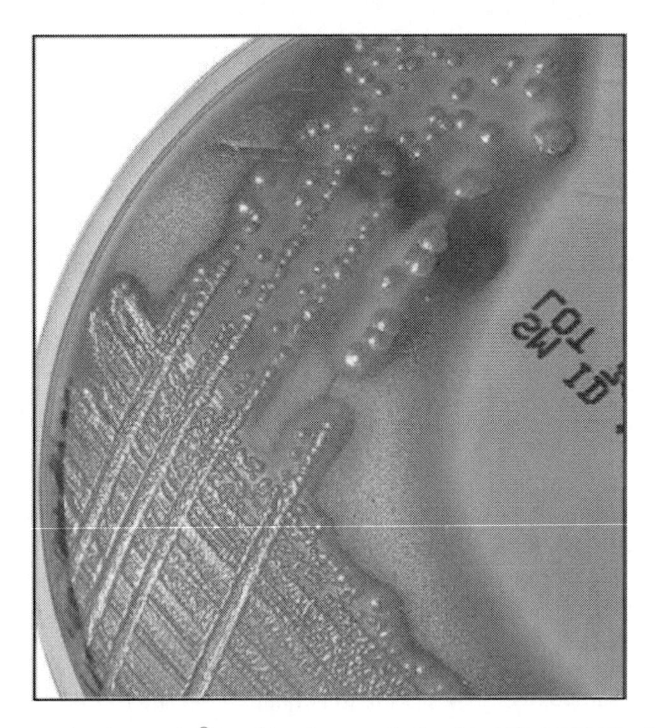

Figure 1.*Salmonella enterica* in SM ID® medium (image obtained in VISAVET Health Surveillance Centre, Universidad Complutense de Madrid, Spain).

The current view of *Salmonella* taxonomy [4] assigns the members of this genus to two species: *S. enterica* and *S. bongori*, in addition to a new species "*Salmonella subterranea*", which was recognized in 2005[5]. *S. enterica* itself is divided into six subspecies, *enteric* (I), *salamae*(II), *arizonae*(IIIa), *diarizonae*(IIIb), *houtenae* (IV) and *indica* (VI) [2, 4, 6].

The taxonomic classification of *Salmonella* has been continually revised over the years[4, 6, 7]. Beyond the level of subspecies, serotyping is used for differentiation. The serotypes have been described on the basis of the Kauffmann-White scheme according to somatic (O), flagellar (H), and capsular (Vi) antigens [2, 7]. A total of 2,501 different *Salmonella* serotypes have been identified up to 2004 [6, 8]. The antigenic formulae of *Salmonella* serotypes are defined and maintained by the World Health Organization (WHO) Collaborating Centre for Reference and Research on *Salmonella* at the Pasteur Institute, Paris, France (WHO Collaborating Centre), and new serotypes are listed in annual updates of the Kauffmann-White scheme [8, 9].

Salmonella nomenclature is complicated and still evolving. However, uniformity in *Salmonella* nomenclature is necessary for communication between scientists, health officials, and the public. Currently, the nomenclature system used is based on recommendations from the WHO Collaborating Centre and is summarized in Tables 1and 2 [6, 7, 8].

Table 1. *Salmonella* species, subspecies, serotypes, and their usual habitats (Adapted from Lin-Hui *et al.,* 2007)

Salmonella species and subspecies	No. of serotypes within subspecies	Usual habitat
S. enterica subsp. *enterica* (I)	1,478	Warm-blooded animals
S. enterica subsp. *enterica* (II)	498	Cold-blooded animals and environment
S. enterica subsp. *enterica* (IIIa)	94	Cold-blooded animals and environment
S. enterica subsp. *enterica* (IIIb)	327	Cold-blooded animals and environment
S. enterica subsp. *enterica* (IV)	71	Cold-blooded animals and environment
S. enterica subsp. *enterica* (VI)	12	Cold-blooded animals and environment
S. bongori	21	Cold-blooded animals and environment
S. subterranea	1	Cold-blooded animals and environment
Total	2,502	

Table 2. *Salmonella* nomenclature in use at Centers for Disease Control and Prevention (CDC) (Adapted from Lin-Hui *et al.,* 2007)

Taxonomic position	Nomenclature
Genus (italics)	*Salmonella*
Species (italics)	*enterica* *bongori* *subterranea*
Subspecies *S. enterica* (italics)	*enterica* (I), *salamae* (II), *arizonae* (IIIa), *diarizonae* (IIIb), *houtenae* (IV), *indica* (VI)
Serotype (capitalized, not italicized)	Serotypes are named in subspecies I and designated by antigenic formulae in subspecies II to VI and *S. bongori* and *S. subterranea.*

2. Epidemiology of *Salmonella*

The epidemiology of *Salmonella* is dynamic, complex and evolving. *Salmonella* is capable to survive inside the cells of host immune system, in the intestinal lumen and even in the environment. That proves that they are true survivors [10, 27].

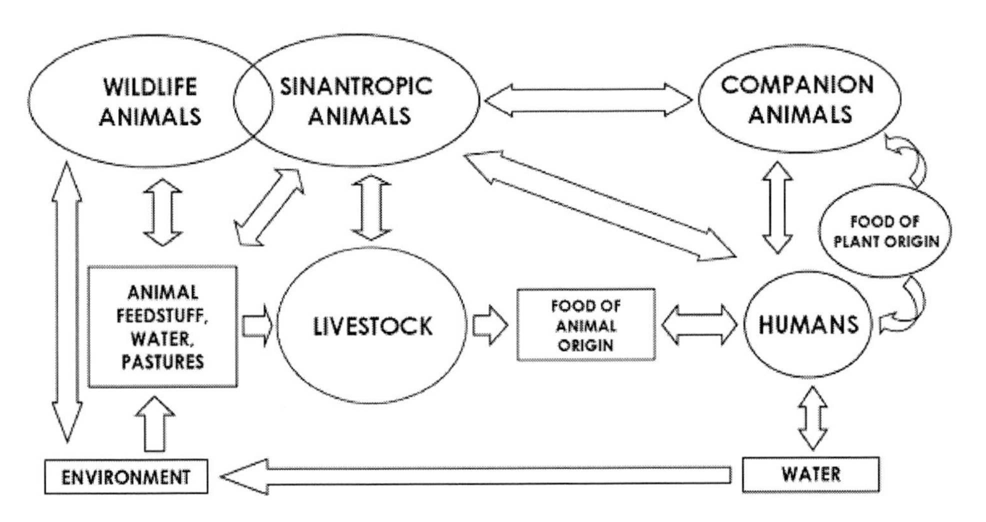

Figure 2. Epidemiological cycle of *Salmonella* spp. (Adapted from WHO, 1988).

While all serotypes can cause disease in humans and/or animals, they are often classified according to their adaptation to animal hosts. From a clinical perspective, these may be broadly grouped on the basis of host range and disease presentation [2, 11].

Salmonella serotypes are normally divided into three groups on the basis of host range; host-restricted, host-adapted and ubiquitous. Host-restricted serotypes are exclusively associated with one particular host in which these cause systemic disease (e.g., Typhi and Paratyphi in humans, Abortusovis in sheep, Abortusequi in horses, Gallinarum/Pullorum in poultry and Typhisuis in swine) [11, 47].

Host-adapted serotypes are associated with one particular host species, but able to cause disease in other host species: Dublin and Choleraesuis, for example, are generally associated with severe systemic disease in cattle and pigs, respectively but may also infrequently cause disease in other mammalian hosts including humans.

Ubiquitous serotypes (e.g. Typhimurium and Enteritidis) are capable of causing systemic disease in a wide range of host animals, but usually induce a self-limiting gastroenteritis in a broad range of unrelated host species [11, 47]. These serotypes are also the most important cause of human salmonellosis [38, 44].

In all cases, infectedanimals and humans frequently become clinically asymptomatic carriers. These infected animals becomesource of spread and represent a health hazard since they contaminate their environment and increase the number of infected individuals [11, 47]. Environmental contamination plays a role in colonization, persistence, and transmission within animals and man [1, 2, 10, 11].

3. Pathogenicity and Clinical Signs in Human and Animal Species

Salmonella is an invasive, facultative intracellular pathogen of animals and man with the ability to colonize various niches in diverse host organisms [2, 10].The nature of the pathogenic action of *Salmonella* varies with the serotype, the strain, the infectious dose, the nature of the contaminated food, and the host status [1, 2, 10].*Salmonella* virulence requires the coordinated expression of complex arrays of virulence factors that allow the bacterium to evade the host immune system. All *Salmonella* serotypes share the ability to invade the host by inducing their own internalization into cells of the intestinal epithelium and, in addition,serotypes associated with gastroenteritis are able to develop an intestinal inflammatory and secretory response. Those serotypes that are capable of causing systemic disease, establish infection through their ability to survive and replicate in mononuclear phagocytes (Figure 3) [10, 12, 13].

The colonization and invasion of hosts by *Salmonella* depends on the function of a large number of virulence determinants [10].The knowledge of the whole genome sequence has revealed many virulence genes involved in pathogenesis of *Salmonella*. Both invasion and intracellular survival is mediated by at least 60 virulence genes that have a plasmid or chromosomal origin and are located on several pathogenicity islands which are acquired by horizontal gene transfer [15, 16].

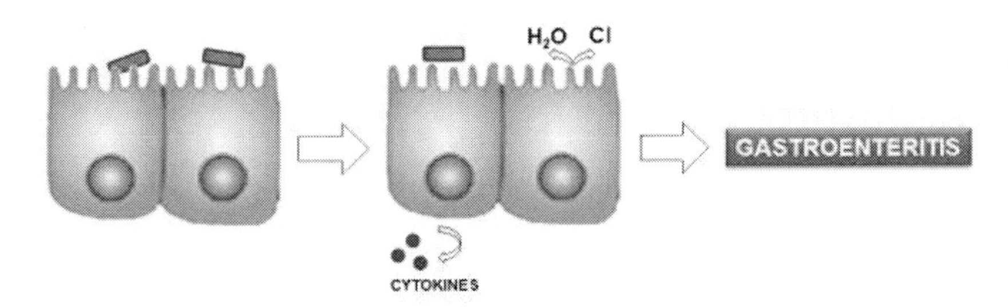

B. Invasion of mucosal epithelial cells

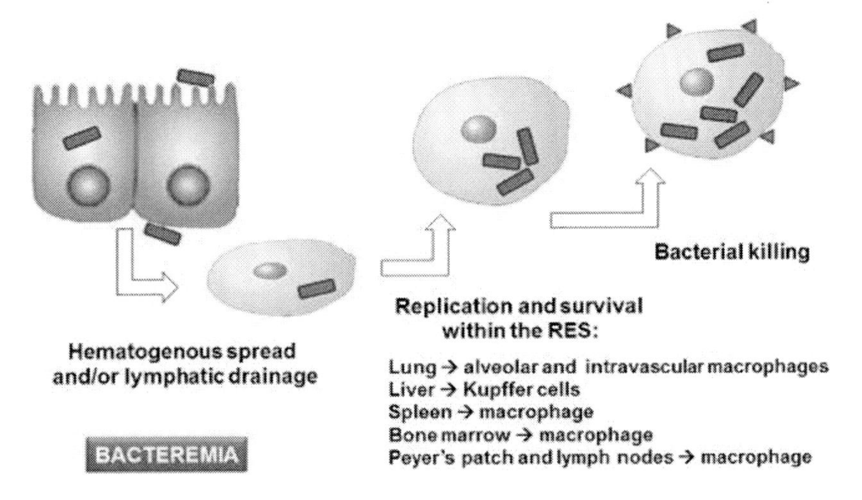

Figure 3. Principal steps in *Salmonella* pathogenesis (Adapted from Uzzau *et al.*, 2000).

To date, 12 pathogenicity islands have been identified in *Salmonella*, which help bacteria to invade the host system and in intracellular survival as well as in proliferation. The pathogenicity islands (SPIs) 1 and 2 are the two major virulence determinants of *Salmonella* because they encode type III secretion system (T3SSI and T3SSII, respectively)that form syringe-like organelles on the surface of bacteria that enable them to inject effector proteins directly inside the eukaryotic cells [15,16]. These virulence proteins sabotage the host processes, manipulate the cellular functions of the infected host and facilitate the progression of the infection. SPI1 is present in all the subspecies and serotypes of *Salmonella* and is required for the invasion of non-phagocytic cells and SPI2 is required for the intracellular survival and systemic pathogenesis [16, 17]. But even considerable differences in virulence can be found within a serovar those are strain dependent [10, 18].

In animals and humans, the outcome of infection depents on both serotype- and host-specific factors and may involve enteric or systemic phases (Table 3) [14].

Table 3.Diseases caused by *Salmonella* subspecies I serotypes in humans and higher vertebrates (Adapted from Bäumler *et al.*, 1998)

Host species	Disease	*Salmonella* serotype most frequently encountered	Typical symptoms or signs
Humans	Salmonellosis	Typhimurium, Enteritidis, Hadar, Virchow, Infantis	Diarrhoea, dysentery, fever
	Typhoid fever	Typhi	Septicemia, fever
	Paratyphoid fever	Paratyphi A, B and C, Sendai	Septicemia, fever
Cattle	Salmonellosis	Typhimurium	Diarrhoea, fever
		Dublin	Diarrhoea, dysentery, fever, septicemia, abortion
Poultry	Pullorum disease	Pullorum	Septicemia, diarrhea
	Fowl typhoid	Gallinarum	Septicemia, diarrhea
	Avian paratyphoid	Typhimurium, Enteritidis	Septicemia, diarrhea
Sheep	Salmonellosis	Abortusovis	Septicemia, abortion, vaginal discharge
		Typhimurium	Septicemia, diarrhea
Pigs	Pig paratyphoid	Choleraesuis	Septicemia, fever, skin discoloration
	Salmonellosis	Typhimurium	Diarrhoea
	Chronic paratyphoid	Typhisuis	Intermittent diarrhea
Horses	Salmonellosis	Abortusequi	Abortion, septicemia
		Typhimurium	Septicemia, diarrhea
Rodents	Murine typhoid	Typhimurium, Enteritidis	Septicemia, fever

3.1. Clinical Sings in Humans

In humans, the clinical pattern of salmonellosis can be divided into three disease patterns: enteric fever, gastroenteritis andbacteremia and other complications of non-typhoidal salmonellosis [2, 3].

3.1.1. Enteric Fever

Typhoid fever is caused by serovar Typhi whereas serovars Paratyphi A, B and C cause paratyphoid fever. The term enteric fever refers to both typhoid and paratyphoid infections. Both serotypes are restricted human pathogens and are transmitted through the ingestion of food or water contaminated by the feces or urine of infected people [2, 3].

Symptoms usually develop in one to three weeks after exposure; those may be mild to severe. They include headache, high fever, dehydration, constipation or diarrhea, rose-colored spots on the chest, and enlarged spleen and liver. Roughly 10percent of patients may relapse, die or encounter serious complications such as typhoid encephalopathy, gastrointestinal bleeding, metastatic abscesses and intestinal perforation [2, 3]. Nowadays, typhoid and paratyphoid fever continue to be important causes of illness and death, particularly among children and adolescents in South-central and Southeast Asia, where enteric fever is associated with poor sanitation and unsafe food and water [19].

3.1.2. Gastroenteritis

Enterocolitis is caused by at least 150 *Salmonella* serotypes (non-typhoidal serotypes) with Typhimurium and Enteritidis being the most common serotypes worldwide. However, other serovars are often more prevalent in specific countries, and result in more severe infections and outcome.Infection always occurs via ingestion of water or food contaminated with the feces of infected animals or humans [2, 20, 21].

Patients typically present with an acute onset of fever, diarrhea, vomiting and abdominal cramping; there is a wide spectrum of severity of illness. The incubation period is dependent upon the host and inoculum, but is generally 6 to72 hours [22].

Treatment of patients with symptoms of infectious diarrhea with antimicrobial agents remains controversial. The decision is complicated by the fact that (i) presenting patients could have any one of a number of enteric pathogens, so treatment for severe illness "up front" is usually empiric; and (ii) the long-appreciated but counterintuitive finding that antibiotic treatment of patients with non-typhoidal salmonellosis may actually prolong rather than limit the fecal shedding of these organisms [22, 23].

3.1.3. Bacteremia and Other Complications of Non-Typhoidal Salmonellosis

Bacteremia and other forms of extra-intestinal *Salmonella* infections are serious complications of mild primary infection. In these cases*Salmonella* is able to enter the bloodstream after passing through the intestinal barrier [2, 3, 22]. About 8 percent of the untreated cases of salmonellosis result in bacteremia. It has been associated with immunocompromisedstatus and with highly invasive serotypes.Complications of bacteremia include endocarditis, mycotic aneurysms and osteomyelitis. Patients with bacteremia and other complications should be treated with antibiotics [24].

3.2. Clinical Sings in Animals

In most animal species, *Salmonella* results in a clinically inapparent infection of variable duration, which is significant as a potential zoonosis. However, under various stress conditions, some serovars may also cause disease in animal species.

3.2.1. Livestock

3.2.1.1. Cattle and Small Ruminants

Clinically, *Salmonella* infection in cattle is typically manifested as watery or bloody diarrhea, and often associated with fever, depression, anorexia, dehydration and endotoxemia. Less common clinical manifestations include abortion and respiratory disease. The mortality rates can be high. In adult animals, *Salmonella* frequently causes subclinical disease and is known to persist on infected farms for months or years.

Individual animals shed *Salmonella* intermittently over variable periods of time and infections with host adapted serotypes such as Dublin may potentially result more frequently in the development of asymptomatic shedders than infections with broad host range serotypes [26, 27].

The severity and clinical manifestation of *Salmonella* infection in small ruminants differ by age group and serotype. Acute enteric salmonellosis is common in adult sheep leading to fever, anorexia, depression and diarrhea, while septicemia is common in young animals. However, asymptomatic carriage, chronic gastro-enteritis and abortion have also been described [26].

Abortion due to infection with serotypes such as Typhimurium or Dublin has been reported, but abortion is most frequently caused by serotype Abortusovis, an ovine-adapted serotype that also occasionally infect goats, and abortion generally occurs during the last weeks before parturition. Infections of ewes with Abortusovis can also lead to stillbirth, metritis, placental retention, or peritonitis, and infected ewes may present with fever, anorexia and depression prior to abortion [26, 27].

3.2.1.2. Pigs

A variety of clinical manifestations have been observed in *Salmonella* infected pigs, ranging from asymptomatic to peracute disease. Infections with certain serotypes such as Typhimurium usually cause mild or no disease and infected animals may shed *Salmonella* for considerable periods of time.In contrast, infection with host adapted serotype Choleraesuis generally causes severe systemic disease with high mortality.All age groups are susceptible to *Salmonella* infection but disease is more commonly observed in weaned pigs of more than eight weeks of age and asymptomatic carriers are thought to represent the most important source of *Salmonella* introduction onto the pig farms[26, 28].

A variety of clinical manifestations have been documented among *Salmonella* infected pigs including enteritis, septicemia, pneumonia, meningitis and arthritis. Fever, diarrhea, in-appetence, depression, respiratory distress, lameness, edema, and hypoxia in the extremities are common symptoms in clinically sick pigs and mortality rates in such instances are high [26, 28].

3.2.1.3. Poultry

The clinical symptoms associated with *Salmonella*infection vary considerably by age group and serotype. Infections with general serotypes rarely causeclinical disease in galliform birds and most animalsbecome asymptomatic carriers, even though severe clinical disease with high mortality has been observed insome cases, particularly during infections of youngbirds. Infections with the host adaptedserotype Gallinarum biovars Gallinarum and Pullorum, however, cause severe disease with high mortalityand immense economic losses on chicken and turkeyfarms [26].

Serotype Gallinarum biovar Pullorum causes "Pullorum disease"in young birs, which is associated with septicemiaand high mortality that can exceed 85 percent in some cases. Pulloruminfections of adult birds are generally mild or asymptomatic, although decrease in fertility and egg productionas well as increased mortality has been observed insome cases. Adult animals can develop a carrierstate, and transovarian transmission is thought to be theprimary rout of transmission to young birds, even though rodents and other vectors are also thought toplay an important epidemiological role. Clinical symptoms include anorexia, diarrhea, dehydration, decreased hatching and high mortality [29].

Serotype Gallinarum biovar Gallinarum causes "fowl typhoid" in young and particularly in adult birds. Clinical symptoms are very similar to those observedduring infections with biovarPullorum and economic losses during outbreaks can be very high [26, 29].

3.2.2. Companion Animals

3.2.2.1. Horses

Salmonellosis is an important disease of horses. Mortality ratesin equine vary depending on the host age, predisposing factors andthe *Salmonella* serotype involved. Mortality can be as high as 40 to 60 percentbut in general it appears to be considerably low. In most cases, animals present withprofuse watery and malodorous diarrhea, frequentlyassociated with abdominal pain and endotoxemia. Fever, dehydration and depression are common and insevere cases these symptoms are accompanied by colic, gastric reflux, cardiovascular shock or coagulopathies.However, the severity of disease can vary considerablyand in animals of the same age group, it may range from severe to asymptomatic. Both peracute andchronic forms of disease are common, and convalescentcarriers may shed *Salmonella* for months but acarrier state does not appear to develop in all cases [26, 30]. Disease may also manifest without gastrointestinalsigns. Some serotypes appear to resultmore frequently in systemic disease than others but the underlying mechanisms are still incompletely understood. Respiratory forms are comparablyfrequent and systemic forms of infection are commonly associated with arthritis, osteomyelitis or softtissueabscesses. Foals, pregnant mares, andimmune-compromised horses are at higher riskof infection. In foals, *Salmonella*-associated meningo-encephalitis has been described. Abortions due to *Salmonella* infection with host adapted serotype Abortusequi cause important economic losses on the stud farms [26, 30].

3.2.2.2. Dogs and Cats

A considerable numberof *Salmonella* serotypes have been isolated from domestic dogs and cats around the world. The majority of infections are asymptomatic. However, gastrointestinal disease manifestedas enterocolitis and endotoxemia can occur and is often associated with fever, vomiting, anorexia, dehydration and depression. Abortion, still birth, meningo-encephalitis, respiratory distress and conjunctivitishave have also been described [26, 31].

3.2.2.3. Rabbits and Rodents

Salmonella infection can cause severe disease in rabbits, which is sometimes associated with high mortality. Clinical symptoms include enteritis, metritis andabortion, but striking differences in pathogenic potentialseem to exist among different *Salmonella* serotypes [26]. In contrast, the majority of infections in mice andrats are asymptomatic. However, clinical disease amongrodents has also been described, for instance duringlarge outbreaks among laboratory rodents it wasassociated with high mortality rates [26].

3.2.2.4. Birds

Numerous *Salmonella* serotypes have been isolatedfrom a variety of captive birds kept as pets. Acute and chronic infectionshave been reported, which range from asymptomatic to clinically severe and can manifest as diarrhea, anorexia, dehydration, depression, crop stasis, septicemiaor osteomyelitis [26, 32].

3.2.2.5. Reptiles and Amphibians

Salmonella occurs naturally in the gastrointestinal tractof many reptiles, is commonly shed by these animalsand around the world a large number of serotypes havebeen isolated from feral and captive reptiles as well as from their eggs [26]. Clinicaldisease including septicemia, osteomyelitis, salpingitis, nephritis, dermatitis and abscesses seems to be occasionallyassociated with *Salmonella* infection in snakes, turtles and lizards but the overwhelming majority ofinfections in reptiles are undoubtedly asymptomatic. Clinical salmonellosis in reptiles is rare, appears to beassociated with underlying disease or other stressorsand a causal relationship between *Salmonella* infectionand disease is generally difficult to establish conclusively [26, 33].

Figure 4.Chronic osteomyelitis in hip articulation in a *Chamaeleo calyptratus*due to a *Salmonella* infection (image obtained in VISAVET Health Surveillance Centre, Universidad Complutense de Madrid, Spain).

3.3. Chronic Carrier State in Animals and Humans

A feature of some *Salmonella* infections in humans and animals is the development of a carrier state after primary challenge.

- Active carriers: animals or humans that excrete high levels of *Salmonella*, even in the absence of clinical signs during recovery from enteric or systemic disease. In some cases (e.g. Dublin infection of cattle), this state may persist for life.
- Passive carriers: animals or humans with no active pathology that excrete *Salmonella* acquired from a contaminated environment. Such individuals will clear the organism if removed to a clean environment.
- Latent carriers: *Salmonella* can persist asymptomatically in the gallbladder, mesenteric lymph nodes, etc. and be excreted intermittently, mainly in response to stress [1, 2, 34].

In humans, factors contributing to the chronic carrier state have not been fully explained.

- On average, non-typhoidal serotypes persist in the gastrointestinal tract from six weeks to three months, depending on the serotype. Only about 0.1 percent of these cases are shed in stool samples for more than one year.
- About two to five percent of untreated typhoid infections result in a chronic carrier state. Up to 10 percent of untreated convalescent typhoid cases will excrete serotype Typhi in feces for one to three months and between one and four percent become chronic carriers excreting the microorganism for periods exceeding one year [2, 35, 36].

In animals, carrier status is not easy to identify. As in humans, gallbladder and mesenteric lymph nodes are favorite places of multiplication and bacterial survival[36, 37]. *Salmonella* can further form biofilms on gallstones during persistent infection and this might be the mechanism of the pathogen to allow its host to become an asymptotic carrier and spread the disease [10, 36]. Fecal shedding coincides with stressful situations such as feed withdrawal, loading, transportation from farm to slaughterhouse, etc. [1, 2, 10].

Among livestock animal species, pigs and poultry are those that reflect a higher prevalence of *Salmonella* carriers [26, 38]. *Salmonella* prevalence estimates for pig farms seem to differ considerably by production and management type [26, 38]. In some countries, up to 60 percent of the pigs at slaughterhouse are shedding*Salmonella*. In poultry, the prevalence of *Salmonella*carriers can be as high as 50 to 70 percent [38]. *Salmonella* prevalence varies considerably by poultry type, differs between serotypes and biovars. The intestinal carriage often appears to be lower than isolation rates from egg shells, dead birds, and environmental samples [26].

In cattle, *Salmonella* prevalence estimates vary considerably, ranging from two to forty two percent. Usually they are around one to two percent, perhaps somewhat higher in young animals [26, 38, 39]. Large herd size represents an important risk factor for salmonellosis and the risk of *Salmonella* shedding seems to vary by production system, housing type, hygiene level, management type and animal age [26, 39].

In all cases, there is intermittent excretion of *Salmonella*during peaks of elimination, which makes serial testing of several fecal samples necessary to verify carrier status [1, 2, 34].

In summary, improved understanding of the biology of the carrier state is vital to control the *Salmonella* in the environment and its influence in food safety and public health [1, 2, 34].

4. Treatment and Antimicrobial Resistance

Diagnosis of clinic salmonellosis should be confirmed by isolation, identification, and serotyping of *Salmonella* strains by classical methods or molecular techniques. *Salmonella* can be detected in stool, blood and in affected organs at necropsy [1, 2]. Difficulties in detecting asymptomatic carriers by culture techniques could be due to sporadic and intermittent shedding of the bacteria. *Salmonella* can be detected inconsecutive samples of stool or rectal/cloacal swabs [1, 2].

4.1. Treatment of Salmonellosis in Animals

In companion animals, the treatment for salmonellosis includes fluid therapy and replacing lost electrolytes, non-steroidalanti-inflammatory drugs, gastrointestinal protectants and probiotics to help replace the microbiota in the gastrointestinal tract. The use of antibiotics in the treatment of salmonellosis is controversial because it has been found that they may lead to antibioticresistance and also because the use of oral antibiotics may prolong the presence of bacteria in the stool. In production animals, treatment is usually contraindicated but when necessary, can be given via injection with several treatment alternatives based on considerations such as withdrawal time [40, 41].

4.2. Treatment of Salmonellosis in Humans

Antibiotic treatment is usually not advised except for rare cases because it can prolong the shedding of the organism [25]. Fluoroquinolones are the antimicrobials most widely regarded as optimal for the treatment of salmonellosis in adults because they are well tolerated, relatively inexpensive, have good oral absorption and are rapidly and reliably effective. Third-generation cephalosporins (which need to be given by injection) are widely used in children with serious infections. The earlier drugs chloramphenicol, ampicillin and amoxicillin and trimethoprim-sulfamethoxazole are occasionally used as alternatives [35, 36].

4.3. Antimicrobial Resistance

Since the early 1990s, antimicrobial-resistant *Salmonella* strains have emerged and are threatening to become a serious animal and human health problem. This resistance results from the use of antimicrobials, both in humans and animal husbandry [16, 23, 42].

The development of resistance in *Salmonella* toward antimicrobial agents is attributable to one of the multiple mechanisms: (i) production of enzymes that inactivate antimicrobial agents through degradation or structural modification, (ii) reduction of bacterial cell permeability to antibiotics, (iii) activation of antimicrobial efflux pumps, and (iv) modification of the cellular target for drug [43].

Nowadays, the emergence of multidrug resistant (MDR) *Salmonella* strains with resistance to fluoroquinolones and third-generation cephalosporins is a serious human and animal health problem. The consequencesdue to the occurrence of resistant microorganisms results in severe limitation of the possibilities for effective treatment of human and animal infections and can be divided into two categories: (i) infections that would otherwise not have occurred; and (ii) increased frequency of treatment failures and increased severity of infections [44]. An example of this emergence is the global spread of a multidrug-resistant serotype Typhimurium phage type DT104 in animals and humans [16, 42, 44].

These alarms have led to the development of different measures that have improved substantially by the use of antimicrobial agents in veterinary medicine. Basically, these measures have to do with the implementation of a new legislation on drugs and on a hygiene package [23, 45].

Even though the level of awareness of regulatory organisms is very satisfactory and implemented surveillance systems and the promotion of a sensible use start blooming. It is necessary to go deeper than that while promoting alternate actions to the use of antimicrobials, especially those addressed to the prevention of bacterial diseases, either specifically, through vaccination and improvement of the biosafety (using insecticides, disinfectants and insect repellents, or generally), improving cattle facilities and animal comfort. All these alternatives would lower the need to use antimicrobial agents in animals. These actions need to be complemented with training involving all agents (veterinarians, farmers, the pharmaceutical industry, drug suppliers and wholesale food distributors) and establishing more effective ways of cooperation between human and veterinary medicine specialists [45].

5. *SALMONELLA* AS A FOOD SAFETY PROBLEM AND ITS CONTROL

Salmonella is an important human foodborne pathogen worldwide. A recent study estimated that approximately 93.8 million human cases of gastroenteritis and 155,000 deaths occur due to *Salmonella*worldwide each year [21].*Salmonella* serotypes Enteritidis and Typhimurium are responsible for 95 percent of outbreaks with known serotype [46].

The importance of food-producing animals as reservoirs of non-typhoidal serovars affecting humans is well-established.*Salmonella* spp. colonize a wide range of hosts and all the major livestock species (poultry, cattle, and pigs) can become colonized, frequently asymptomatically, eventually producing contaminated meat and other food products [38, 44].

Although consumption of contaminated poultry meat and eggs is the primary risk factor for human non-typhoidal salmonellosis, other food-producing animals also pose a threat of zoonotic transmission [44].

The control of *Salmonella* in livestock is a main objective in developed countries. Different countries have adopted different control strategies for *Salmonella* with varying degrees of success[38, 44].For example, the European Union is developing control measures to prevent and reduce the prevalence of *Salmonella* in animal populations at the level of primary production and, where necessary, at other appropriate stages of the food chain as part of a farm to fork approach to food safety [46].

From 2003 to date, control measures have been planned for the following animal populations (Regulation (EC) No 2160/2003 of the European Parliament):

- poultry (breeding flocks of *Gallus gallus*, laying hens, broilers and turkeys)
- swine (herds of slaughter and breeding pigs).

For each population a similar approach is followed:

- Baseline studies on the prevalence of *Salmonella* in each Member State have been carried out.
- These baseline studies provide the reference prevalence for setting a reduction target for *Salmonella*. They also allow analysis of risk factors with a view to developingcontrol programs.
- National control programs are submitted to the Commission by MemberStates and third countries for approval each time a target for reduction has been agreed on.
- Control programs, including a harmonized monitoring of the population apply after approval and in any case within 18 months after setting a target for reduction.

This Regulation does not apply to primary production for private domestic use or for direct supply in small quantities, by the producer to the final consumer, or to local retail establishments directly supplying the final consumer. The Member States are to establish their own specific national legislation on these activities.

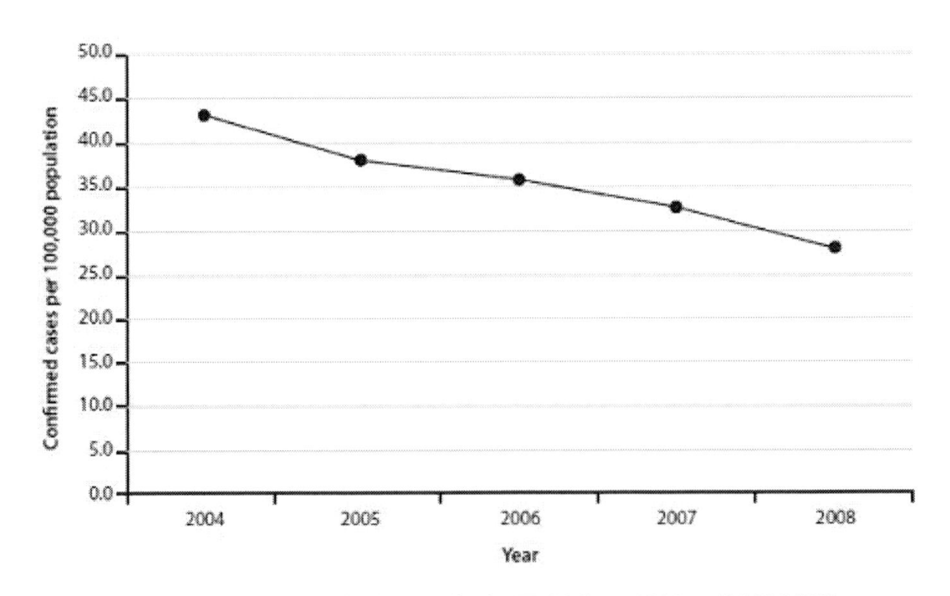

Figure 4. Number of reported salmonellosis cases in the EU24 from 2004 until 2008 [38].

Since Regulation (EC) No 2160/2003 on zoonosis control was adopted, the awareness and motivation of competent authorities and stakeholders to tackle *Salmonella* has increased significantly. The effect on public health can already be seen in the trend of salmonellosis in humans from 2004 until 2007 in Figure 4 [38].

ACKNOWLEDGMENTS

We are grateful to Sabrina Rodriguez for the careful revision of the manuscript.

REFERENCES

[1] Goyache J, Briones V. Géneros *Salmonella* y Shigella. In *Manual de Microbiología Veterinaria*. Eds. Vadillo S, Píriz S, Mateos E, McGraw-Hill/Interamericana de España SAU, 2002.

[2] D'Aoust J, Maurer J. *Salmonella* species. In *Doyle MP and Beuchat LR (Eds) Food microbiology: Fundamentals and frontiers*. Washington, D. C: ASM Press.,2007; 187-219,.

[3] Pui CF, Wong WC, Chai LC, Tunung R, Jeyaletchumi P, Noor Hidayah MS, Ubong A, Farinazleen MG, Cheah YK, Son R. *Int. Food Res. J.,*2011;18: 465-473.

[4] Euzéby JP. Revised *Salmonella* nomenclature: designation of *Salmonella* enterica (ex Kauffmann and Edwards 1952) Le Minor and Popoff 1987 sp. nov., nom. rev. as the neotype species of the genus *Salmonella* Lignieres 1900 (Approved Lists 1980), rejection of the name *Salmonella* choleraesuis (Smith 1894) Weldin 1927 (Approved lists 1980), and conservation of the name *Salmonella* Typhi (Schroeter 1886) Warren and scott 1930 (Approved Lists 1980). Request for an opinion.*Int. J. Syst. Bacteriol.*, 1999; 49: 927-930.

[5] Shelobolina ES, Sullivan SA, O'Neill KR, Nevin KP, Lovley DR. Isolation, characterization, and U(VI)-reducing potential of a facultatively anaerobic, acid-resistant Bacterium from Low-pH, nitrate- and U(VI)-contaminated subsurface sediment and description of *Salmonella* subterranea sp. *Nov. Appl. Environ. Microbiol.*, 2004;70:2959-2965.

[6] Lin-Hui Su MS, Cheng-Hsun Chiu MD.*Salmonella*: Clinical Importance and Evolution of Nomenclature. *Chang Gung Med. J.*, 2007; 30:210-219.

[7] Brenner FW, Villar RG, Angulo FJ, Tauxe R, Swaminathan B. *Salmonella* Nomenclature. *J. Clin. Microbiol.,*2000; 38: 2465-2467.

[8] Popoff MY.*Formules antigeniques des serovars de Salmonella*. WHO Collaborating Centre for Reference and Research on *Salmonella*, Paris, France. 2001.

[9] Langley J. WHO and CDC nomenclature. *Inj. Prev.*,2008; 14:342.

[10] Lahiri A, Lahiri A, Iyer N, Das P, Chakravortty D. Visiting the cell biology of *Salmonella* infection. *Microb. Infect.*, 2010; 12:809-818.

[11] Uzzau S, Brown DJ, Wallis T, Rubino S, Leori G, Bernard S, Casadesús J, Platt DJ, Olsen JE. Host adapted serotypes of *Salmonella* enterica. *Epidemiol. Infect.*,2000; 125: 229-255.

[12] Ohl ME, Miler SI. *Salmonella*: A Model for Bacterial Pathogenesis.*Annu. Rev. Med.*, 2001; 52: 259-274.

[13] Groisman EA, Ochman H. The path to *Salmonella*.*ASM News* 66 (1): 21-27, 2000.

[14] Bäumler AJ, Tsolis RM, Ficht TA, Adams LG. Evolution of host adaptation in *Salmonella*enterica.*Infect Immun.*, 1998; 66:4579-4587.

[15] Marcus SL, Brumell JH, Pfeifer CG, Finlay BB. *Salmonella* pathogenicity islands: big virulence in small packages. *Microb. Infect.*, 2000; 2:145-156.

[16] Foley SL, Lynne AM. Food animal-associated *Salmonella* challenges: Pathogenicity and antimicrobial resistance. *J. Anim. Sci.*,2008; 86:173-87.

[17] Winstanley C, Hart CA. Type III secretion systems and pathogenicity islands. *J. Med. Microbiol.*,2001; 5:116-126.

[18] Hall RM. *Salmonella* genomic islands and antibiotic resistance in *Salmonella* enteric.*Future Microbiol.*, 2010; 5:1525-38.

[19] Crump JA, Mintz ED. Global trends in typhoid and paratyphoid Fever. *Clin. Infect. Dis.*, 2010; 50:241-246.

[20] Crum-Cianflone NF. Salmonellosis and the gastrointestinal tract: more than just peanut butter.*Curr. Gastroenterol Rep.*, 2008; 10:424-431.

[21] Majowicz SE, Musto J, Scallan E, Angulo FJ, Kirk M, O'Brien SJ, Jones TF, Fazil A, Hoekstra RM. The global burden of nontyphoidal *Salmonella* gastroenteritis.*Clin. Infect. Dis.*, 2010; 50: 882-889.

[22] Hohmann EL. Nontyphoidal salmonellosis.*Clin. Infect. Dis.*, 2001; 32: 263-269.

[23] McEwen SA, Fedorka-Cray P. Antimicrobial use and resistance in animals. *J. Clin. Infect. Dis.*, 2002;34: 93-106.

[24] Mastroeni P, Grant AJ. The spread of *Salmonella* enterica in the body during systemic infection: unravelling host and pathogen determinants. *Expert. Rev. Mol. Med.. Apr.*, 2011;11;13:e12.

[25] Torres C, Moreno MA, Zarazaga M. Uso prudente de los agentes antimicrobianos: no solo para las personas. *Enferm. Infecc. Microbiol. Clin.*, 2010; 28:669-671.

[26] Hoelzer K, Moreno Switt AI, Wiedmann M. Animal contact as a source of human non-typhoidal salmonellosis.*Vet. Res.*, 2011; 42:34.

[27] Aceto H, Miller SA, Smith G. Onset of diarrhea and pyrexia and time to detection of *Salmonella* enterica subsp enterica in feces in experimental studies of cattle, horses, goats, and sheep after infection per os. *J. Am. Vet. Med. Assoc.*,2011; 238:1333-9.

[28] Xiong N, Brewer MT, Day TA, Kimber MJ, Barnhill AE, Carlson SA. Evaluation of the pathogenicity and virulence of three strains of *Salmonella* organisms in calves and pigs.*Am. J. Vet. Res.*, 2010; 71:1170-7.

[29] Shivaprasad HL. Fowl typhoid and pullorum disease. *Rev. Sci. Tech.*, 2000; 19:405-24.

[30] Chapman AM. Acute diarrhea in hospitalized horses.*Vet Clin North Am Equine Pract.*, 2009; 25:363-80.

[31] Enriquez C, Nwachuku N, Gerba CP. Direct exposure to animal enteric pathogens. *Rev. Environ. Health.*, 2001; 16: 117-31.

[32] Grimes JE. Zoonoses acquired from pet birds. *Vet. Clin. North Am. Small Anim. Pract.*, 1987; 17:209-18.

[33] Chiodini RJ, Sundberg JP. Salmonellosis in reptiles: a review. *Am. J. Epidemiol.*, 1981; 113: 494-499.

[34] Stevens MP, Humphrey TJ, Maskell DJ. Molecular insights into farm animal and zoonotic *Salmonella* infections.*Phil. Trans. R. Soc. B.,* 2009; 364:2709-2723.

[35] Parry CM. Epidemiological and clinical aspects of human typhoid fever. In Matroeni P and Maskell D (Eds.).*Salmonella infections: Clinical, immunological and molecular aspects,* New York: Cambridge University Press,2006; pp.1-18.

[36] Sirinavin S, Pokawattana L, Bangtrakulnondh A. Duration of nontyphoidal *Salmonella* carriage in asymptomatic adults. *Clin. Infect. Dis.*, 2004; 38:1644-1645.

[37] He GZ, Tian WY, Qian N, Cheng AC, Deng SX. Quantitative studies of the distribution pattern for *Salmonella* Enteritidis in the internal organs of chicken after oral challenge by a real-time PCR. *Vet. Res. Comm.*, 2010; 34:669-676.

[38] The Community Summary Report on trends and sources of zoonoses, zoonotic agents and food-borne outbreaks in the European Union in 2008, 2010; 8:1496.

[39] Dargatz DA, Fedorka-Cray PJ, Ladely SR, Ferris KE. Survey of *Salmonella* serotypes shed in feces of beef cows and their antimicrobial susceptibility. *J. Food Prot.,* 2002; 63:1648-1653.

[40] Hardy B. The issue of antibiotic use in the livestock industry: What have we learned?.*Anim Biotechnol.*, 2002; 13:129-147.

[41] FAO/OIE/WHO. *Workshop on Non-human Antimicrobial Usage and Antimicrobial Resistance: Management Options*. Oslo, Norway, 15-18 March, 2004.

[42] Alcaine SD, Warnick LD, Wiedmann M. Antimicrobial resistance in nontyphoidal *Salmonella. J. Food Prot.*, 2007; 70:780-90.

[43] Sefton AM. Mechanisms of antimicrobial resistance: their clinical relevance in the new millennium. *Drugs,* 2002; 62:557-66.

[44] Newell DG, Koopmans M, Verhoef L, Duizer E, Aidara-Kane A, Sprong H, Giessen J, Kruse H. Food-borne diseases-The challenges of 20 years ago still persist while new ones continue to emerge. *Int. J. Food Micro.,* 2010; 139:S3-S15.

[45] Rodríguez LD, Romo MA, Calonge MC, Peña ST. Prudent use of antimicrobial agents and proposals for improvement in veterinary medicine. *Enferm. Infecc. Microbiol. Clin,*2010; 28:40-4.

[46] EU-COMMUNICATION FROM THE COMMISSION TO THE EUROPEAN PARLIAMENT AND TO THE COUNCIL with regard to the state of play on the control of food-borne *Salmonella* in the EU Brussels, 2009. http://ec.europa.eu/food/food/biosafety/*Salmonella*/docs/comm_control_foodborne_*Salmonella*_EU_en.pdf.

[47] Soyer Y, Orsi RH, Rodriguez-Rivera LD, Sun Q, Wiedmann M. Genome wide evolutionary analyses reveal serotype specific patterns of positive selection in selected *Salmonella* serotypes.*BMC Evol. Biol.,* 2009;14;9:264.

In: Livestock: Rearing, Farming Practices and Diseases
Editor: M. Tariq Javed
ISBN 978-1-62100-181-2

Chapter 10

BOVINE TUBERCULOSIS AT THE HUMAN-ANIMAL INTERFACE, SITUATION AND POSSIBLE RISK FACTORS OF DISEASE IN ANIMALS IN PAKISTAN, FUTURE OF DISEASE AND ACTION PLAN

M. Tariq Javed, Farooq A. Farooqi, M. Irfan, M. Usman, Latif Ahmad, Imtiaz Ali, A. Latif Shahid, M. Wasiq, F. Rizvi, and Monica Cagiola

Department of Pathology, Faculty of Veterinary Science,
University of Agriculture, Faisalabad

ABSTRACT

The situation of bovine tuberculosis in Pakistan and the role of *M. bovis* as cause of human tuberculosis have been presented. The study in 1969 indicated a prevalence of 6.72percent in animals in Faisalabad. In 1972, it was 2.9, 1.6 and 8.6percent in buffaloes, Australian and Sahiwal cattle, respectively; in 1974 in Quetta, it was 0.53percent in buffaloes. In 1989, it was 7.3percent in cattle and buffaloes being slaughtered at Lahore Abattoir. In 2001, a study reported 1.76 percent prevalence in buffaloes in Faisalabad. In 2003, the respective prevalence of bovine tuberculosis in buffaloes and cattle was 6.91 and 8.64percent, respectively in Lahore. In 2006, the prevalence in buffaloes at two farms was 2.45 and 8.48percent. A study during 2006 and 2007 reported 2.2percent prevalence of tuberculosis in buffaloes at 11 farms. Another study reported a prevalence of 2.4percent in goats and 0.9percent in sheep at 3 and 7 livestock farms, respectively. A study during same period in buffaloes reported three percent prevalence involving 14percent farms around two cities of Punjab Pakistan. Recently, in cattle a prevalence of 7.6percent in cattle at 100percent farms has been reported. An overall prevalence of 3.3percent has been reported in zoo animals with 3.6percent in Bovidae, 3.2percent in Cervidae and 0percent in Equidae families.It is quite possible that some of the sheep and goat breeds are genetically resistant to tuberculosis. The stronger risk factors in cattle at 11 farms were the age of cattle, number of calving, total milk produced and lactation length. Certain possible risk factors identified in buffaloes were the lactating status of the buffalo, the presence of cattle at the farm, total number of cattle at the farm and the total number of animals at the farm. Results in zoo animals suggested that odds of tuberculosis were 1.2 times higher when animal number was less than 10. Because of the growing trend in developing livestock farms in the private sector in Pakistan, there is likelihood that the situation of bovine tuberculosis will worsen in coming years. Therefore, it is important to develop a strategy to control the disease right at the beginning rather than to think about it when it is already late. Certain strategies are suggested to control the disease in animals and humans.

1. Bovine Tuberculosis at the Human-Animal Interface

Tuberculosis is one of the oldest important zoonotic disease known to mankind and the *Mycobacterium tuberculosis* complex organisms infect a wide range of animals and human [1, 2]. Bovine tuberculosis is a disease caused by *M. bovis* in cattle/buffalo and other animal species. Infected animals can transmit the disease to humans and vice versa. *M. bovis* is one of the important members of *M. tuberculosis* complex organisms and is classified as a Risk 3 pathogen for public health [3]. Tuberculosis in humans is a disease known to mankind since ancient times. Recently, *M. bovis* has been recovered from Iron Age peoples of Siberia [4]. So the link between *M. bovis* and human infection is as old as the history of mankind and *M. bovis* as cause of human infection has been reported from almost all over the world, and where from it is not reported it does not mean that this link does not exist. *M. bovis* has been linked with 25percent cases of human TB in developed countries in the late 19th and early 20th centuries [5]. A study between 1995 and 2005 in United States reported *M. bovis*as cause of human infection in 1.4percent cases of human tuberculosis[6]. In same country, *M. bovis* was isolated from 45percent of TB cases in children and 6percent of adult with a significant ($P<0.001$) rise in *M. bovis* cases and the decrease in *M. tuberculosis* cases in United States [7]. Almost all *M. bovis* cases in US during 2001 to 2005 were in Hispanic ethnic group [7]. A study during 2004 and 2005 in Taiwan reported *M. bovis* as cause of human infection in 0.5percent cases of human tuberculosis [8].

Figure 1. Lungs showing caseous exudate in a cow died of tuberculosis.

In Ireland in 1980s, 4 to 6percent cases of human tuberculosis were associated with *M. bovis*, while it decreased to 3 percent in 2006 [9]. In Nigeria in 1986, 4percent of human cases were due to *M. bovis* and in 2005-06; these were 5percent [10, 11]. *M. bovis* was linked with 17percent cases of TB meningitis in India [12]. In Eastern Germany, in 1982, *M. bovis* was the cause of 4.6percent of newly diagnosed pulmonary cases with a decrease in percentage in the following years (1983 and 1984), i.e., 3.5 and 3.1percent, respectively [13]. In Tanzania, *M. bovis* was isolated in 16percent cases of human lymphadenitis [14]. It has been reported that patients with *M. bovis* were 2.5 times more likely to die during the treatment than those with *M. tuberculosis* [7].

The source for human infection is contaminated milk, meat, saliva of animals, air, soil and water [15, 16, 17, 18, 19]. However, the main source (*M. bovis*) to human transmission is the infected milk or meat causing infection in the gastrointestinal tract in human, although respiratory disease can also occur [15]. The pulmonary and extra-pulmonary form of disease by *M. bovis* varies in different studies. In Taiwan, of the 15 (0.5%) patients having *M. bovis* infection, 13 (87%) had pulmonary TB, one had both pulmonary and extra-pulmonary TB, and one had extra-pulmonary TB [8]. In United States, *M. bovis* was found associated with 48percent of extra-pulmonary, 26percent of pulmonary and 26percent of pulmonary and extra-pulmonary infections in one study and 54percent of pulmonary and 46percent of extra-pulmonary in another study with an overall infection of 1.5and 10percent by *M. bovis* in tuberculosis cases, respectively [6,7]. The risk of human infection in these studies has been linked withunpasteurized, un-boiled milk, using raw milk for producing cream, butter, or curd [15]. Even pasteurization does not kill certain Mycobacterium species [20]. The infection from animal to animal occurs through colostrum/milk to calves, ingestion of infected flies (*Muscadomesticus*), droppings of birds, aerosol, contact with each other and wildlife, excreta of flies, etc. The causative agent of tuberculosis can remain viable in the soil for up to two years [21, 22, 23, 24, 25, 26]. This highlights the reason that why this disease could not be eliminated/eradicated from the world and did not vanish on its own. It was believed that human pathogen;*M. tuberculosis* is evolved from *M. bovis* and then developed into a phylogenetically distant species. However, recent studies suggest that this is probably not the case and *M. tuberculosis* and *M. bovis* are independent. Instead, it is supposed that *M. tuberculosis* appears to be more closely related to the ancestral tubercle bacilli and has undergone fewer DNA deletions than *M. bovis* [27]. This data indicate the significance of *M. bovis* as cause of human disease over times in different regions of the world.

2. A Brief Account of Prevalence of Bovine Tuberculosis in Pakistan

Bovine tuberculosis is an important zoonotic disease of animals characterized by progressive emaciation and formation of tuberculous nodules, mainly in the lungs and regional lymph nodes in animals. The OIE has classified this disease as a list B disease, a disease of public health and socioeconomic importance [3]. The disease is present all over the world but the prevalence of bovine tuberculosis is higher in developing countries. Some studies in Pakistan on bovine tuberculosis in cattle, buffalo, sheep and goat have been carried out. The study in 1969 indicated a prevalence of 6.72 percent in animals in Faisalabad [28]. In

1972, it was 2.9, 1.6 and 8.6 percent in buffalo, Australian and Sahiwal cattle, respectively; in 1974 in Quetta, it was 0.53percent in buffaloes [29, 30]. In 1989, it was 7.3percent in cattle and buffaloes being slaughtered at Lahore abattoir [31]. In 2001, one study reported 1.8percent prevalence in buffaloes in Faisalabad [32]. In 2003, the respective prevalence of bovine tuberculosis in buffaloes and cattle in Lahore was 6.9and 8.6percent, respectively [33]. In 2006, the prevalence in buffaloes at two farms was 2.45 and 8.5percent [34]. A study recent during reported 3 percent prevalence of tuberculosis in buffaloes around two cities of Pakistan [35].

3. Prevalence and Risk Factors Identified in Small Ruminants [36]

The goat farms showed 100 percent herd prevalence, while sheep farms showed 86 percent of herd prevalence. Relatively higher percentage of positive animals in goats than sheep has been related with probable resistance to tuberculosis by local sheep breeds. A higher prevalence of tuberculosis in small ruminants at farms where large ruminants were also present than farms where only small ruminants were present and has been linked with spread of infection by large ruminants to small ruminants. However, molecular studies are suggested to clarify these facts. A very low prevalence has been reported in sheep kept at homes in cities in small groups (1/254; 0.4%). The only positive sheep for tuberculosis was of Lohi breed more than five years old and weighing more than 40 kg. These results of very low prevalence in small ruminants were surprising as no eradication program in Pakistan is going on. The lower prevalence around two cities was linked with keeping very low number of these animals at home or Deras/farms [37]. Very few studies in Pakistan have been carried out to assess the prevalence of tuberculosis in sheep and goats, and 2percent prevalence in sheep and 0percent in goats has been documented [38]. That shows a strong reason that small ruminants probably rarely contract this disease and are probably slightly resistant to Mycobacterium. The published work from other parts of the world also suggests a lower prevalence of bovine tuberculosis in small ruminants than in cattle or buffaloes. Previously, 0.5 and 3.5percent prevalence in goats has been reported [39, 40]. It has been reported that small ruminants are probably incapable of maintaining the infection and are not involved in the maintenance of tuberculosis in livestock [41]. However, contrary reports are also published those suggest high susceptibility of goats and they can maintain infection in the herds [42].

An association of age and live weight with tuberculosis in small ruminants was found. Tuberculosis was about 6 times higher in older goats (>4 years). It has been established that the chances of tuberculosis increases with the increase in age [43, 44]. An association of live weight with tuberculosis in small ruminant is interesting as heavy and lightweight animals were more affected. It may be speculated as that the infection occurs in healthier animals and with the passage of time, the effected animals become week. Other studies also indicated an association of low body weight with tuberculosis [45].

Tuberculosis in sheep and goats also showed an association with breed of these animals in Pakistan. The prevalence of tuberculosis was higher in Thalli and Kajli breeds of sheep and was low in Pak-karakul and Kachhi breeds.

Figure 2. Sheep Farm in Pakistan, the herd is taken for grazing in the morning to the adjoining harvested agriculture Fields.

Figure 3.Fresh water being offered to a herd of goats.

Further, it was found that tuberculosis in Thalli breed could occur despite the fact that they are kept with goats, cattle/buffaloes or with multiple species of animals and under different management conditions.However, chances of occurrence of tuberculosis are higher in the Kajli breed when it is kept with multiple animal species including buffaloes, cattle and sheep and it probably does not contract the disease when kept alone. These results suggest some kind of interplay of genetics, epigenetic/environmental and Mycobacterium. Two breeds of sheep including Pak-karakul and Kachhi did not show tuberculous animals; may be they are genetically resistant but that need experimental studies to prove it. This is similar for goat breeds including Nachhi and Dera Din Pinah. The Teddy, Beetal and Pak-Angora breeds of goats are susceptible to tuberculosis irrespective of keeping them with cattle, buffaloes or sheep.

As for the usefulness of comparative intradermal tuberculin test in small ruminants is concerned both kind of reports are available, some indicating a higher sensitivity and specificity (83.7%, 100%, respectively) than the IFN-gamma assay (83.7%, 96%, respectively) and the anamnestic ELISA (88.6%, 95.8%, respectively) [46], while others indicating a low sensitivity (44.6%) [42]. Thus, it can be inferred that the results of SCID Test vary under different geno-environmental conditions. Probably genetics of animals and/or the environmental factors plays some kind of role in immunoreactivity in small ruminants in different regions. There may be other factors involved as one study indicated that paratuberculosis interferes with the results of CID Test in animals [47]. Now, we can say that we need to be sure and conduct certain genetic studies to relate the resistance of animals to certain diseases, ruling out all other possible contributory factors such as interference of certain other diseases or other environmental factors including some dietary factors. Our experience show there may be some genetic factors involved in resistance to disease, especially for certain breeds, as we were unable to find positive reaction to avian PPD during our investigations on small ruminants. However, it does not mean that paratuberculosis is not present in Pakistan in small ruminants and thus we suggest a comprehensive study in Pakistan and other parts of the world to claim the genetic resistance of certain breeds of animals.

4. PREVALENCEAND RISK FACTORS IDENTIFIED IN CATTLE AT 11 EXPERIMENT STATIONS [48]

In Pakistan, some work on bovine tuberculosis in animals has already been done but the sample size was small. A study was carried out on a considerably large sample size (on 11 large cattle Farms of more than 40 years old) in Punjab, Pakistan to have a better picture of the disease. At the farms, animal-based prevalence recorded was 8percent, while herd based prevalence was 100percent. The latter is definitely a worrying factor. The possible explanation given for relatively low animals-based prevalence is the normal annual culling of animals, which are low producers and/or identified with certain disease problems where treatment failed. The other contributory factors in lower prevalence may be of introduction of new young healthy animals and the open/semi-close housing system that allow natural disinfection by direct sunlight as the country mostly has dry and sunny environmental conditions. The other management practices probably also help to keep the infection rate low including grazing practices adopted on vast cultivated/harvested fields from early morning to late afternoon. The animals are brought directly to milking parlors where hand milking is done with one attendant milking about 10 animals (though vary from farm to farm). After milking, the animals are taken to capacious open houses in summer and semi-close houses in winter with availability of water round the clock those were already exposed to direct sunlight for about more than 6 hours. The prevalence recorded at these 11 farms varied with < 5 percent at 3 farms, 5 to 10percent at 6 farms and > 10percent at 2 farms.

This study reported an association of tuberculosis with age and number of calving in cattle at these 11 farms, which is quite understandable as more time spent on the farm by an animal, there, are more chances that the animal can contract the infection. There was also an association found with milk production with the drop in later along with shortening of lactation length. The age association has been reported from world-over and the susceptibility to *M. bovis* is related with the increase in age of cattle [49] as the disease in older cattle was higher than the yearlings and calves [14] and it was twice as high in cows as in heifers [50]. Similar findings were from India, where an increase in prevalence with age (1.7% in cattle of 0.5-1 years to 22.3% of >five years of age) has been reported [51].

Figure 4.A large ruminant farm where cattle and buffaloes are kept in separate sheds.

The association of bovine tuberculosis with different breeds at these 11 farms of cattle was not found, although it was numerically higher in crossbred cattle than indigenous breeds of cattle. However, in Bangladesh and India, such an association has been reported with prevalence of tuberculosis as high as 7.8 percent in crossbred than 2.1 percent indigenous cattle in Bangladesh [52] and 26 percent in introduced breeds, 9.7 percent crossbred (9.7%) and 7 percent in local breeds [51] suggest some kind of resistance in local breeds of cattle. There might also be an association of presence of buffaloes at the farm with tuberculosis as odds of cattle being infected were 2.1 times higher when they were kept with buffaloes. The results of logistic analysis after controlling for farm, farm + breed revealed significant association of tuberculosis with age, number of calving, milk production and lactation length. Similarly, the results of multivariate logistic analysis including all variables in the model revealed a significant association of tuberculosis in cattle with the increase in age (OR=1.2), numbers of calving (OR = 1.5), per day milk (OR = 1.2) and lactation length (OR= 1.01). There was an interesting finding that the presence of sheep at the farm has shown a protective effect on the occurrence of bovine tuberculosis in cattle (log odd = 0.646). The latter has been related with grazing habit of sheep that they clean the field well and do not leave grass on the field and thus may also pick the causative agent of the disease (Mycobacterium) and kills it in its digestive system (needs to be proved).The role of sheep identified in Pakistan is that of protection to the animals in contracting the disease, while it has been reported that the risk of tuberculosis in cattle increases if they are kept with sheep (OR: 1.7) [53].

Figure 5. The milking area inside the shed, after milking the animals are shifted to open or semi-open areas at the farm.

5. Prevalence and Risk Factors Identified in Buffaloes around Two Cities [35]

There were some interesting data as at 57percent of the farms, total number of animals kept were less than 20; at 78percent of the farms total number of both buffaloes and cattle were less than 25and at 80percent of the farms total number of buffaloes were <20. We can see from here that the farming system in Pakistan in private sector is not fully developed and dairy animals are kept in small numbers with buffalo as the main dairy animal. The main reason of keeping buffalo as main dairy animals is that the buffalo milk (white in color) is preferred over cow milk (yellow in color) in local population along with the taste of buffalo milk is liked by the people.

The results of the recent study are quite encouraging in buffalo where 14percent herds had tuberculous animals and the overall prevalence of disease was 2.6percent in buffaloes around two cities of Pakistan. A study from Ethiopia reported herd prevalence of 19 percent with 1.6percent animal-based prevalence by using a comparative tuberculin test, while it was 51percent at herd-level and 4.1percent at animal level by using single intradermal tuberculin test [54]. Another study from Ethiopia, carried out by using comparative tuberculin test reported a much higher prevalence at herd-level (48%) and animal-level (19%) [55]. Similarly, herd prevalence of 47percent has been reported from Uganda by using comparative tuberculin test [56]. The low herd prevalence around two cities in Pakistan has been related with small herd size, as small as single animal. However, the herd prevalence increased to 18percent when herds with less than 10 dairy animals were excluded from the analysis and it became 19percent when herds with less than 10 buffaloes were excluded from the analysis.

The overall animal-level prevalence was 2.3 and 2.7percent at Okara and Faisalabad, respectively. The higher herd prevalence and lower animal-level prevalence are similar to studies reported from other countries in dairy animals [5, 55, 57]. The animal-level prevalence in Pakistan was even lower in positive herds suggesting that the transmission of disease within herds is not serious. This lower animal-based prevalence has been linked with keeping animals in Pakistan in an open environment and are offered fresh water (mostly hand pump water). Thus the source of disease transmission is suspected to be the manger and contaminated air as about two to four animals use one manger (mostly wooden) for fodder although many workers stated that the chances of disease transmission through direct contact from the infected animals is low [5, 58, 59]. In Pakistan, the dairy farming is under the process of development and it is supposed to be the next big industry in Pakistan in near future. Therefore, if we look at the data of tuberculosis at well-established farms where animal number is in hundreds we see much higher prevalence of tuberculosis and with the development of private farming system in future, the prevalence of disease is going to increase. In most other countries, where herd prevalence is high, the wild animals are acting as source of spread ofdisease to cattle. However, in Pakistan, grazing is very rarely practiced in private settingsmostly due to non-availability of cultivated land or the land is at far distance from where the animals are kept. Further,animals are fed in small numbers from two to four, are chained, and kept in open mainly around villages at Deras (huts near in the cultivated lands away from villages) those are without a boundary wall. The indoor keeping of animals is only practiced during winter rains and during nights of winter (about three months) and are fed the chopped green fodder with mixing of wheat straw and concentrate. These husbandry practices are probably strong factors to keep the prevalence of disease low, which is less than 5percent at animal-level, even though there are no test and slaughter programs running in Pakistan. The lower than 5 percent prevalence of bovine tuberculosis in Pakistan as indicated by many studies is ideal to start with test and slaughter policy to control/eradicate the disease from Pakistan [33, 34, 60, 61].

The crude or adjusted bivariate and multivariate logistic analysis revealed that the chances of infection with Mycobacterium are 1.8 times higher in dry, lactating and pregnant buffaloes (in that order). Similarly, crude or adjusted logistic analysis revealed 2 to 4 times higher risk of tuberculosis in buffaloes when they are kept with cattle, while it is 1.1 times higher with the increase in number of cattle at the farm. From these results, one can infer that the risk of tuberculosis in buffaloes is higher during pregnancy, if the cattle are present at the farm and if the number of cattle at the farm increases. The frequency analysis confirmed these results and in addition, the odds of buffalo having tuberculosis were 2.24 times higher if cattle+other animals were present at the farm. The presence of cattle at the farm and the higher chances of tuberculosis in buffaloes was also shown by stratified analysis after controlling for the breed. However, the results showed that the presence of small ruminants, dogs and equine at the farm are not associated with tuberculosis in buffaloes. The other factors including age, sex, breed, weight, milk production, number of calves produced and lactation length did not show association with tuberculosis in buffaloes around cities. Although, it has been well perceived that the older animals have more chances of being positive for tuberculosis than young animals [49, 53, 62]. From the results in buffaloes around two cities, an age association could not be proved. It may be because buffaloes around these two cities were kept in small numbers, mostly the young animals are kept, old animals are culled along with the contributory factors of keeping cattle with buffalo, and pregnancy stress in young animals predispose to infection with Mycobacterium. A study from Ethiopia reported an insignificant role of herd size in the prevalence of tuberculosis [53] but reports advocating the role of herd size are also published [57, 62]. The study in Pakistan indicated 0.9 times the risk of having tuberculosis at the farm with the increase ofone animal at the farm. So, the role of herd size depends on circumstances as what is the minimum herd size taken into consideration while carrying out the study as in the latter study majority of the herds were having <20 animals.

Figure 6. A wooden manger used for offering fodder to the animals, a total of four animals can share this type of manger. This cow was found positive for tuberculosis by tuberculin test.

Figure 7.A herd of buffalo being kept in private setting under shade of a large tree outside the village.

There was another interesting finding from the study under report, the chances of tuberculosis is 1.5 times higher in buffaloes if they are kept on west sides of the village. This is linked with the airflow (which in summer is mostly from east to west, the dust particles are easily suspended in the air in the scorching heat oftropical summer) and the village positioned in the way. The study in buffaloes showed higher chances of tuberculosis in Nili Ravi breed than Kundi breed of buffaloes (OR= 2.1). Further studies will clarify the predisposition of certain breeds of buffaloes to tuberculosis. The water as source of spread of tuberculosis may be out of question in study from Pakistan as the water source is mainly the hand pump, less so is the tube well or canal's water. The use of stagnant water in spread of disease has been reported in other reports [56]. There were some shortcomings in the study as the role of backyard poultry which finds their food in the mangers and the role of human as source of spread of disease to buffalo, especially those who are in contact with animals was not addressed. A study from neighboring country India speculated human-to-cattle transmission [63].

6. Prevalence of Bovine Tuberculosis in Zoo Animals in Pakistan [64]

There were few postmortem reports available indicating the presence of disease in zoo animals supported with laboratory results. This is why there was a need to carry out a study with use of intradermal tuberculin to investigate the magnitude of the problem. The only study conducted in Pakistan to investigate the disease in zoo animals reported a prevalence of around 3.3percent [64]. The disease in zoo animals is not only a threat to the workers and veterinarians working in the zoo but also to the general public that visits the zoo. There are reports of bovine tuberculosis in zoo animals from other countries [65, 66]. The disease in wild deer has been known that is acting as a reservoir for infection in cattle in many European countries [67, 68, 69, 70]. The tuberculous animals identified were one spotted deer, one Chinkaragazella and one Blackbuck gazelle [64]. All other animal species including hog deer, grey gorals, urial, mouflon, nilgai and zebra were found negative. This does not mean that these animals are resistant to tuberculosis. The prevalence of tuberculosis in animals of Bovidae (3.6%) and Cervidae (3.2%) family was almost similar [64]. Results revealed that there are 32percent higher chances for females to contract the disease than males, while in Cervidae family there are more chances of a positive test in males. We should interpret these results with little caution as the data is small and it requires further studies. The results in zoo animals also revealed that there are 19percent higher chances of contracting tuberculosis if animal number is less than 10. Now this is something we do not often see as the larger the herd the higher are the chances of infection. The possible explanation about these findings can be that the social setup among animals might be a contributory factor, e.g., when the animals are in small number, the fodder offered to them is also in small quantity and that brings social dominance into play for feeding. Contrary, if the animal number is large, the fodder offered to the animals is also in large quantity that gives more chances to weaker animals to get some share. This is just a presumptive explanation, there may be many other factors involved, and as the data is also from small sample size, there is a need to have further studies to clarify the real interplay of social dominance of fodder offering. The possible explanation for females to contract the disease is the pregnancy and parturition stresses those bringing the immune system of the animal down. This study could not find an association of age with tuberculosis in zoo animals. However, another study from Spain reported a significant association of age with tuberculosis in deer [67]. Further, it has been reported in a study from New Zealand that factors including age, environment, population density, exposure and genetics play role in deer susceptibility [71]. Another study reported the prevalence of disease in deer between the ages from 1.5 to 5.5 years [72]. It was quite interesting not to find any positive animals among the animals of more than eight years old [64]. Another finding of animals having lower live weight had higher prevalence of tuberculosis. Therefore, as previously explained, the social setup among the deer might have played some role in causing general body weakness and lowering the immune system. Another possibility is that the animals might have become weaker with the passage of time for being infected. It may be possible that both the factors were involved. Tuberculosis in zoo animals has been reported in monkeys and many other animal species including elephants [73]. An outbreak of tuberculosis in elephants, giraffes, rhinoceroses and buffaloes has been reported in a Swedish zoo [74].

7. Future of Diseasein Pakistan and Action Plan

These results show the presence of tuberculosis in different animal species in Pakistan. Although it is not more than 10percent in animals but is present more or less at 100percent of public farms. This is quite an alarming situation as the milk from these farms is sold untreated in the open market. The consumers buy it and some use it without boiling although most of the consumers use milk after boiling. The under-boiling conditions do exist and majority of people are unaware of the threat of tuberculosis from consumption of un-boiled or under-boiled milk. As the Government in Pakistan in recent times has paid more attention to the Livestock and thus the Livestock farming is on the increase. It is quite obvious that the situation of bovine tuberculosis in coming years is speculated to become worse. The above data about bovine tuberculosis in Pakistan is from the time when the private farming was almost non-existent and livestock were mainly kept at homes in less than 10 animalsgroup or is at the stage of early development. Further, the most of the old studies are conducted with the use of slightly low quality PPDs, the test was conducted only once, and the doubtful reactors were not followed in most studies to rule out the real situation of tuberculosis in animals. However, the negative tuberculin test does not mean that the animals are real-negative to bovine tuberculosis as a recent study reported that 19percent of tuberculin negative (apparently healthy) cattle were culture positive for *M. bovis* and 15 percent of milk samples were also positive for *M. bovis* [75]. Thus it can be said that the situation of bovine tuberculosis in near future in Pakistan is perhaps going to worsen and it is the time to take some concrete steps in the right direction, not only to protect Pakistani people but also people of other nations living in Pakistan and further to have export of safer animal products.

One of the most important steps required in this direction is to have test and cull programs in livestock in place as is the case in most of the industrialized countries; there the eradication programs of bovine tuberculosis have significantly reduced the prevalence of this disease in animals and thus in humans. Secondly, the sale of fresh milk in the open market should be banned and Government should provide subsidized pasteurized milk to the people. There is also need to establish a zoonotic disease control center/lab/institute in the country and to make a special slaughter/postmortem block to carry out slaughtering/postmortem of animals confirmed or suspected for zoonotic diseases. These programs are need of the time and Pakistan cannot afford to further delay such vital programs. The culling of diseased animals is not only going to boost the animal production but with supply of pasteurized milk to people will make a healthier nation. Bovine tuberculosis is one of the important zoonotic diseases that is prevalent in livestock and zoo animals in Pakistan [35, 36, 48, 64]. Looking at the prevalence of disease in zoo animals in Pakistan it is emphasized that there should be routine testing of zoo animals for tuberculosis. The tuberculin testing is still regarded as an efficient test for various animal species.

ACKNOWLEDGMENTS

Funding provided by Pakistan Science Foundation under project grant PSF/Res/P-AU/AGR (283) and bovine PPD donated by InstitutoZooprofilattico, Perugia, Italy for this project study on bovine tuberculosis in small and large ruminants is highly acknowledged.

REFERENCES

[1] Griffith AS. Infections of wild animals with tubercule bacilli and other acid-fast bacilli.*Proc. R. Soc. Med.*, 1939; 32: 1405–1412.

[2] Francis J. *Tuberculosis in animals and man: a study in comparative pathology*. London: Cassel and Co., 1958;pp: 16.

[3] OIE. *Bovine Tuberculosis*, 2005. http://www.cfsph.iastate.edu/Factsheets/pdfs/bovine_tuberculosis.pdf (accessed 5 April 2005).

[4] Taylor GM, Murphy E, Hopkins R, Rutland P, Chistov Y. First report of M. bovis DNA in human remains from the Iron Age.*Microbiology,* 2007; 153: 1243–1249.

[5] O'Reilly LM, Daborn CJ. The epidemiology of M. bovis infections in animals and man: a review. *Tuber. Lung Dis.,*1995; 76(Suppl1):S1–46.

[6] Hlavsa Michele C, Moonan PK, Cowan LS., Navin TR, Kammerer JS, Morlock GP, Crawford JT, Philip AL.Human Tuberculosis due to Mycobacterium bovis in the United States, 1995–2005.*Clinical Infectious Diseases,*2008; 47:168–75.

[7] Rodwell TC, Moore M, Moser KS, Brodine SK, Strathdee SA. Tuberculosis from M. bovis in Binational Communities, United States.*Emerging Infectious Dis.,*2008; 14: 909-916.

[8] Jou R, Huang WL, Chiang CY. Human tuberculosis caused by M. bovis, Taiwan. *Emerg. Infect. Dis.*, 2008;14: 515–7.

[9] OjoOlabisi, Sheehan S, Corcoran GD, Okker M, Gover K, Nikolayevsky V, Brown T, Dale J, Gordon SV, Drobniewski F, Prentice MB. M. bovis Strains Causing Smear-Positive Human Tuberculosis, Southwest Ireland.*Emerging Infectious Dis.*, 2008; 14: 1931-1934

[10] Idigbe EO, Anyiwo CE, Onwujekwe DI. Human pulmonary infections with bovine and atypical mycobacteria in Lagos, Nigeria.*J. Trop. Med. Hyg*.,1986; 89: 143–148.

[11] Cadmus S, Palmer S, Okker M, Dale J, Gover K, Smith N, et al. Molecular analysis of human and bovine tubercle bacilli from a local setting in Nigeria. *J. Clin. Microbiol.*, 2006;44:29–34. DOI: 10.1128/JCM.44.1.29-34.2006

[12] Shah NP, Singhal A, Jain A, Kumar P, Uppal SS, SrivastavaMVP, et al. Occurrence of overlooked zoonotic tuberculosis:detection of M. bovis in human cerebrospinal fluid. *J. Clin. Microbiol.*,2006; 44: 1352-8.

[13] Kappler W, Kalich R, Fischer P. Incidence and significance of lung diseases caused by tuberculosis bacteria and atypical mycobacteria in East Germany. *Z. Erkr.Atmungsorgane*, 1986; 167: 42–46.

[14] Kazwala RR, Kambarage DM, Daborn CJ, Nyange J, Jiwa SF, Sharp JM. Risk factors associated with the occurrence of bovine tuberculosis in cattle in the Southern Highlands of Tanzania. *Vet. Res. Commun.*,2001; 25: 609-614.

[15] Cotter TP, O'Shaughnessy E, Sheehan S, Cryan B, Bredin CP. Human M. bovis infection in the southwest of Ireland 1983–1992: a comparison with M. tuberculosis. *Ir. Med. J.,* 1996; 89: 62–63.

[16] Marshal CJ. Progress in controlling bovine tuberculosis.*J. Am. Vet .Med. Assoc*., 1932; 33: 625.

[17] Wendt SL, George KL, Parker BC, Gruft H, Falkinham JO. Epidemiology of infection by nontuberculous Mycobacteria- III.Isolation of potentially pathogenic Mycobacteria from aerosols.*Am. Rev. Respir.Disease*,1980; 122: 259-263.

[18] Brooks RW, Parker BC, Gruft H, Falkinham JD. Epidemiology of infection by nontuberculous Mycobacteria - V. numbers in eastern United States soils and correlation with soil characteristics.*Am. Rev. Respir. Disease,*1984; 130: 630-633.

[19] Falkinham JD, Parker BC, Gruft CB, Epidemiology of infection by non-tuberculous Mycobacteria-I. Geographic distribution in the eastern United States.*Am. Rev. Respir. Disease*,1980; 121: 931-937.

[20] Grant IR, Ball HJ, Rowe MT. Thermal inactivation of several Mycobacterium spp. in milk by pasteurization. *Letter Appl. Microbiol.*,1996; 22: 253-256.

[21] Evangelista TBR,AndaJHD. Tuberculosis in Dairy Calves: risk of Mycobacterium spp. exposure associated with management of colostrum and milk. *Prevent. Vet. Med.,* 1996; 27: 23-27.

[22] Polyakov VA, Ishekenov MS, Kosenko VI. Flies - one of the links in the transmission mechanism of tuberculosis in animals. *ProblemyVeterinarnoiSanitariEkdojii.*, 1994; 93: 84-92.

[23] Hejlicek K, Trend F. Epidemiology of Mycobacterium avium infection in cattle. *Veterinarstvi,* 1995; 45: 351-354.

[24] Sauter CM, Morris RS.Behavioral studies on the potential for direct transmission of tuberculosis from feral ferrets and possums to farmed livestock. *New Zealand Vet. J.,* 1995; 43: 294-300.

[25] Barlow ND, Kean JM, Hickling G, Livingstone PG, Robson AB. A simulation model for the spread of bovine tuberculosis within New Zealand cattle herds. *Prevent. Vet. Med.,*1997; 32: 57-75.

[26] Hutchings MR, Harris S. Effects of farm management practices on cattle grazing behaviour and the potential for transmission of bovine tuberculosis from badgers to cattle. *Vet. J.*, 1997; 153: 149-162.

[27] Brosch R, Gordon SV, Marmiesse M, Brodin P, Buchrieser C, Eiglmeier K, et al. A new revolutionary scenario for M. tuberculosis complex.*Proc. Natl. Acad. Sci.USA,*2002; 99: 3684-9.

[28] Barya M. Incidence of bovine tuberculosis in West Pakistan. Pros. National conf.On control of communicable disease, Lahore. *Pub. Ass. Pak.*, 1969; Nov. 6-8.

[29] Akhtar S, Khan MI, Anjum AD.Comparative delayed cutaneous hypersensitivity in buffaloes and cattle; Reaction to Tuberculin Purified Protein Derivatives.*Buffalo J.*, 1972; 8: 39-45

[30] Khilji IA. Incidence of tuebrculosis amongst Kundi buffaloes.*Pak J Anim. Scie.*,1974; 13: 27-31.

[31] Amin S. Prevalence of buffalo tuberculosis by using short thermal test and identification of organism from lymph nodes. MSc (Hons) Thesis, CVS, *Univ. Agric. Faisalabad.* 1989.

[32] Ifrahim. *Epidemiological studies on tuberculosis in cattle and buffalo population in villages around Faisalabad.* Msc (Hons) thesis, Department of Veterinary Microbiology, UAF. 2001.

[33] Jalil H, Das P, Suleman A. *Bovine tuberculosis in dairy animals at Lahore, threat to the public health.* Metropolitan Corporation Lahore, Pakistan. 2003; http://priory.com/vet/bovinetb.htm.

[34] Javed MT, Usman M, Irfan M, Cagiola M. A study on tuberculosis in buffaloes: some epidemiological aspects, along with haematological and serum protein changes.*VeterinarskiArhiv,*2006; 76: 193-206.

[35] Javed M. Tariq Javed, A. Latif Shahid, Farooq A. Farooqi, M. Akhtar, Gabriel A. Cardenas, M. Wasiq and Monica Cagiola. Association of some of the possible risk factors with tuberculosis in water buffalo around two cities of Punjab Pakistan. *ActaTropica.,*2010; 115: 242-247.

[36] Javed MT , Munir A, Shahid M, Severi G, Irfan M, Aranaz A, Cagiola M. Percentage of reactor animals to SCCIT in small ruminants in Punjab Pakistan. *ActaTropica,*2010; 113: 88-91.

[37] DeLisle GW, Mackintosh CG,Bengis, RG.Mycobacterium bovis in free-living and captive wildlife, including farmed deer.*Revue scientifiqueet technique (International Office of Epizootics),*2001; 20: 86–111.

[38] Ashraf M, Khan MZ,Chishti MA. Incidence and pathology of lungs affected with tuberculosis and hydatidosis in sheep and goats. *Pakistan Vet. J.*, 1986; 6: 119-120.

[39] Arellano RB, RamirezCIC, Diaz AE, Valero EG,Santillan FMA. Diagnosis of tuberculosis in goat flocks using the double intradermal test and bacteriology. *TecnicaPecuaria en Mexico.*,1999; 37: 55-58.

[40] Sanson RL. *Tuberculosis in goats.Surveillance*, 1998; 15: 7-8.

[41] Coleman JD, Cooke MM.Mycobacterium bovis infection in wildlife in New Zealand. *Tuberculosis,*2001; 81: 191-202.

[42] Liebana E, Aranaz A, Urquia JJ, Mateos A,Domínguez L. Evaluation of the gamma-interferon assay for eradication of tuberculosis in a goat herd. *Aust .Vet. J.,*1998; 76: 850-53.

[43] Guindi SM, Lofty O,Awad WM. Some observations regarding the infectivity and sensitivity for tuberculosis in buffaloes in Arab Republic of Egypt.*J. Egyp. Vet. Med. Assoc.,*1975; 125-138.

[44] Siva R, William BJ, Rao GD, David WPA,Balasubramanian NN. Radiological diagnosis of respiratory diseases in large animals - a survey report.*Cheiron*,1997; 26: 9-11.

[45] Rao VNA, Ramadas P,Dhinakran M. A study on the effect of tuberculosis on body weight and haemogram values in cattle. *Cheiron,*1992; 21: 1.

[46] Gutiérrez M, Tellechea J,García MJF. Evaluation of cellular and serological diagnostic tests for the detection of Mycobacterium bovis infected goats. *Vet. Micro.*, 1998; 62: 281-90.

[47] Alvarez J, deJuan L, Bezos J, Romero B, Saez JL, Reviriego GFJ, Briones V, Moreno MA, Mateos A, Dominguez L,Aranaz A. Interference of paratuberculosis with the diagnosis of tuberculosis in a goat flock with a natural mixed infection. *Veterinary Microbiology*,2008; 128: 72-80.

[48] Javed MT, Irfan M, Ali I, Farooqi FA, Wasiq M,CagiolaM. Risk factors identified associated with tuberculosis in cattle at 11 livestock experiment stations of Punjab Pakistan. *ActaTropica*,2011; 117: 109-113.

[49] Phillips CJ, Foster CR, Morris PA, Teverson.Genetic and management factors that influence the susceptibility of cattle to Mycobacterium bovis infection.*Anim. Hlth. Res Rev.,*2002; 3: 3-13.

[50] Bonsu OA, Laing E, Akanmori BD. Prevalence of tuberculosis in cattle in the Dangme-west district of Ghana, Public Health Implication. *ActaTriopica*,2000; 76: 9-14.

[51] Sharma AK, Vanamayya PR, Parihar NS. Tuberculosis in cattle: a retrospective study based on necropsy. *Indian J. Vet. Path.*, 1985; 9, 14-18.

[52] Samad MA, Rahman MS.Incidence of bovine tuberculosis and its effect on certain blood indices in dairy cattle of Bangladesh. *Indian J. Dairy Sci.*, 1986; 39: 231-234.

[53] Tschopp R, Schelling E, Hattendorf J, Aseffa A, Zinsstag J. Risk factors of bovine tuberculosis in cattle in rural livestock production systems of Ethiopia. *Prev. Vet. Med.*, 2009; 89: 205–211.

[54] Laval G, Ameni G. Prevalence of bovine tuberculosis in zebu cattle under traditional animal husbandry in Boji district of western Ethiopia. *Revue De MedecineVeterinaire,* 2004; 155: 494-499

[55] Shitaye JE, Getahun B, Alemayehu T, Skoric M, Treml F, Fictum P, Vrbas V, Pavlik I. A prevalence study of bovine tuberculosis by using abattoir meat inspection and tuberculin skin testing data, histopathological and IS6110 PCR examination of tissues with tuberculous lesions in cattle in Ethiopia.*Vet. Med.*, 2006; 51: 512–522.

[56] Oloya J, Muma JB, Opuda-Asibo J, Djonne B, Kazwala R, Skjerve E. Risk factors for herd-level bovine-tuberculosis seropositivity in transhuman cattle in Uganda. *Prev. Vet. Med.,* 2007; 80: 318–329.

[57] Ameni G, Amenum K, Tibbo M. Bovine tuberculosis: prevalence and risk factors assessment in cattle and cattle owners in Wuchale-Jida district, Central Ethiopia. *Int. J. Appl. Res. Vet. Med.,*2003; 1: 17–25.

[58] Flangan PA, Kelly G. A study of tuberculosis breakdowns in herds in which some purchased animals were identified as reactors. *Ir. Vet. J.,*1996; 49: 704-6.

[59] Morrison WI, Bourne FJ, Cox DR, Donnelly CA, Gettinby G, McInerney JP, Woodroffe R. Pathogenesis and diagnosis of infections with Mycobacterium bovis in cattle. *Vet. Rec.,* 2000; 146: 236–242.
[60] Javed MT, Latif M, Irfan M, Ali I, Khan A, Wasiq M, Farooqi FA, Shahid AL, Cagiola M. Haematological and serum proteins values in tuberculin reactor and non-reactor water buffaloes, cattle, sheep and goats. *Pakistan Vet. J.,* 2010; 30: 100-104.
[61] Usman M. Prevalence and Pathological Studies on Tuberculosis in Buffaloes At Various Livestock Farms. M. Sc. (Hons) Thesis. University of Agriculture, *Faisalabad,* Pakistan. 2003.
[62] Cleavelan S, Shaw DJ, Mfinanga SG, Shirima G, Kazwala RR, Eblate E, Sharp M. Mycobacterium bovis in rural Tanzania: risk factors for infection in human and cattle populations. *Tub.,*2007; 87: 30–43.
[63] Srivastava K, Chauhan DS, Gupta P, Singh HB, Sharma VD, Yadav VS, Sreekumaran TSS, Dharamdheeran JS, Nigam P, Prasad HK, Katoch VM. Isolation of Mycobacterium bovisand M. tuberculosis from cattle of some farms in north India - Possible relevance in human health.*Indian J. Med. Res.,* 2008; 128: 26-31.
[64] Shahid AL, Javed MT, Khan MN,Cagiola M. Prevalence of Bovine Tuberculosis in Zoo Animals in Pakistan on the Basis of Single Comparative Cervical Intradermal Tuberculin Test (SCCIT). *Iranian J. Vet. Res.*, (accepted) 2011.
[65] Sternberg S, Bernodt K, Holmström A, Röken B. Survey of tuberculin testing in Swedish zoos. *J. Zoo Wildl Med.*, 2002; 33: 378-80.
[66] Kiers A, Klarenbeek A, Mendelts B, Van Soolingen D, Koëter G. Transmission of Mycobacterium pinnipedii to humans in a zoo with marine mammals. *Int. J. Tuberc. Lung Dis.,* 2008; 12: 1469-73.
[67] Jaroso R, Vicente J, Marta-n-Hernando MP, Aranaz A, Lyashchenko K, Greenwald R, Esfandiari J, Gortazar C. Ante-mortem testing wild fallow deer for bovine tuberculosis. *Vet. Microbiol.,*2010; 146: 285-289.
[68] Ward AI, Smith GC, Etherington TR, Delahay RJ. Estimating the risk of cattle exposure to tuberculosis posed by wild deer relative to badgers in England and Wales.*J. Wildl Dis.,* 2009; 45: 1104-20.
[69] Surujballi O, Lutze-Wallace C, Turcotte C, Savic M, Stevenson D, Romanowska A, Monagle W, Berlie-Surujballi G, Tangorra E, (2009). Sensitive diagnosis of bovine tuberculosis in a farmed cervid herd with use of an MPB70 protein fluorescence polarization assay. *Can. J. Vet. Res.,* 2009; 73: 161-6.
[70] Aranaz A, de Juan L, Montero N, Sanchez C, Galka M, Delso C, Alvarez J, Romero B, Bezos J, Vela AI, Briones V, Mateos A, Domínguez L. Bovine tuberculosis (Mycobacterium bovis) in wildlife in Spain. *J. Clin. Microbio.*,2004; 42: 2602–2608.
[71] Mackintosh CG, de Lisle GW, Collins DM, Griffin JF. Mycobacterial diseases of deer.*New Z. Vet. J.,* 2004; 52: 163-74.
[72] Schmitt SM, Fitzgerald SD, Cooley TM, Bruning-Fann CS, Sullivan L, Berry D, Carlson T, Minnis RB, Payeur JB, Sikarskie J. Bovine tuberculosis in free-ranging white-tailed deer from Michigan. *J. Wildl Dis.,* 1997; 33: 749-58.
[73] Une Y, Mori T. Tuberculosis as a zoonosis from a veterinary perspective. *Comp. Immunol. Microbiol. Infect. Dis.,* 2007; 30: 415-25.

[74] Lewerin SS, Olsson SL, Eld K, Roken B, Ghebremichael S, Koivula T, Kallenius G, Bolske G. Outbreak of Mycobacterium tuberculosis infection among captive Asian elephants in a Swedish zoo. *Vet. Rec.,*2005; 156: 171–175.

[75] Bose M. Natural reservoir, zoonotic tuberculosis and interface with human tuberculosis: An unsolved question. *Indian J. Med. Res*., 2008; 128: 4-6.

In: Livestock: Rearing, Farming Practices and Diseases
Editor: M. Tariq Javed
ISBN 978-1-62100-181-2

Chapter 11

PARATUBERCULOSIS (JOHNE´S DISEASE): CLINICAL SIGNS, DIAGNOSIS, LESIONS, PROPHYLAXIS/ TREATMENT/CONTROL AND ZOONOTIC POTENTIAL

Elena Castellanos,[1] Lucas Domínguez,[1,2] Lucía de Juan,[1,2] and Alicia Aranaz[1]

[1]Department of Animal Health. Faculty of Veterinary Sciences, Universidad Complutense de Madrid. Spain

[2]Animal Health Surveillance Center (VISAVET), Universidad Complutense de Madrid, Spain

ABSTRACT

Paratuberculosis (Johne's disease) is a chronic inflammatory disease of the gastrointestinal tract caused by *Mycobacterium avium* subspecies *paratuberculosis*(*M. a. paratuberculosis*). Paratuberculosis affects mainly ruminants (domestic and wild), but it has also been reported in monogastric animals and the main route of transmission is the fecal-oral.

There are no clinical signs that are pathognomonic of the disease, but it is very characteristic to observe progressive weight loss in the infected animals. The disease hasbeen divided into four stages: silent infection, subclinical, clinical and advanced clinical stage, where in the latter stage animals present intermandibularedema due to hypoproteinemia, emaciation, profuse diarrhea, generalized muscle atrophy and in some cases, death.

The diagnosis of the disease can be made by the direct detection of the agent, *M. a. paratuberculosis*(staining, culture –conventional, radiometric, non-radiometric, or direct DNA extraction), the detection of the host immune response by serological tests (complement fixation test, agar-gel immunodifusion or enzyme-linked immunosorbent assay) or by tests that measure the cell mediated immune response (gamma interferon assay or delayed-type hypersensitivity).Moreover, some studies have reported *M. a. paratuberculosis*as the etiological agent of Crohn's disease in humans. This hypothesis shows the possible zoonotic potential of this organism.

1. Introduction

Paratuberculosis (Johne's disease) is a chronic inflammatory disease of the gastrointestinal tract that leads to a progressive and debilitating condition and is caused by *Mycobacterium avium* subspecies *paratuberculosis* (*M. a. paratuberculosis*). This microorganism is clustered into the Genus *Mycobacterium* and the *Mycobacterium avium* complex (MAC), a group of acid-fast Gram-positive bacilli that include human pathogens as well. Paratuberculosis is found worldwide and affects mainly ruminants (both domestic and wild ruminants), but it has also been described in monogastric animals (horses, pigs, dogs, primates, etc.) [1].

The main route of transmission is the fecal-oral via contaminated pasture, milk or colostrum. However, other routes of infection have been reported such as the intravenous [2], intramammary[3] or the intrauterine [4], among others. Other than these, another possibleroute of infection that has been proposed is predation in the case of carnivores. In one study the isolation of *M. a. paratuberculosis* in predators was 62percent in comparison to 10percent isolation of *M. a. paratuberculosis* in prey [5, 6]. This fact supports the existence of a possible role of wild animals in the epidemiology of the disease [7]. *M. a. paratuberculosis* has been isolated from an extensive number of domestic hosts and their products. Therefore, the consumption of these products or the direct contact with these animals could act as a potential source of zoonosis [7, 8] (Figure 1).

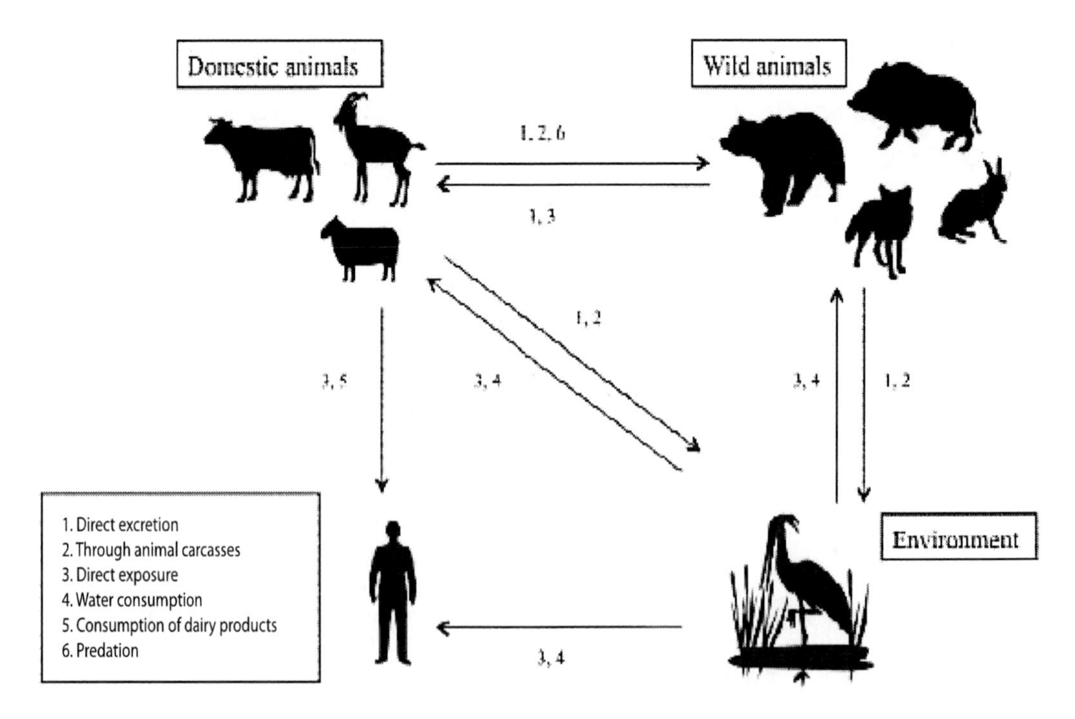

Figure 1. Possible routes of transmission of paratuberculosis between domestic animals, wild animals, environmental reservoirs and humans (adaptation of Biet et al. 2005 [7]).

2. Clinical Signs and Progression of the Disease in Livestock

The progression of the disease has been divided into four stages. There are no pathognomonic clinical signs of the disease but there are clinical signs that are variable and characteristic of each phase [9]:

- Stage I: Silent infection. This phase is characterized by the absence of clinical signs and it normally occurs in young animals (calves, heifers and young stock), but can also occur in adult animals that are exposed to *M. a. paratuberculosis*. Animals in this stage of the disease are difficult to diagnose with the current methods.
- Stage II: Subclinical stage. This occurs mainly in adult animals and older heifers. Animals do not show clinical signs but diagnostic tests can sometimes detect circulating antibodies or altered immune responses. The majority of the animals are fecal culture negative, but there are also animals that act as intermittent fecal shedders.
- Stage III: Clinical stage. This only occurs after a few years of infection with *M. a. paratuberculosis*. During this stage the main clinical signs are a reduction in milk production, weight loss, diarrhea and decreased hair quality (Figure 2).
- Stage IV: Advanced clinical stage. The onset of this phase occurs a few weeks after the establishment of the clinical stage. The animals in this stage present lethargy, weakness, emaciation, intermandibularedema due to the presence of hypo-proteinemia, cachexia and profuse diarrhea. Frequently, animals are slaughtered earlier due to decreased milk production and weight loss; sometime animals die due to severe dehydration and cachexia.

It is generally considered that in herds with paratuberculosis for every positive animal detected, there are 25 animals in the herd that are not diagnosed and clinicaly diseased animals only represent the "peak of the iceberg" (Figure 3). Current diagnostic tests are only capable of detecting animals in the clinical phase of the disease (which represent only 15-25percent of the infected animals in the herd).

*M. a. paratuberculosis*can also remain in the gastrointestinal tract of sub-clinically infected animals but these animals can become clinically apparent during stress periods such as lactation or high animal density, etc. [10].

Although young animals are very susceptible to the disease (which could take from months to years until its clinical onset), the disease could also occur in adult animals if they are exposed to high bacterial loads [11]. There are also different factors that are associated to the host disease susceptibility such as species or breed. Thus, guinea pigs and rats are particularly resistant to the disease, whereas young deer and goats are highly susceptible. There are certain breeds of cattle such as Channel Island, Limousine and Welsh Black that have been shown to be highly susceptible [1].

Figure 2. Cow in the clinical stage of the disease (image obtained in Animal Health Surveillance Centre, VISAVET and Faculty of Veterinary Medicine, Universidad Complutense de Madrid, Spain).

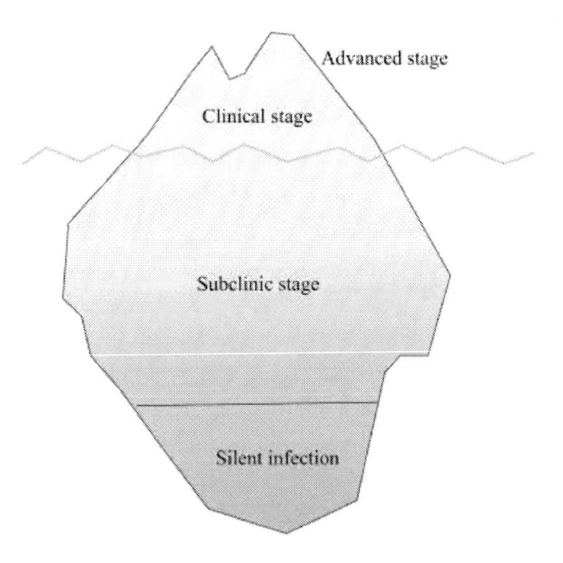

Figure 3. Simile between the capacity of tests to diagnose infected animals in a herd and an iceberg. Clinical cases that are detectable correspond to the visible part of the iceberg, whereas there are many animals that are not detected (false negatives animals) and maintain the infection within the herd.

3. Clinical Signs and Progression of the Disease in Small Animals and Wild Ruminants

Clinical signs of the disease in sheep and goats are very similar to those described in cattle although the presence of diarrhea is less frequent than in cattle and the onset of the disease occurs in younger animals [12]. In case of non-domestic ruminants, clinical signs are more variable but more similar to those described in cattle.

4. DIAGNOSIS OF THE DISEASE

4.1. Direct Detection of *M. A. Paratuberculosis*

4.1.1.Macroscopic Diagnosis

There is no correlation betweenthe severity of clinical signs and the presence of macroscopic intestinal lesions. Nevertheless, there are gross lesions that are characteristic of the disease such as thickening of the intestinal mucosa and ileocecal valve, edema of the intestine and lymph nodes and lymphangiectasia (Figure 4). However, to get an accurate diagnosis of the disease,it is necessary to perform histology of the different intestinal segments and the mesenteric lymph nodes along with culture and isolation of*M. a. paratuberculosis*[13].

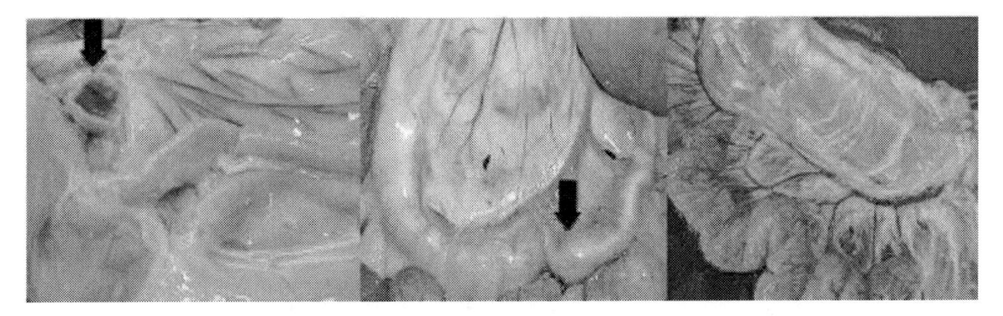

Figure 4. From left to right: edema and enlargement of the ileocecal valve (black arrow), mesenteric lymphnode (black arrow) and lymphangiectasia of an animal in an advanced status of paratuberculosis (image obtained in Animal Health Surveillance Centre, VISAVET and Faculty of Veterinary Medicine, Universidad Complutense de Madrid, Spain).

4.1.2.Direct Staining

This method uses a Ziehl-Neelsen staining of smears of the intestinal mucosa or feces. The test is considered preliminary positive if bacterial aggregates of three or more microorganisms are observed.

4.1.3.Conventional Culture

It is considered the gold standard method in the diagnosis of the disease. The main disadvantage of this method is related to the slow growth rate of *M. a. paratuberculosis* (up to one year) and the presence of non-cultivable forms of *M. a. paratuberculosis*[14, 15]. The principal advantage of this assay is that it presents 100percent specificity in the detection of *M. a. paratuberculosis* [17].

Fecal culture in solid media can detect between 30-40percent of the infected cattle as it only detects animals that excrete more than 100 colony-forming units of organism per gram of feces[13, 16]. Therefore, this method only identifies animals in the advanced clinical phase of the disease (with a 100% sensitivity) but only detects a few animals in the initial stages [16].

The most widespread culture media used for the isolation of *M. a. paratuberculosis* are Herrold's egg yolk medium (HEYM), Middlebrook 7H9, 7H10 and 7H11, Dubos medium and Löwestein-Jensen [17]. One of the components that is present in all these media is mycobactin J (an iron chelated cell extract), which is necessary for *M. a. paratuberculosis* growth.

The method of sample processing has been well described by Greig and collaborators [6]. Briefly, tissues are homogenized in 10 ml of distilled water and then 10 ml of 1.5 percenthexadecylpyridinium chloride (HPC) is added. The mixture is incubated at room temperature for about 18 hours for a proper decontamination. Subsequently, supernatant is centrifuged at 3800 g for 30 minutes and the precipitate is homogenized in 10 ml of distilled water. Again, under the same conditions, the former solution is centrifuged and the resulting sediment is re-suspended in 0.5 ml of distilled water. Finally, 0.1 ml is inoculated in the solid culture media and incubated at 37ºC for at least three months until visible growth (Figure 5).Then, the identification of *M. a. paratuberculosis* can be performed directly from colonies by PCR targeting insertion sequences (IS*900*, ISMav2, IS*Mpa1,* IS*MAP0*, and IS*1311*) or other loci such as F57, *HspX,* Genes 251 and 255.

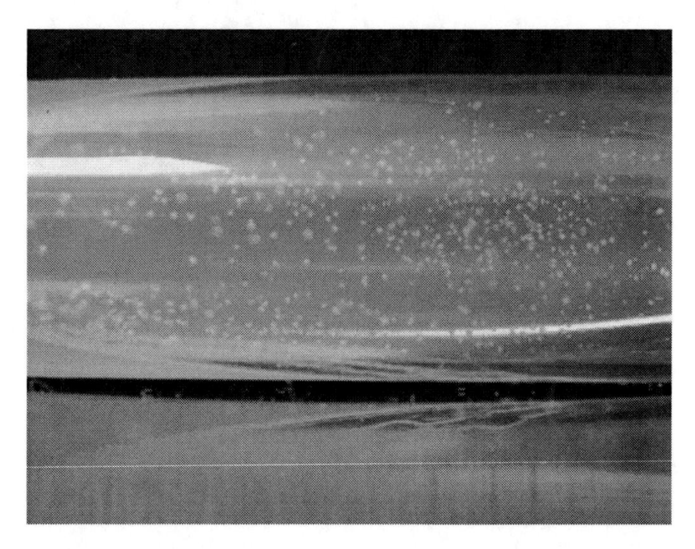

Figure 5. Colonies of *M. a. paratuberculosis* isolated in Middlebrook 7H11 medium with mycobactin J (image obtained in Animal Health Surveillance Centre, VISAVET and Faculty of Veterinary Medicine, Universidad Complutense de Madrid, Spain).

4.1.4.Radiometric Culture

Radiometric culture is based on the detection of isotope CO_2^{14} released from palmitic acid because of bacterial metabolism. This method is capable of detecting *M. a. paratuberculosis*growth faster than culture in solid media (in just nine days after inoculation of culture media) [18]. However, the drawback to this method is that the use of radioactive isotopes is restricted to certain laboratories and it is not cost effective.

4.1.5.Non-Radiometric Culture

This method employs culture media that incorporates a detection system that reacts when there is an alteration of oxygen concentration, carbon dioxide or pressure in each culture vial [13]. Results are automatically saved in a database. This method has been compared to conventional culture from cattle feces, where radiometric culture was capable of detecting *M.*

a. paratuberculosis faster than conventional culture method, constituting a good alternative for conventional culture [19].

4.1.6.Direct Extraction of DNA

DNA extraction can be performed directly from tissue samples or biological fluids, leading to rapid results in comparison to the culture method [20].

The main disadvantage of this method is the presence of PCR inhibitors, especially in fecal samples [20, 21]. Nonetheless, the determination of PCR inhibitors in the sample could be solved with the incorporation of internal controls [22], elimination ofthese inhibitors with a DNA purification stage [23] or by diluting the extracted DNA [20].

Direct extraction of *M. a. paratuberculosis* DNA has been used in samples directly obtained from feces [20, 24-28], milk [29-32], blood and tissue [33-37].

To date, this type of assay is not standardized so it is considered a complementary method to increasethe sensitivity for detecting positive animals.

4.2. Detection of Host Immune Response

4.2.1.Serologic Tests

4.2.1.1. Complement Fixation Test (CFT)

This test has been a standard practice in the diagnosis of Johne's disease in cattle. It is recommended for the detection of clinical animals, but its application as a control tool in the eradication of the disease is very limited. This test is capable of detecting lightly infected animals, but it has a very low sensitivity, which frequently leads to false negative results [16].

4.2.1.2. Agar-Gel Immunodiffusion (AGID)

It is mainly indicated as aconfirmatory test in cattle, goats and sheep with clinical signs of the disease [38]. The usefulness of this test in the diagnosis of sub-clinically infected animals is very limited; however, in terms of sensitivity and specificity, it is better than CFT [39].

4.2.1.3. Enzyme-Linked Immunosorbent Assay, (ELISA)

This technique presents a better sensitivity compared to CFT and AGID while detecting sub-clinically infected animals [17]. This method detects 30-40percent of cattle identified as positive by fecal culture [16]. However, the sensitivity of this assay depends on the age, estimating that the overall value for every age group is 15percent [16, 40]. Nevertheless, in small ruminants, commercial ELISA presents a greater sensitivity (between 98.2 and 99.5%) in those animals that present macroscopic lesions [41].

The main advantage of this technique is the automatization, reproducibility, high throughput and objectivity of the results [10, 42]. To date, it is widely used in terms for the determination of the infection status within herds [43]. Additionally, novel ELISAs with greater sensitivity and specificity have been developed with the description of new antigens and antibodies [44-46].

4.2.2. Test Based on Cellular Immune Response

4.2.2.1. Gamma Interferon Assay (γ-IFN)

This assay is based on the production of γ-IFN by the lymphocytes, which are sensitized with specific antigens (avian tuberculin purified protein derivate- avian PPD, bovine PPD or johnin) during an incubation period of 18-36 hours [47]. Then, the quantitative detection of the γ-IFN produced is done by a sandwich ELISA using monoclonal antibodies anti-γ-IFN. In the literature, the specificity of this assay ranges between 67-94percent [48] depending on the interpretation criteria as this method was originally designed for the diagnosis of bovine tuberculosis [17].

4.2.2.2. Delayed-Type Hypersensitivity (DTH)

This test measures the delayed cellular immune hypersensitivity produced after the intradermic inoculation of an antigen by measuring the induration atthe inoculation site after 72 hours. Initially, the antigens utilized were avian PPD and johnin until a report described the sensitization of some animals against other mycobacterial species members of MAC, demonstrating that these antigens did not have a high specificity for the diagnosis of para-tuberculosis[49].

The interpretation of this test is difficult due to the lack of standardization in the interpretation of the results. In one study, it was observed that the antigen johnin presented a specificity of 88.8percent in cattle if the threshold for the dermal induration was greater than 2 mm, 91.3percent if more than 3 mm and 93.5percent if larger than 4 mm [48].

5. Histopathology of Johne's Disease

In small ruminants, paratuberculosis-associated lessions have been classified on the basis of localization, intensity of the lesions, cell types that are found in the lesion or number of mycobacteria present [50, 51].

Diffuse lesions are present in animals with clinical signs in two principal pathological forms, paucibacillary and multibacillary. The former is characterized by the presence of an inflammatory infiltrate composed of lymphocytes with low number of macrophages that haveno mycobacteria in their cytoplasm and later by the presence of macrophages with mycobacteria in their cytoplasm [50-52]. The intestinal focal lesions with defined granuloma are seen in sheep and goats that did not show clinical signs of the disease[50, 51]. In cattle, focal lesions in the lymphoid intestinal tissue or mesenteric lymph nodes in the first stage of experimental infection in calves have been reported [53].

The lesions produced by *M. a. paratuberculosis* have been classified into five categories on the basis of different criteria (presence of granulomatous lesions, their localization, intensity and distribution of lesions, cell types present in the inflammatory infiltrate and the presence/absence of mycobacteria):

5.1. Focal Lesions

Focal lesions are constituted by small granulomas that present macrophages with pale cytoplasm, enlarged nucleus and a small amount of chromatin. The number of macrophages per granuloma varies (from 5 to 30) and in cattle, there arealso Langhans giant cells inside or in isolated localization. In sheep (type 1 lesion) and goats, this type of lesion has been observed in 24 and 12percent of animals, respectively [50, 51].

5.2. Multifocal Lesions

These lesions are described for cattle, present small granulomas with Langhans giant cells in the lamina propria of the intestinal mucosa. The difference from focal lesions is that it does not appear only in the lymphoid tissue but also in the lamina propria. In the small intestine, this lesion is in the formofsmall granulomas or Langhans giant cells surrounded by an infiltrate of lymphocytes and plasma cells, localized in some intestinal villus [54]. This type of lesion has also been reported in sheep (types 2 and 3a) [50], which corresponds to the mild and moderate forms of the disease initially described in cattle [55].

5.3. Diffuse Multibacillary Lesion

It is associated with severe granulomatous enteritis that affects different intestinal localizations such as lymph nodes. In these types of lesions, macrophages (principal inflammatory cells) contain a high number of acid-fast microorganisms and there are also epithelioid cells that infiltrate towards the intestinal wall. The histopathology that is frequently observed consists of the fusion of intestinal villus and the presence of granulomas in the body of those intestinal villi that are not fused. In these cases, the mucosa appears thickened with lymphocytes and Langhans giant cells in the epithelioid infiltrate. In addition, in many cases, the intestinal glands present necrotic debris inside and they are normally positive in the Ziehl-Neelsen staining, indicating the presence of *M. a. paratuberculosis* [51, 54]. Diffuse lesions are comparable to the advanced stage of the disease reported in cattle [55]. Thistype of lesionhasbeen reported as characteristic of paratuberculosis in goats [51].

5.4. Diffuse Lymphocytic Lesions (Paucibacillary)

It is characterized by the presence of severe and diffuse granulomatous enteritis. The cellular types that are predominant are different from those observed in other lesions; lymphocytes are the principal inflammatory cells that infiltrate towards the lamina propriaand in occasions some macrophages and Langhans giant cells are also seen. The mucosal architecture is altered and there is a large amount of inflammatory infiltrate in the basal part of the lamina propria of the intestinal villus. However, the submucosa only presents giant cells and small granulomas. In this case, the number of acid-fast bacilli is very low or even absent in several parts of different intestinal segments.

5.5. Intermediate Diffuse Lesions

In this case, there is an inflammatory infiltrate formed mainly by abundant lymphocytes and macrophages [51, 54].

6. Prophylaxis, Therapy and Control of Johne's Disease

6.1. Vaccination

To date, vaccines developed for the prevention of paratuberculosis have been obtained from attenuated strains, inactivated strains or fractions of *M. a. paratuberculosis* [56, 57].

The first vaccine available against paratuberculosis was developed by Vallée and Rinjard in 1926 and produced with an attenuated strain of *M. a. paratuberculosis,* reconstituted in liquid paraffin and olive oil. Initially, this vaccine was successfully applied subcutaneously in France [57, 58]. Then, another attenuated strain of *M. a. paratuberculosis*recovered from a culture media without mycobactin J was used in the vaccination of animals in France and United Kingdom [58]. In this study, they concluded that this type of vaccine was capable of eradicating the disease in a four-year period.

There are different commercial vaccines such as the one with an attenuated strain of *M. a. paratuberculosis* (316F) reconstituted in an oleous adjuvant (Neoparasec™, Lab. Rhone-Mérieux), which has been extensively applied in France. Nevertheless, the principal disadvantage of this vaccine is the production of anti-mycobacterial antibodies in those animals inoculated subcutaneously with the vaccine, possibly leading to positive reactions in serological test used for the diagnosis of the disease [59].

Later, another study utilized this vaccine orally for 10 weeks and then challenged the animals with a clinical isolate of *M. a. paratuberculosis*onceper week for 10 weeks. As a result, necropsies of infected animals did not present differences among challenged animals and the control group, indicating the low protection of this oral vaccine against a pathogenic strain of *M. a. paratuberculosis* [60, 61]. However, some authors have questioned the relevance of these results because differences between vaccinated and unvaccinated animals should only be evaluated six months after the last challenge [34, 58, 62].

Another commercial vaccine against paratuberculosis is Gudair™ (CZ Veterinaria S.A. Spain). This vaccine is composed of *M. a. paratuberculosis* strain 316F heat inactivated in mineral oil adjuvant. The vaccine was applied in sheep in Australia (200 vaccinated and 200 control animals) and they observed positive effects that included a 90percent reduction in mortality at herd level in the vaccinated group compared to non-vaccinated and decreased fecal excretion of *M. a. paratuberculosis*in the infected animals (up to 90%). The drawback was that the vaccinated animals excreted high bacterial loads to the environment along with vaccinated animals with subclinical infection and animals that died because of the infection showed multibacillary lesions. These facts highlight the potential risk of moving vaccinated animals to areas of low prevalence and demonstrate the usefulnessof applying this vaccine in the control of the disease in herds with high level of prevalence [63].Also, the Gudair™ vaccine was compared to two vaccines obtained from inactivated and attenuated *M. a. paratuberculosis* strains reconstituted in a lipid formulation denominated matrix-K [64].

These vaccines were tested on 35 sheep per vaccine against a control group of 30 sheep (analyzing the intra-peritoneal and subcutaneous route). They observed that the lipidic formulation produced less inflammatory reaction at the inoculation site, a reduction in the prevalence of the disease within the herd and produced fewer macroscopic lesions (especially in intraperitoneal route) compared to the Gudair™ vaccine. Nonetheless, histopathology revealed fewer lesions in the Gudair™ vaccine (multibacillary lesions were less frequent in the lipidic formulation). In this study, they concluded that none of the vaccines or routes of administration presented variations in the morphometric indexes (weight loss after the challenge or weight gain during chronic infection), thus the ultimate impact in the disease was the same.

In cattle, another vaccine inactivated in an oleous adjuvant has recently been used. This vaccine was capable of reducing clinical signs of the disease in infected herds and producing a statistically significant response against *M. a. paratuberculosis* [65]. Additionally, in astudyconducted on 12 calves that were administered an inactivated strain of *M. a. paratuberculosis*orally and after challenge with a virulent strain of *M. a. paratuberculosis*did not show positive responses in serological tests performed. However, there was less colony forming units in the cultures obtained from the tissues of vaccinated animals [66].

Overall, in those herds with vaccinated animals the number of infected animals and individuals in the advanced clinical stage of the disease decreased, the number offecal shedders was reduced and the proportion of mycobacteria excreted to the environment diminished [57]. However, in some cases vaccination produced cutaneous lesions at the site of injection along with hypersensitivity and positivity to serological diagnostic tests thuscomplicatingthe estimation of the real prevalence within the herd [57, 59, 67]. Moreover, vaccination also interfered with diagnostic tests includingγ-IFN or intradermal tuberculin test (IDTB); official tests for the detection of tuberculosis [68-70]. Recentlya novel vaccine of subunits of the heat shock protein 70 that does not interfere with diagnostics tests for tuberculosis has been reported. Nevertheless, these results have to be confirmed with a higher number of animals (in this case, only 30 animals were tested) [71].

6.2. Treatment of the Disease

It is not common to treat infected animals and it is just done in pets or animals with high genetic value. Animals have to be treated daily and it only helps to decrease the signs of the disease but it does not cure it.

Although treated animals show a decrease in clinical signs, they continue excreting mycobacteria to the environment and their derived products for human consumption are not acceptable [72, 73]. The most commonly used drugs for the treatment of the disease are different anti-mycobacterial agents such as cycloserine, ethambutol, ethionamide, isoniazid, para-aminosalicylic acid, thiocarlide and thiosemicarbazones, fenamides, pyrazinamides, etc.; clinically active antibiotics aminoglycosides (gentamicin, kanamycin, neomycin, strepto-mycin, etc.), capreomycin, rifabutin, rifampicin and viomycin. However, the results obtained after the treatment with these drugs using *in vivo* models are not comparable to those obtained in *in vitro* models. The main reason is that mycobacteria can multiply in the interior of phagocytes and thus they remain inaccessible to anti-mycobacterial drugs [57].

In a recent study [74], the agent rodhamine (D157070) was described as a good option for the treatment against both the replicative and latent phenotypes of *M. a. paratuberculosis*. Nevertheless, these results were obtained using an *in vitro* model; therefore, it would be necessary to use *in vivo* studies to demonstrate the efficacy of this agent against the disease.

7. *Mycobacterium Avium* Subspecies *Paratuberculosis*and Crohn's Disease

The disease was initially reported in 1932, when Burril B. Crohn observed the presence of a chronic granulomatous ileitis in humans [61]. The clinical signs of Crohn's disease are enteralgia, diarrhea and vomiting, which are manifested in remissions and relapses. The disease commonly affects individuals between the ages of 16 and 25 years, although it has also been reported (with less frequency) in individuals between 55 and 65 [75].

The target organs are the terminal ileum, the caecum or colon (commonly accompanied with perforation, formation of abscess and fistulas, constipation and intestinal obstruction among others).

The cause of the disease is unknown; however, there are several hypotheses as the presence of a host aberrant autoimmune response against unknown antigens or against infectious agents including *M. a. paratuberculosis*, *Escherichia coli,* chronic viral infection, etc. Other possible etiologies that have been described include chronic ischemia or micro-infarction, abnormal response against a dietary component, altered inflammatory response against normal intestinal microflora or genetic predisposition (due to the presence of single nucleotide polymorphisms in NOD2 gene), etc.[13, 76].

The initial description of *M. a. paratuberculosis* as the etiological agent of the disease in the chronic inflammation of the gastrointestinal tract in humans took place in 1913 in Glasgow by T.K. Dalziel[1]. However, due to the difficulty in isolation and slow growth rate of *M. a. paratuberculosis,* it was not until 1984 when Dr. Rod Chiodini and collaborators characterized the presence of a mycobacteria for the first time, mycobactin-dependent obtained from three Crohn's disease patients [77]. These isolates were pathogenic for mice, other laboratory animals and young goats, producing granulomatous infection in the distal intestine.

To date, there are numerous reports those support or depict the relationship of *M. a. paratuberculosis* as the etiological agent of Crohn's disease in humans. The different aspects that support or reject this theory are [75, 76]:

- *M. a. paratuberculosis* has been demonstrated to cause a severe and terminal gastrointestinal disease in ruminants, which present several similarities with Crohn's disease in humans, both at the clinical and pathological level [61, 75].
- Due to the high genetic similarity between *M. a. paratuberculosis* and *M. avium* (opportunistic pathogen in immunodepressed humans) and the pathogenic capacity in a wide range of host for both microorganisms, we could hypothesize that *M. a. paratuberculosis*could act as opportunistic pathogens in humans [75].

- *M. a. paratuberculosis* has been detected in dairy products such as milk and cheese but also in meat and water for human consumption whichcould potentially constitute a source of infection in humans [78-85].
- Genes associated with Crohn's disease suggest that the presence of an inadequate immune response to an intracellular pathogen could be the cause of the disease. *M. a. paratuberculosis* is an intracellular pathogen and it could be feasible that this could be the cause of the disease [86, 87].
- *M. a. paratuberculosis* has been detected in tissue, blood, and milk of Crohn's disease patients by molecular techniques, histopathology and antibodies against *M. a. paratuberculosis* antigens [36, 88-91].
- The agent *M. a. paratuberculosis* has also been observed in the tissues of Crohn's disease patients in the Ziehl-Neelsen staining [92].
- A study that reviewed reports that considered the relationship between *M. a. paratuberculosis*and Crohn's disease patients estimated that those patients presented a probability seven times higher of having *M. a. paratuberculosis* in their tissues compared to the rest of the population [93].

However, there are other reports those consider the lack of a link between *M. a. paratuberculosis* and Crohn's disease:

- Farmers and personnel more exposed to *M. a. paratuberculosis*did not present higher prevalence of Crohn's disease compared with other sectors of the population [94]. Although it has been shown that population migration from areas without disease to areas of high prevalence ended up in the presentation of the disease in these individuals [95]. On the other hand, Sweden has the higher prevalence of Crohn's disease but the lowest prevalence of paratuberculosis [96, 97].
- To date there is no evidence of aggravation in Crohn's disease patients due to immune-deficiency such as in case of infections due to *Mycobacterium tuberculosis* [98].
- In a controlled assay where patients were treated with antibiotics they did not show a continuous response and the relapse observed in these patients after two years of treatment was similar to that obtained from non-treated patients (control) [99].

In summary, the role of *M. a. paratuberculosis*in the development of Crohn's disease is not confirmed and it could only be accepted or ruled out after improvingthe isolation methods for the organism or the detection of *M. a. paratuberculosis*bacilli, the creation of animal models of Crohn's disease (to date there are only animal models for Johne's disease) or the improvement of clinical assays with chemotherapeutic agents, etc. [75].

REFERENCES

[1] Hermon-Taylor J, El-Zaatari FAK. *Pathogenic mycobacteria in water.* ed. London, UK: World Health Organization (WHO), IWA Publishing. 2004;74-93

[2] Kluge JP, Merkal RS, Monlux WS, Larsen AB, Kopecky KE, Ramsey FK, Lehmann RP.Experimental paratuberculosis in sheep after oral, intratracheal, or intravenous inoculation lesions and demonstration of etiologic agent. *Am. J. Vet. Res.*, 1968; 29:953-962.

[3] Larsen AB, Miller JM. Mammary gland exposure of cows to *Mycobacteriumparatuberculosis. Am. J. Vet. Res.,* 1978; 39:1972-1974.

[4] Merkal RS, Miller JM, Hintz AM, Bryner JH.Intrauterine inoculation of *Mycobacterium paratuberculosis* into guinea pigs and cattle.*Am. J. Vet. Res*., 1982; 43:676-678.

[5] Greig A, Stevenson K, Perez V, Pirie AA, Grant JM, Sharp JM. Paratuberculosis in wild rabbits (*Oryctolagus cuniculus*). *Vet. Rec*., 1997;140:141-143.

[6] Greig A, Stevenson K, Henderson D, Perez V, Hughes V, Pavlik I, Hines ME, McKendrick I, Sharp JM. Epidemiological study of paratuberculosis in wild rabbits in Scotland. *J. Clin. Microbiol*., 1999; 37:1746-1751.

[7] Biet F, Boschiroli ML, Thorel MF, Guilloteau LA. Zoonotic aspects of *Mycobacterium bovis* and *Mycobacterium avium-intracellulare* complex (MAC).*Vet. Res*., 2005; 36:411-436.

[8] Thorel MF, Huchzermeyer H, Weiss R, Fontaine JJ. *Mycobacterium avium* infections in animals. Literature review.*Vet. Res*., 1997; 28:439-447.

[9] Whitlock RH, Buergelt C C. Buergelt. Preclinical and clinical manifestations of paratuberculosis (including pathology). *Vet. Clin. North Am. Food Anim. Pract*., 1996; 12:345-356.

[10] Manning EJ, Collins MT s. *Mycobacterium avium* subsp. *paratuberculosis*: pathogen, pathogenesis and diagnosis. *Rev. Sci. Tech*., 2001; 20:133-150.

[11] Rankin JD. The experimental infection of cattle with *Mycobacterium johnei*. II. Adult cattle inoculated intravenously.*J. Comp. Pathol*., 1961; 71:6-9.

[12] Stehman SM. Paratuberculosis in small ruminants, deer, and South American camelids.*Vet. Clin. North Am. Food Anim. Pract*., 1996; 12:441-455.

[13] Committee on Diagnosis and Control of Johne's Disease, Board on Agriculture and Natural Resources ed. *Diagnosis and control of Johne's disease*. Washington, D.C: The National Academies Press, 2003.

[14] Gunnarsson E, Fodstad FH. Cultural and biochemical characteristics of *Mycobacterium paratuberculosis* isolated from goats in Norway. *Acta Vet. Scand*., 1979; 20:122-134.

[15] Whittington RJ, Hope AF, Marshall DJ, Taragel CA, Marsh I. Molecular epidemiology of *Mycobacterium avium* subsp. *paratuberculosis*: IS*900* restriction fragment length polymorphism and IS*1311* polymorphism analyses of isolates from animals and a human in Australia.*J. Clin. Microbiol*., 2000; 38:3240-3248.

[16] Whitlock RH, Wells SJ, Sweeney RW, Van Tiem J. ELISA and fecal culture for paratuberculosis (Johne's disease): sensitivity and specificity of each method. *Vet. Microbiol.,*2000; 77:387-398.

[17] OIE: *Manual of diagnostic tests and vaccines for terrestrial animals* (2008).

[18] Damato JJ, Collins MT. Growth of *Mycobacterium paratuberculosis* in radiometric, Middlebrook and egg-based media. *Vet. Microbiol.,*1990; 22:31-42.

[19] Stich RW, Byrum B, Love B, Theus N, Barber L, Shulaw WP. Evaluation of an automated system for non-radiometric detection of *Mycobacterium avium paratuberculosis* in bovine feces.*J. Microbiol. Methods,*2004; 56:267-275.

[20] Garrido JM, Cortabarria N, Oguiza JA, Aduriz G, Juste RA. Use of a PCR method on fecal samples for diagnosis of sheep paratuberculosis. *Vet. Microbiol.,* 2000; 77:379-386.

[21] Vary PH, Andersen PR, Green E, Hermon-Taylor J, McFadden JJ. Use of highly specific DNA probes and the polymerase chain reaction to detect *Mycobacterium paratuberculosis* in Johne's disease.*J. Clin. Microbiol.*, 1990; 28:933-937.

[22] Englund S, Ballagi-Pordany A, Bolske G, Johansson KE. Single PCR and nested PCR with a mimic molecule for detection of *Mycobacterium avium* subsp.*paratuberculosis.Diagn. Microbiol. Infect. Dis.,* 1999; 33:163-171.

[23] Whittington RJ, Marsh I, McAllister S, Turner MJ, Marshall DJ, Fraser CA. Evaluation of modified BACTEC 12B radiometric medium and solid media for culture of *Mycobacterium avium* subsp. *paratuberculosis* from sheep. *J. Clin. Microbiol.,* 1999; 37: 1077-1083.

[24] Marsh IB, Whittington RJ. Progress towards a rapid polymerase chain reaction diagnostic test for the identification of *Mycobacterium avium* subsp. *paratuberculosis* in faeces *Mol. Cell Probes,* 2001; 15:105-118.

[25] Eamens GJ, Whittington RJ, Turner MJ, Austin SL, Fell SA, Marsh IB. Evaluation of radiometric faecal culture and direct PCR on pooled faeces for detection of *Mycobacterium avium* subsp. *paratuberculosis* in cattle.*Vet. Microbiol.,* 2007; 125:22-35.

[26] Kawaji S, Taylor DL, Mori Y, Whittington RJ. Detection of *Mycobacterium avium* subsp. *paratuberculosis* in ovine faeces by direct quantitative PCR has similar or greater sensitivity compared to radiometric culture. *Vet. Microbiol.*, 2007; 125:36-48.

[27] Scott HM, Fosgate GT, Libal MC, Sneed LW, Erol E, Angulo AB, Jordan ER.Field testing of an enhanced direct-fecal polymerase chain reaction procedure, bacterial culture of feces, and a serum enzyme-linked immunosorbent assay for detecting *Mycobacterium avium* subsp *paratuberculosis* infection in adult dairy cattle. *Am. J. Vet. Res.,*2007; 68:236-245.

[28] Irenge LM, Walravens K, Govaerts M, Godfroid J, Rosseels V, Huygen K, Gala JL.Development and validation of a triplex real-time PCR for rapid detection and specific identification of *M. avium* subsp. *paratuberculosis* in faecal samples.*Vet. Microbiol.,* 2009; 136:166-172.

[29] Grant IR, O'Riordan LM, Ball HJ, Rowe MT. Incidence of *Mycobacterium paratuberculosis* in raw sheep and goats' milk in England, Wales and Northern Ireland. *Vet. Microbiol.*, 2001; 79:123-131.

[30] Odumeru J, Gao A, Chen S, Raymond M, Mutharia L. Use of the bead beater for preparation of *Mycobacterium paratuberculosis* template DNA in milk.*Can. J. Vet. Res.,*2001; 65:201-205.

[31] Djonne B, Jensen MR, Grant IR, Holstad G. Detection by immunomagnetic PCR of *Mycobacterium avium* subsp. *paratuberculosis* in milk from dairy goats in Norway. *Vet. Microbiol.,* 2003; 92:135-143.

[32] Kaur P, Filia G, Singh SV, Patil PK, Sandhu KS. Molecular detection and typing of *Mycobacterium avium* subspecies *paratuberculosis* from milk samples of dairy animals. *Trop. Anim Health Prod.*, 2010; 42: 1031-1035.

[33] Gwozdz JM, Reichel MP, Murray A, Manktelow W, West DM, Thompson KG. Detection of *Mycobacterium avium* subsp. *paratuberculosis* in ovine tissues and blood by the polymerase chain reaction.*Vet. Microbiol.,* 1997; 57:233-244.

[34] Gwozdz JM, Thompson KG, Manktelow BW, Murray A, West DM.. Vaccination against paratuberculosis of lambs already infected experimentally with *Mycobacteriumavium* subspecies *paratuberculosis. Aust. Vet. J.,*2000; 78:560-566.

[35] Bull TJ, McMinn EJ, Sidi-Boumedine K, Skull A, Durkin D, Neild P, Rhodes G, Pickup R, Hermon-Taylor J.Detection and verification of *Mycobacterium avium* subsp. *paratuberculosis* in fresh ileocolonic mucosal biopsy specimens from individuals with and without Crohn's disease. *J. Clin. Microbiol*., 2003;41:2915-2923.

[36] Naser SA, Ghobrial G, Romero C, Valentine JF. Culture of *Mycobacterium avium* subspecies *paratuberculosis* from the blood of patients with Crohn's disease. *Lancet,*2004; 364: 1039-1044.

[37] Juste RA, Garrido JM, Geijo M, Elguezabal N, Aduriz G, Atxaerandio R, Sevilla I.Comparison of blood polymerase chain reaction and enzyme-linked immunosorbent assay for detection of *Mycobacterium avium* subsp. *paratuberculosis* infection in cattle and sheep.*J. Vet. Diagn. Invest.,* 2005; 17:354-359.

[38] Sherman DM, Markham RJ, Bates F. Agar gel immunodiffusion test for diagnosis of clinical paratuberculosis in cattle. *J. Am. Vet. Med. Assoc*., 1984; 185:179-182.

[39] Sherman DM, Gay JM, Bouley DS, Nelson GH. Comparison of the complement-fixation and agar gel immunodiffusion tests for diagnosis of subclinical bovine paratuberculosis. *Am. J. Vet. Res*., 1990; 51:461-465.

[40] Jubb TF, Sergeant ES, Callinan AP, Galvin J. Estimate of the sensitivity of an ELISA used to detect Johne's disease in Victorian dairy cattle herds. *Aust. Vet. J*., 2004; 82:569-573.

[41] Hope AF, Kluver PF, Jones SL, Condron RJ. Sensitivity and specificity of two serological tests for the detection of ovine paratuberculosis. *Aust. Vet. J.,* 2000; 78:850-856.

[42] Sergeant ES, Marshall DJ, Eamens GJ, Kearns C, Whittington RJ. Evaluation of an absorbed ELISA and an agar-gel immuno-diffusion test for ovine paratuberculosis in sheep in Australia. *Prev. Vet. Med*., 2003; 61:235-248.

[43] Nielsen SS, Houe H, Thamsborg SM, Bitsch V. Comparison of two enzyme-linked immunosorbent assays for serologic diagnosis of paratuberculosis (Johne's disease) in cattle using different subspecies strains of *Mycobacterium avium. J. Vet. Diagn. Invest.,*2001; 13:164-166.

[44] Griffin JF, Spittle E, Rodgers CR, Liggett S, Cooper M, Bakker D, Bannantine JP.Immunoglobulin G1 enzyme-linked immunosorbent assay for diagnosis of Johne's Disease in red deer (*Cervus elaphus*). *Clin. Diagn. Lab Immunol*., 2005; 12:1401-1409.

[45] Eda S, Bannantine JP, Waters WR, Mori Y, Whitlock RH, Scott MC, Speer CA.A highly sensitive and subspecies-specific surface antigen enzyme- linked immunosorbent assay for diagnosis of Johne's disease. *Clin. Vaccine Immunol*., 2006; 13:837-844.

[46] Speer CA, Scott MC, Bannantine JP, Waters WR, Mori Y, Whitlock RH, Eda S 1.A novel enzyme-linked immunosorbent assay for diagnosis of *Mycobacterium avium* subsp. *paratuberculosis* infections (Johne's Disease) in cattle.*Clin. Vaccine Immunol.,* 2006; 13:535-54.

[47] Wood PR, Rothel JS, McWaters PG, Jones SL. Production and characterization of monoclonal antibodies specific for bovine gamma-interferon.*Vet. Immunol. Immunopathol.,*1990; 25:37-46.
[48] Kalis CH, Collins MT, Hesselink JW, Barkema HW. Specificity of two tests for the early diagnosis of bovine paratuberculosis based on cell-mediated immunity: the Johnin skin test and the gamma interferon assay. *Vet. Microbiol.*, 2003; 97:73-86.
[49] Inderlied CB, Kemper CA, Bermudez LE. The *Mycobacterium avium* complex. *Clin.Microbiol. Rev.*, 1993; 6:266-310.
[50] Pérez V, García Marín JF, Badiola JJ. Description and classification of different types of lesion associated with natural paratuberculosis infection in sheep.*J. Comp Pathol.,* 1996; 114:107-122.
[51] Corpa JM, Garrido J, Garcia Marin JF, Perez V. Classification of lesions observed in natural cases of paratuberculosis in goats. *J. Comp. Pathol.*,2000; 122:255-265.
[52] Carrigan MJ, Seaman JT. The pathology of Johne's disease in sheep.*Aust. Vet. J.,* 1990; 67:47-50.
[53] Larsen AB, Merkal RS, Cutlip RC. Age of cattle as related to resistance to infection with *Mycobacterium paratuberculosis*. *Am. J. Vet. Res.*, 1975; 36:255-257.
[54] González J, Geijo MV, García-Pariente C, Verna A, Corpa JM, Reyes LE, Ferreras MC, Juste RA, García Marín JF, Pérez V.Histopathological classification of lesions associated with natural paratuberculosis infection in cattle. *J. Comp Pathol.*, 2005; 133:184-196.
[55] Buergelt CD, Hall C, McEntee K, Duncan JR. Pathological evaluation of paratuberculosis in naturally infected cattle. *Vet. Pathol.*, 1978; 15:196-207.
[56] Saxegaard F, Fodstad FH. Control of paratuberculosis (Johne's disease) in goats by vaccination. *Vet. Rec.*, 1985; 116:439-441.
[57] Cocito C, Gilot P, Coene M, de Kesel M., Poupart P, Vannuffel P. Paratuberculosis. *Clin. Microbiol. Rev.,* 1994; 7:328-345.
[58] Stuart P. Vaccination against Johne's disease in cattle exposed to experimental infection.*Br. Vet. J.,* 1965; 121:289-318.
[59] Hilbink F, West D. The antibody response of sheep to vaccination against Johne's disease. *N. Z. Vet. J.,* 1990; 38:168-169.
[60] Gilmour NJ, Angus KW. Absence of immunogenicity of an oral vaccine against *Mycobacterium johnei* in sheep. *Res. Vet. Sci.,* 1974; 16:269-270.
[61] Harris NB, Barletta RG. *Mycobacterium avium* subsp. *paratuberculosis* in Veterinary Medicine. *Clin. Microbiol. Rev.,* 2001; 14:489-512.
[62] Juste RA, Garcia Marin JF, Peris B, Saez de Ocariz CS, Badiola JJ. Experimental infection of vaccinated and non-vaccinated lambs with *Mycobacteriumparatuberculosis*. *J. Comp Pathol.*, 1994; 110:185-194.
[63] Reddacliff L, Eppleston J, Windsor P, Whittington R, Jones S S. Efficacy of a killed vaccine for the control of paratuberculosis in Australian sheep flocks. *Vet. Microbiol.*, 2006; 115:77-90.
[64] Griffin JF, Hughes AD, Liggett S, Farquhar PA, Mackintosh CG, Bakker D. Efficacy of novel lipid-formulated whole bacterial cell vaccines against *Mycobacterium avium* subsp. *paratuberculosis* in sheep.*Vaccine,*2009; 27:911-918.

[65] Platt R, Thoen CO, Stalberger RJ, Chiang YW, Roth JA. Evaluation of the cell-mediated immune response to reduced doses of *Mycobacterium avium* ssp. *paratuberculosis* vaccine in cattle.*Vet. Immunol. Immunopathol.*, 2010; 136:122-126.
[66] Sweeney RW, Whitlock RH, Bowersock TL, Cleary DL, Meinert TR, Habecker PL, Pruitt GW.Effect of subcutaneous administration of a killed *Mycobacterium avium* subsp *paratuberculosis* vaccine on colonization of tissues following oral exposure to the organism in calves. *Am. J. Vet. Res.,* 2009; 70:493-497.
[67] Chiodini RJ, Van Kruiningen HJ, Merkal RS. Ruminant paratuberculosis (Johne's disease): the current status and future prospects. *Cornell. Vet.,* 1984; 48:185-195.
[68] Kohler H, Gyra H, Zimmer K, Drager KG, Burkert B, Lemser B, Hausleithner D, Cubler K, Klawonn W, Hess RG. Immune reactions in cattle after immunization with a *Mycobacterium paratuberculosis* vaccine and implications for the diagnosis of *M. paratuberculosis* and *M. bovis* infections. *J. Vet. Med. B Infect. Dis. Vet. Public Health,* 2001; 48:185-195.
[69] Muskens J, van Zijderveld F, Eger A, Bakker D. Evaluation of the long-term immune response in cattle after vaccination against paratuberculosis in two Dutch dairy herds. *Vet. Microbiol.,* 2002; 86:269-278.
[70] Mackintosh CG, Labes RE, Griffin JF. The effect of Johne's vaccination on tuberculin testing in farmed red deer (*Cervus elaphus*). *N. Z. Vet. J.,* 2005; 53:216-222.
[71] Santema W, Hensen S, Rutten V, Koets A. Heat shock protein 70 subunit vaccination against bovine paratuberculosis does not interfere with current immunodiagnostic assays for bovine tuberculosis.*Vaccine,*2009; 27:2312-2319.
[72] Larsen AB, Vardaman TH. The effect of isonicotinic acid hydrazide on *Mycobacterium paratuberculosis*. *J. Am. Vet. Med. Assoc*., 1953; 122:309-310.
[73] Merkal RS, Larsen AB. Clofazimine treatment of cows naturally infected with *Mycobacterium paratuberculosis*.*Am. J. Vet. Res.,* 1973; 34:27-28.
[74] Bull TJ, Linedale R, Hinds J, Hermon-Taylor J. A rhodanine agent active against non-replicating intracellular *Mycobacterium avium* subspecies *paratuberculosis*. *Gut. Pathog.,*2009; 1: 25.
[75] Chacon O, Bermudez LE, Barletta RG. Johne's disease, inflammatory bowel disease, and *Mycobacterium paratuberculosis*.*Annu. Rev. Microbiol.*, 2004; 58: 329-63.
[76] Nacy C, Buckley M. *Mycobacterium avium paratuberculosis*: Infrequent human pathogen or public health threat?*Report from the American Academy of Microbiology,Colloquium,* 15-17 June, Salem, Massachusetts, 2008.
[77] Chiodini RJ, Van Kruiningen HJ, Thayer WR, Merkal RS, Coutu JA. Possible role of mycobacteria in inflammatory bowel disease. I. An unclassified *Mycobacterium* species isolated from patients with Crohn's disease. *Dig. Dis. Sci.,* 1984; 29:1073-1079.
[78] Corti S, Stephan R. Detection of *Mycobacterium avium* subspecies *paratuberculosis* specific IS*900* insertion sequences in bulk-tank milk samples obtained from different regions throughout Switzerland. *BMC. Microbiol*., 2002; 2:15.
[79] Donaghy JA, Totton NL, Rowe MT. Persistence of *Mycobacterium paratuberculosis* during manufacture and ripening of cheddar cheese. *Appl. Environ. Microbiol*., 2004; 70:4899-4905.
[80] Pickup RW, Rhodes G, Arnott S, Sidi-Boumedine K, Bull TJ, Weightman A, Hurley M, Hermon-Taylor J. *Mycobacterium avium* subsp. *paratuberculosis* in the catchment area and water of the River Taff in South Wales, United Kingdom, and its potential

relationship to clustering of Crohn's disease cases in the city of Cardiff. *Appl. Environ. Microbiol.,*2005; 71:2130-2139.

[81] Clark DL, Jr., Anderson JL, Koziczkowski JJ, Ellingson JL. Detection of *Mycobacterium avium* subspecies *paratuberculosis* genetic components in retail cheese curds purchased in Wisconsin and Minnesota by PCR. *Mol. Cell Probes,* 2006; 20:197-202.

[82] Pickup RW, Rhodes G, Bull TJ, Arnott S, Sidi-Boumedine K, Hurley M, Hermon-Taylor J. *Mycobacterium avium* subsp. *paratuberculosis* in lake catchments, in river water abstracted for domestic use, and in effluent from domestic sewage treatment works: diverse opportunities for environmental cycling and human exposure. *Appl. Environ. Microbiol.,* 2006; 72:4067-4077.

[83] Abubakar I, Myhill DJ, Hart AR, Lake IR, Harvey I, Rhodes JM, Robinson R, Lobo AJ, Probert CS, Hunter PR.A case-control study of drinking water and dairy products in Crohn's Disease--further investigation of the possible role of *Mycobacterium avium paratuberculosis.Am. J. Epidemiol.*, 2007; 165:776-783.

[84] Jaravata CV, Smith WL, Rensen GJ, Ruzante J, Cullor JS. Survey of ground beef for the detection of *Mycobacterium avium paratuberculosis. Foodborne. Pathog. Dis.*, 2007; 4:103-106.

[85] Alonso-Hearn M, Molina E, Geijo M, Vazquez P, Sevilla I, Garrido JM, Juste RA.Isolation of *Mycobacteriumavium* subsp. *paratuberculosis* from muscle tissue of naturally infected cattle. *Foodborne. Pathog. Dis.,* 2009; 6:513-518.

[86] Ferweda G, Kuliberg DJ, deJong DJ. *Mycobacterium paratuberculosis* is recognized by Toll-like receptors and NOD2. *J. Leukoc. Biol.,*2007; 82:1011-1018.

[87] Parkes M, Barrett JC, Prescott NJ, Tremelling M, Anderson CA, Fisher SA, Roberts RG, Nimmo ER, Cummings FR, Soars D, Drummond H, Lees CW, Khawaja SA, Bagnall R, Burke DA, Todhunter CE, Ahmad T, Onnie CM, McArdle W, Strachan D, Bethel G, Bryan C, Lewis CM, Deloukas P, Forbes A, Sanderson J, Jewell DP, Satsangi J, Mansfield JC, Cardon L, Mathew CG.Sequence variants in the autophagy gene IRGM and multiple other replicating loci contribute to Crohn's disease susceptibility. *Nat. Genet.,* 2007; 39:830-832.

[88] Hulten K, Karttunen TJ, El-Zimaity HM, Naser SA, Almashhrawi A, Graham DY, el-Zaatari FA.In situ hybridization method for studies of cell wall deficient *M. paratuberculosis* in tissue samples. *Vet. Microbiol.,* 2000; 77:513-518.

[89] Naser SA, Schwartz D, Shafran I. Isolation of *Mycobacterium avium* subsp *paratuberculosis* from breast milk of Crohn's disease patients. *Am. J. Gastroenterol.,* 2000; 95:1094-1095.

[90] Olsen I, Wiker HG, Johnson E, Langeggen H, Reitan LJ. Elevated antibody responses in patients with Crohn's disease against a 14-kDa secreted protein purified from *Mycobacterium avium* subsp. *paratuberculosis.Scand. J. Immunol.*, 2001; 53:198-203.

[91] Chamberlin W, Ghobrial G, Chehtane M, Naser SA. Successful treatment of a Crohn's disease patient infected with bacteremic *Mycobacterium paratuberculosis. Am. J. Gastroenterol.,* 2007; 102:689-691.

[92] Jeyanathan M, Boutros-Tadros O, Radhi J, Semret M, Bitton A, Behr MA. Visualization of *Mycobacterium avium* in Crohn's tissue by oil-immersion microscopy. *Microbes. Infect.,* 2007; 9:1567-1573.

[93] Feller M, Huwiler K, Stephan R, Altpeter E, Shang A, Furrer H, Pfyffer GE, Jemmi T, Baumgartner A, Egger M. *Mycobacterium avium* subspecies *paratuberculosis* and Crohn's disease: a systematic review and meta-analysis.*Lancet Infect. Dis.,* 2007; 7:607-613.

[94] Jones PH, Farver TB, Beaman B, Cetinkaya B, Morgan KL. Crohn's disease in people exposed to clinical cases of bovine paratuberculosis. *Epidemiol. Infect.*, 2006; 134:49-56.

[95] Probert CS, Jayanthi V, Rampton DS, Mayberry JF. Epidemiology of inflammatory bowel disease in different ethnic and religious groups: limitations and aetiological clues. *Int. J. Colorectal Dis.*, 1996; 11:25-28.

[96] Ekbom A, Helmick C, Zack M, Adami HO. The epidemiology of inflammatory bowel disease: a large, population-based study in Sweden. *Gastroenterology,* 1991; 100:350-358.

[97] Sternberg S, Viske D. Control strategies for paratuberculosis in Sweden. *Acta Vet. Scand.,*2003; 44:247-249.

[98] Sartor RB. Does *Mycobacterium avium* subspecies *paratuberculosis* cause Crohn's disease? *Gut.,* 2005; 54:896-898.

[99] Selby W, Pavli P, Crotty B, Florin T, Radford-Smith G, Gibson P, Mitchell B, Connell W, Read R, Merrett M, Ee H, Hetzel D l.Two-year combination antibiotic therapy with clarithromycin, rifabutin, and clofazimine for Crohn's disease. *Gastroenterology,* 2007; 132:2313-2319.

In: Livestock: Rearing, Farming Practices and Diseases
Editor: M. Tariq Javed
ISBN 978-1-62100-181-2

Chapter 12

CHANGES IN CONSUMERS' FOOD PURCHASES DUE TO NEW LEGISLATION ON FOOD LABELING MAY AFFECT LIVESTOCK PRODUCTION PRACTICES IN THE UNITED STATES

Terence J. Centner and Steven M. Gower
College of Agricultural and Environmental Sciences,
The University of Georgia, Athens, GA 30602, US

ABSTRACT

Consumer demand for additional information on food labels has been accompanied by legislative and judicial edicts that are expected to affect livestock production in the United States.One development involves legislation enacted by the state of Ohio that limited labels on milk products concerning recombinant bovine somatotropin, commonly called rbST.Dairy processors challenged the regulations because they wanted to be able to tell consumers more about whether products contained milk from cows are treated with rbST.Because of potential concerns for dangers to human health, some consumers are willing to pay more for milk produced from cows that were not treated with rbST.An appellate court found some of the Ohio labeling restrictions to be unconstitutional.This decision should facilitate more labeling, a reduction in market share for milk from cows treated with rbST, and a corresponding need for more dairy animals and high yielders.

The second issue involves the labeling of livestock products with attributes claiming that they are "natural" or "naturally raised."Firms have successfully convinced consumers that products with these labels are healthier and more environmentally friendly than their "non-natural" counterparts.Over the past 20 years, consumer demand for "natural" food products has steadily increased.Accompanying labels is consumer confusion of the actual meaning of the terms.In 2007, the U.S.D.A's Agricultural Marketing Service proposed a Naturally Raised Marketing Claim standard for producers that want to identify their livestock as "naturally raised."A voluntary marketing claim standard for naturally raised livestock and meat products would preclude the use of promotants (hormones), antibiotics (except to prevent parasites), and animal by-products as feed.Further regulatory developments may define these terms to reduce the confusion and alter producers' marketing strategies.

Keywords: food labels, rbST, dairy processors, milk, natural, naturally raised

1. INTRODUCTION

Consumer demand for more information on food labels has significantly increased in recent years in response to changes in the marketplace and advances in technology.Product differentiation and the ability to market identifiable products have led to different production processes and compositional differences to gain higher prices.One such product difference that has had a significant effect in the marketplace is the promotion of milk and milk products as "hormone-free."The leading artificial hormone often marketed as not being present in milk is recombinant bovine somatotropin (rbST), also known as rBGH, which is analogous to natural bovine growth hormone.This chapterexaminesrbST and its intended effects on agriculture, looking at the development of legislation that allows informationregardingthe absence of said artificial hormones in milk and milk products on food labels. Labeling requirements of various U.S. state governments are examined as well as asubsequent legal challenge to Ohio's labeling legislation. The second issue involves the identification of attributes for livestock and meat products. Ideas for labeling products "naturally raised" and "natural" are being considered. Vendors seek to use these terms in their marketing programs, but there is disagreement as what the terms mean. Vendors seek broad definitions so more products qualify; consumer groups tend to favor more truthful nomenclatures that would disqualify products that do not meet strict qualifications.

2. RBST AND LABELING REQUIREMENTS

Utilizing recombinant DNA technology, Monsanto developed rbST in 1993 under the trade name Posilac® [1]. Monsanto marketed the product to increasemilk production and extend the lactation periods of cows. The increases in production allow operations that choose to utilize rbST treatment to operate with fewer animals than operations utilizing conventional methods for the same amounts of milk produced. The U.S. Food and Drug Administration (FDA) approved the use of rbST in lactating dairy cows in 1993 with the purpose of increasing the production of marketable milk [2].

The dosage form for rbST is a prolonged-release injectable as an over the counter drug.The route of administration is a subcutaneous injection usually in the neck area, post-scapular region (behind the shoulder), or in the depression on either side of the tail head. The use of rbST in cattle has lead to an increased observance of mastitis, a swelling of the udder that is characterized by the presence of abnormal milk. The use of rbST has also lead to increased incidence of twinning, cystic ovaries, and uterine disorders[2].

In 1994, the FDA issued interim guidance on the labeling of milk and milk products from cows that have not been treated with rbST as a result of multiple requests from different U.S. state governments, from industry, and from consumer representatives [3]. As stated in the guidance, the FDA intended to rely primarily on the enforcement activities of state governments as a result of the state's traditional role in overseeing milk production to determine whether rbST labeling claims are in fact truthful and not misleading. Moreover,

FDA's guidance reflects merely an interpretation and does not bind the FDA or states in making their individual guidelines for labeling products [3].

A food is misbranded if statements on its label or in its labeling are false or misleading.Moreover, the absence of information relevant to the issue may cause labeling to be misleading [4].Thus, misbranding precludes information that without further details might be expected to mislead. Due to the fact that natural bST is in all milk, FDA felt that truthful information means that labels cannot claim that milk is "bST free."Moreover, FDA maintained a claim that milk is "rbST free" might convey the idea that there exists a meaningful distinction between milk from cows that have been treated with rbST and those that have not been treated. Although there is a distinction in the way milk is produced, FDA concluded there was no meaningful distinction in the milk.

To prevent misleading information, differentiation between rbST and bST may be achieved by a statement that the milk comes "from cows not treated with rbST."Standing alone, however, a statement that milk comes from cows not treated with rbST may be misleading by implying that such milk is safer or of higher quality than milk from treated cows.To avoid this problem, FDA suggested that such a statement be placed in a proper context with an accompanying notation that: "No significant difference has been shown between milk derived from rbST-treated and non-rbST-treated cows."

In 2008, the state of Ohio decided to addressrbST labeling when Governor Ted Strickland issued an executive order on behalf of the Ohio Department of Agriculture [5]. This order covered the immediate adoption of a rule concerning the labeling of milk and milk products sold in Ohiowith information on rbST by adding a definition to what would be considered false and misleading labels.The order maintained that producers were using labels to imply a difference in quality or safety between milk from cows treated with rbST and milk from cows not treated with rbST to create an image of a superior product in the milk and milk products from cows not treated with rbST.To best serve consumers, the order prescribed rbST labeling information coupled with labeling that informed customers that the FDA has found no significant difference between milk from cows treated with rbST and cows that have not been treated.

On May 22, 2008, the dairy labeling legislation became law [6]. The Ohio rule requires all milk and milk products making production claims on rbST to have the following contiguous additional statement: "The FDA has determined that no significant difference has been shown between milk derived from rbST-supplemented and non-rbST-supplemented cows"[6].In addition, the Ohio rule prohibits all composition claims such as "no hormones" or "rbST free" because they are inherently false and misleading [6]. The Ohio labeling requirements are very similar to what theFDA had suggested for labeling milk and milk products.

Other U.S.states have produced their own legislation concerning the labeling of milk and milk products produced or sold within their states with emphasis on how to prevent labeling from being false and misleading when concerning claims involving hormone use.Minnesota's legislation delineates labeling requirements for rbST(called rBGH)as to what constitutes acceptable labeling and what is deemed false and misleading [7]. Production claims and composition claims are allowed under Minnesota law without the requirement of a disclosure statement.Furthermore, these claims may be placed on the package and may also be conspicuously placed in advertising for the product.This differs from Ohio's legislation in that production claims do not require a disclosure statement and composition claims are allowed.

Wisconsin's regulations are addressed in its administrative code's provisions on advertising and labeling claims for dairy products [8]. The provisions allow persons to represent a dairy product as "farmer certified rbST-free" if the representation complies with the remaining provisions. Several prohibitions are enumerated that preclude persons from making claims that are false, deceptive, or misleading; are without qualifying statements; without substantiation; or representing that dairy products produced from cows not treated with rbST differ significantly in composition from other dairy products. Wisconsin's rules also forbid representations that dairy products produced with milk from cows treated with rbST are of lower quality, are less safe, or are less wholesome than other dairy products. The subsection on qualifying statements requires processors to make production claims rather than composition claims,and requires a disclosure statement that shows no significant difference has been found between milk from cows treated with rbST and milk from cows not treated with rbST. Thus, Wisconsin's regulations are similar to the Ohio provisions and FDA's interim guidance.

The Vermont general assembly enumerated a policy of allowing consumers to have truthful and non-misleading information on dairy products that have not had rbST used in their production [9]. Rules adopted in the state of Vermont allow persons to make representations that milk or milk products are derived from cows not treated with rbST [10].To best serve this interest, the rules preclude any implication that a dairy product contains no bovine somatotropin, and qualifying statements to acknowledge that there is no significant difference between milk derived from rbST treated and non-rbST-treated cows. The requirements for regulating milk and milk products in these states coincide with the Ohio Rule and the FDA's Interim Guidance.

In 2010, a federal appellate court found some of the Ohio labeling restrictions of rbST to be unconstitutional in *International Dairy Foods Association v. Boggs* [11]. Ohio's restrictions followed FDA's guidance by placing a prophylactic ban on milk composition claims including "rbST free" [6]. The plaintiffs argued this regulation violated their commercial free speech guarantees by the First Amendment of the federal Constitution.A separate provision of the state labeling restrictions mandating contiguous information on a production claim was also challenged[12].

The *International Dairy Foods Association* court decided that there was a compositional difference between milk from untreated cows and conventional milk.By giving credence to evidence of elevated levels of insulin-like growth factor 1 and higher somatic cell counts, the court found a compositional difference. With this distinction, milk from cows not treated with rbST is "rbST free."However, milk from cows treated with rbST might contain rbST, although there exists no way to determine whether this is the case.The court thereby decided that dairy processors should be able to make claims that the milk is "rbST free." This means that the state of Ohio cannot ban claims that milk is "rbST free," but may require a disclaimer to inform consumers that rbST has not been detected in milk from cows treated with rbST.

With respect to the production claim that milk came from cows not supplemented with rbST, the court approved the requirement of a disclosure that the "FDA has determined that no significant difference has been shown between milk derived from rbST-supplemented and non-rbST-supplemented cows" [6]. Ohio could enunciate such a requirement since there was some evidence of possible deception absent in the disclosure. However, the disclosure did not need to be contiguous to the statement that milk is from cows not supplemented with rbST.

3. Attributes for Animals and Their Products

More and more agricultural producers seek to label their products to convey information to consumers about special attributes about quality that may enable sellers to command a price premium. For livestock producers, several labeling options exist, including organic, natural, naturally raised, pasture-raised, hormone-free, rbST-free, grass-fed, free-range, certified humane, and others. The twin purposes of using these labels are to entice buyers to select the products as opposed to others and to be able to charge a higher price. For producers of animal products, labeling animals as "naturally raised" is attractive, and labeling meat products as "natural" may allow the seller to charge a higher price.

Yet, these terms may mean different things to different people.Many people attribute inaccurate connotations with the term "natural," and assume it means improved production and processing attributes [13]. Moreover, there is no third-party verification of naturally raised products [14]. A significant segment of the public feels that because the terms are not certified, they do not mean much [13]. Thus, the U.S. Department of Agriculturehas suggested details for defining these terms and their voluntary use when describing products. "Naturally raised" is intended to address pre-slaughter production practices [15]. A definition of "natural" is being considered to address labeling claims set forth on meat products and would exclude animal production practices. However, the definitions of these terms are controversial so that the U.S.Department of Agriculture has not adopted any final rules for either term. This means that both definitions are still being debated and their meanings remain confusing.

3.1. Naturally Raised Livestock

The Agricultural Marketing Service (AMS), a division of the U.S. Department of Agriculture, delineated a proposal for a voluntary standard for "naturally raised" livestock and the products from such livestock in 2009 [16]. The agency sought to end confusion created by companies and vendors using differing definitions for naturally raised livestock product claims.By reaching an agreement on the definition of the term "naturally raised," the AMS felt that the term would become more meaningful in the marketing of livestock and meat products.Vendors voluntarily using the term would comply with the rules and buyers would know what attributes were associated with the noted claim.

AMS decided that a naturally raised livestock-marketing claim should involve three core criteria: (1) restrictions on byproducts in animal feed, (2) no use of growth promotants, and (3) no use of antibiotics. AMS also considered other issues and decided that the three selected criteria would best describe naturally raised livestock.

The first criteria involved the diet of naturally raised livestock and a requirement that animals never be fed mammalian, avian, or aquatic byproducts[16].AMS described what feed was prohibited, and this included all byproducts connected to the slaughter and harvest processes including meat and fat, animal waste materials, fishmeal, and fish oil. Byproducts not derived from slaughter and harvest processes such as eggs and milk would be allowed to be feed to animals under the naturally raised marketing claim.

Various comments to AMS's proposal to define naturally raised livestock suggested that the term should encompass production practices such as being raised in an environment natural to the species or allowing the animals to exhibit natural behaviors [16]. Some commentators wanted to exclude animals raised at concentrated animal feeding operations or in cages.Others objected to genetic selection, cloning, artificial insemination, tail docking, and surgical mutilation.However, due to a lack of a clear, unified approach to these production issues, AMS felt that none of them was appropriate to include in a definition of naturally raised livestock. Instead, such information could be incorporated into other marketing programs.

The exclusion of growth promotants provided little controversy.Producers claiming their animals were naturally raised cannot administer natural hormones, synthetic hormones, production promotants, estrus suppressants, beta agonists, or other synthetic growth promotants.

Precluding the use of antibiotics in the production of naturally raised livestock was accepted as a criterion for a naturally raised marketing claim[16].However, there was some controversy as how to treat animals that required emergency treatment with an antibiotic.AMS suggested that the use of antibiotics to treat animals when they were sick should disqualify the animals from being called naturally raised.

Turning to the issue of vaccines, AMS decided that they were appropriate and acceptable, so could be administered and the animals would still qualify as naturally raised [16]. Another issue was the use of sulfonamides and ionophores to control coccidiosis and parasites. Coccidiosis is a parasitic disease of the intestinal tract of livestock animals that causes either severe illness with possible death or subtle illness.AMS suggested that ionophores could be used, but the products of treated animals would be identified as not using antibiotics other than ionophores.

Concerns about the use of pesticides and chemicals in the production of animals were considered by AMS but were not incorporated into the standard due to variations among regions of the United States.

3.2. Natural Products

The Food Safety and Inspection Service (FSIS), a division of the U.S. Department of Agriculture, has been charged with defining "natural" for meat and poultry product claims. This marketing claim is completely separate from the term "naturally raised," but an issue arises concerning the compatibility of the two separate marketing claims.To avoid confusion and to not compromise other marketing claims, care is needed in coordinating the definitions of the terms. In September, 2009, FSIS sought comments for defining conditions under which voluntary claims of "natural" may be used in the labeling of meat and poultry products [15].With these comments, FSIS is still considering the issue.

FSIS first issued policy guidance on the use of "natural" to describe products in 1982 [15]. A policy memo defined "natural" to mean the exclusion of artificial flavoring, coloring ingredient, or chemical preservative and a limitation on processing.Minimal processing was allowed on products so they could still use the "natural" claim. The policy memo also provided that an explanation of the claim "natural" should be stated on products. This was rescinded in 2005 and replaced with an entry in FSIS's *Food Standards and Labeling Book*

(*Policy Book*) [17]. The *Policy Book* expanded flavorings that could be added to meat products without disqualifying products from being called natural.

In response to a petition from a food company in 2006, FSIS sought comments for defining "natural" with respect to meat and poultry products.Subsequent public comments showed controversy, considerable desire to maintain flexibility as exists under the *Policy Book*, and highlighted several diverse issues.First, how should multi-functional ingredients added to meat products be treated? For example, if an ingredient such as potassium lactate provides antimicrobial effects as well as flavoring, should it be permitted to be used and the meat qualifies to be labeled "natural"? Some feel that natural preservatives such as sodium lactate should not disqualify a product from being labeled "natural."

Suggestions were also made that FSIS should establish separate criteria for "natural" products and products with "natural ingredients"[15]. Other commentators felt that more direction was needed concerning processing methods that were not in use when the 1982 memo was drafted. This might involve a further definition of what "minimal processing" involves in determining whether a product qualifies for the "natural" claim.Some commentators identifying themselves as members of the Truthful Labeling Coalition objected to allowing poultry products that were enhanced with water, salt, flavorings, seasonings, tenderizing agents, and water-binding ingredients to qualify as "natural" products. These commentators felt that enhanced products were not natural and consumers were being misled when they wanted to purchase a "single-ingredient chicken."

The use of the "natural" claim also is viewed by some people as suggesting that the animals should be naturally raised so meat products from animals raised at intensive confined operations would not qualify as natural products [15]. Related to production practices was some feeling that animals that were genetically altered, treated with hormones, or fed prophylactic antibiotics should not qualify to be labeled as "natural." However, FSIS noted that the "natural" claim was intended to convey information on the finished product, whereas "naturally raised" or other labels are intended to convey information on pre-slaughter conditions.

FSIS claimed that the comments showed a lack of consensus on what natural means focusing on ingredients and processing methods. Therefore, FSIS is seeking additional input on a number of issues prior to taking any action on the further regulation of claims involving natural products [15]. This means that there remains confusion and ambiguity concerning natural marketing claims [13].

4. Conclusion

Farm animals are produced so their products may be sold to consumers. The marketing of these products is tied to consumer demand and rules concerning health and safety.Producers structure their production operations to respond to market prices. To enhance profitability, producers may adopt production practices to reduce costs.Monsanto developed Posilac® for this purpose, and some producers commenced use of rbST to enhance milk production from their herds.However, some consumers are willing to pay more for products that have no association with rbST. Vendors have affixed labels to products with information that the product did not come from cows treated with rbST.Because a subset of consumers object to the use of rbST and have been successful in developing markets with product labels, producers may be better off forgoing the use of rbST.

The Food and Drug Administration issued guidance on the labeling of products from cows not treated with rbST, and felt that "rbST free" might be misleading because there was no test to verify a distinction between milk from treated cows and non-treated cows.However, in *International Dairy Foods Association v. Boggs* [11], a court found a compositional difference between milk from cows treated with rbST and milk from non-treated cows. With this finding, the court decided that a label claiming milk was "rbST free" was not inherently misleading so could be used.Other U.S. state governments may analyze their state food labeling law to discern whether vendors may adopt "rbST free" claims on their labels. The widespread use of these labels may lead to less demand for products from cows treated with rbST and a corresponding decline in the use of Posilac®. In turn, more dairy animals may be needed to meet the market demand for milk and milk products.

Two units of the U.S.Department of Agriculture have considered new rules for labeling animals as "naturally raised" or products as "natural." Considerable controversy exists over production practices and ingredients that should be allowed under labeling rules. Perhaps the most controversial is the addition of natural ingredients to meat and to allow more processing yet allowing these products to qualify for the "natural" claim.Under the U.S. Department of Agriculture's *Policy Book*, vendors are calling their products "natural" despite the fact that up to 15 percent of their weight is an injected solution of ingredients such as salt, both, and seaweed extract [18]. Over one-half of the chicken marketed in the United States is called "natural" [19]. A considerable segment of the public does not understand that natural meats can be enhanced with other ingredients and undergo some processing, and thus various consumer groups would like to limit what may be added to products labeled as "natural." It is unclear whether the FSIS will propose a final rule that continues to cater to the industry or adjust the definition of natural to one that is more aligned with public perception. However, given past experiences, the regulatory definition will probably defer to vendors and allow enhanced products to be called "natural" rather than limiting the term to a single-ingredient meat product.

These labeling developments mean that producers need to stay attuned to consumer demand and consider restructuring their production practices if greater profits are available by qualifying under labels that garner higher prices.Currently, "rbST free," "naturally raised," and "natural" constitute information that may be placed on labels to garner higher prices. If greater numbers of consumers purchase products with these labels, prices may raise and more producers may elect to change production practices so that their products quality for higher prices.

REFERENCES

[1] U.S. Food and Drug Administration. Freedom of Information Summary: Supplemental New Animal Drug Application 140-872 Posilac (1993), www.fda.gov/ohrms/dockets/98fr/140-872-fois001.pdf.

[2] U.S. Food and Drug Administration. Animal drugs, feeds, and related products; Sterile sometribove zinc suspension.*Fed. Register,*1993; 58: 59946-47.

[3] U.S. Food and Drug Administration. Interim guidance on the voluntary labeling of milk and milk products from cowsthat have not been treated with recombinant bovine somatotropin.*Fed. Register,*1994: 59:6279-80.

[4] Centner TJ,Lathrop KW. Labeling rbST-derived milk products: State responses to federal law.*Univ. Kansas Law Rev.*, 1997;45: 511-56.

[5] Ohio Executive Order. *Immediate Adoption of Rule to Define What Constitutes False and Misleading Labels on Milk and Milk Products*, No. 2008-03S,2008.

[6] Ohio Administrative Code. *Dairy Labeling rule* 901:11-8-01,2011.

[7] Minnesota Annotated Statutes. *Dairy Trade Practices: Recombinant Bovine Growth Hormone Labeling, section* 32.75,2010.

[8] Wisconsin Administrative Code. Agriculture, Trade and Consumer Protection: *Dairy Product Advertising and Labeling*, rule 83.02,2011.

[9] Vermont Statutes Annotated. *RBST Labeling: Substantial State Interest*, title 6, section 2760,2007.

[10] Vermont Code Rules.*Rule Relating to rbST Labeling of Milk and Dairy Products,* rule 20-021-005,2011.

[11] International Dairy Foods Association v. *Boggs.*622 F.3d 628-50, 2010.

[12] Rodriguez C,Papoulias E.Recent developments in health law: Ban on milk labeling violates First Amendment -- International Dairy Foods Ass'n v. *Boggs.J. Law MedicineEthics.,*2011; 39: 96-97.

[13] Abrams KM, Meyers CA, Irani TA. Naturally confused: Consumers perceptions of all-natural and organic pork products.*Agric. Human Values.*,2010; 27: 365-74.

[14] Cole L. What's your beef? Grass-fed?Grain-finished? Organic: Free-range: How do you know, and what does it mean? *Oregonian* (Portland). Feb. 17, 2009.

[15] U.S.D.A. Product labeling: Use of the voluntary claim "natural" in the labeling of meat and poultry products. *Fed. Register.*,2009; 74: 46951-57.

[16] U.S.D.A. United States standards for livestock and meat marketing claims, naturally raised claim for livestock and the meat and meat products derived from such livestock. *Fed. Register.*, 2009;74: 3541-45.

[17] U.S.D.A.Food Standards and Labeling Policy Book. Food Safety and Inspection Service, August 2005, http://www.fsis.usda.gov/OPPDE/larc/Policies/Labeling_Policy_Book_082005.pdf.

[18] Skrzycki C. *Crying foul in debate over "natural" chicken.* Washington Post. Nov. 6, 2007.

[19] Reicks AL, Brooks JC, Kelly JM, Kuecker WG, Boillot K, Irion R, Miller MF. National meat case study 2004: Product labeling information, branding, and packaging trends. *J. Animal Sci.*, 2008; 86: 3593-99.

INDEX

D

E

F

G

H

I

M

N

O

P

Q

R

S

T

U

V

W

Y

Z